ACCIDENT PREVENTION MANUAL
FOR BUSINESS & INDUSTRY

SECURITY MANAGEMENT

2ND EDITION

EDWARD J. KRUEGER
EDITOR

National Safety Council
Itasca, IL

Project Editor: Phyllis Crittenden
Interior Design, Layout and Composition, and Cover Design: Jennifer Villarreal
Senior Director of Publishing: Suzanne Powills
Cover Art: ©2010 Shutterstock Images

Printed in the United States of America
24 23 22 21 10 9 8 7 6

Library of Congress Cataloging-in-Publication Data

Accident prevention manual for business & industry : security management / editor, Edward J. Krueger. -- 2nd ed.

p. ; cm.

Includes bibliographical references and index.

ISBN 978-0-87912-260-7

1. Industrial safety--United States--Handbooks, manuals, etc. 2. Accidents--United States--Prevention--Handbooks, manuals, etc. 3. Industries--Security measures--United States--Handbooks, manuals, etc. I. Krueger, Edward J. II. National Safety Council. III. Title: Accident prevention manual for business and industry. IV. Title: Security management.

[DNLM: 1. Accident Prevention--methods--United States--Handbooks. 2. Workplace--United States--Handbooks. 3. Crime--prevention & control--United States--Handbooks. 4. Industry--standards--United States--Handbooks. 5. Safety Management--United States--Handbooks. 6. Security Measures--United States--Handbooks. WA 39 A171 2010]

T55.A334 2010
658.4'7--dc22

2010012839

2C011921 Product Number: 12185-0000

CONTENTS

TABLE OF CONTENTS

FOREWORD

One notion that comes to mind when producing a book on security management is the amount of funding required supporting its recommendations. There is a long-standing adage that suggests "it is better to prevent than to cure." While understandable to those of us in the business, this axiom may not be easily digested by the "penny-wise, pound-foolish" crowd, who are neither schooled in empiricism, nor, as a matter of course, study the dangers of an inadequate security program. This is particularly troublesome as we distance ourselves from the major tragedies of the last decade. Some may falsely conclude that less investment is now okay. Prevention requires up-front funding; and curing a problem before it occurs may be perceived by budget managers as a risk-management issue—an expense to be ignored. This is always a mistake. The instincts of the safety and security professional should be to implement the suggested best practices as described in *Accident Prevention Manual for Business & Industry: Security Management* to help render crisis management unnecessary while, at the same time, always planning for it. Of course, there is a cost. This book provides the rationale and the talking points to convince the financial officers and budget managers why it is necessary to fund an effective safety and security program.

This revised edition of the *Accident Prevention Manual for Business & Industry: Security Management* provides a refinement of topics previously presented. These chapters frame commonsense prescriptions to reduce a company's exposure to threats against an established security system. This book explores where we were, where we are, and where we may be trending as new social and technical vulnerabilities emerge that may change our approach to safety and security issues. Often, the safety and security community responds to new trends with expensive and cumbersome counter trends. This can be avoided to some degree as the security practitioner becomes well-versed in the latest nuances of policy, procedure, and management as described in this edition of the *Accident Prevention Manual for Business & Industry: Security Management*. This edition allows for that to happen. *Security Management*, second edition, is a book that has evolved over time, and its authors are sufficiently flexible to recognize that which works and that which does not. Enjoy this new edition.

Thomas D. Sloan
Special Agent in Charge
U.S. Secret Service (retired)

PREFACE

Security and safety needs for business and industry have dramatically changed since the events of September 11, 2001. Previously, security was often a legal mandate or a necessary response to events to prevent recurrence. Lessons learned have led to the realization that business and industry must develop a security initiative that focuses on prevention, early intervention, loss mitigation, event reaction, and stabilization. This approach has been necessitated by the globalization of criminal activities such as terrorism, piracy, industrial espionage, political kidnapping, cyber crime, and so on. Domestically, it has been estimated that crime costs businesses more than $2 trillion annually. Added to the challenges posed by criminal activity is the need for businesses to prepare for workplace violence, pandemic events, and employee safety when working abroad. These societal issues must be viewed as active and ongoing. No business can afford to have a passive security program.

The purpose of this manual is to provide for security directors, safety directors, facility managers, and students of security a reference tool for assuming the responsibility of security in the workplace. Authored by highly-respected professionals in the safety and security field, *Security Management* is designed to provide a broad overview of the occupational aspects of security management from the perspectives of security practitioners.

Security Management is also designed to prepare students for a career in the security and safety management profession. It also serves as a reference tool for people in management positions (human resource management, facility managers, and operations managers) who have security related responsibilities.

Chapters in this book are arranged into eight major sections as follows:

1. **Security Overview**—Provides historical perspectives regarding the development of security and commentary on future security trends.
2. **Security and the Law**—Focuses on legal aspects affecting an organization's responsibility to provide a safe environment for employees and other property users.
3. **Organizational Management of Security**—Outlines the principles and procedures concerned with defining, organizing, and managing an effective security management program.
4. **Personnel Security Management**—Addresses the various issues, policies, and procedures for securing employees and visitors in the workplace.
5. **Management of Proprietary Information**—Explores measures and models designed to minimize loss from internal and external threats to information assets.
6. **Physical Security Concepts**—Examines security surveys, target hardening, access controls, and other accepted industry standards and regulations for providing a physically secure environment. This section also addresses crime prevention through environmental design (CPTED).
7. **Critical Incident Management**—Discusses plans and preparations necessary for handling security emergencies. The chapters within this section address approaches to developing an emergency operations plan (EOP) as well as the value and goals of establishing a partnership between private security and public law enforcement.
8. **Special Topics**—Looks at the effects of terrorism on global business, some unique preparedness requirements for dealing with a pandemic, the future of security management, and provides a comprehensive list of security resources for those seeking more specific information.

ACKNOWLEDGMENTS

Singling out any one contributor to this book is impossible. It is important to thank each collaborator for his or her time, effort, and support of this endeavor. All of the authors and reviewers had an interest in making a contribution to the field of security by sharing their expertise. This manual is a testimonial to their years of great work.

I wish to acknowledge my family members for their great support: my mother and father, Isabella and Ed; my wonderful wife, Cindy; our five phenomenal daughters, Christine, Jennifer, Allisen, Angella, and Sarah; and our remarkable Ella. This book is for you.

A special acknowledgment to my teacher, mentor, collaborator, colleague, and great friend Arthur A. Kingsbury is in order. If the security field identifies any icons, Art Kingsbury is among them. Thanks to Art for years of friendship and his willingness to share his expertise on our national and international work. Thanks to Tom Sloan for his government and private-sector expertise. To Richard W. Lack, editor of the first edition of the *Security Management*, thank you for your guidance in the early stages of this manual. It was greatly appreciated.

The National Safety Council is gratefully acknowledged for its commitment to the security profession and its willingness to update this manual. There will be many beneficiaries of its publication. Projects such as this never reach the completion stage without that one person, behind the scenes, guiding the work forward. That significant person is Phyllis Crittenden, NSC Press editor. It is with great admiration and respect that I thank Phyllis for her kindness, patience, wisdom, and guidance throughout this project. On behalf of all of the authors, we were fortunate to work with you and we are indebted to you.

As Richard Lack so eloquently stated in the previous edition, this book is dedicated to those who are out there every day ensuring a safe and secure workplace. Your comments and ideas for the continuing improvement of the *Accident Prevention Manual for Business & Industry: Security Management* are solicited. Such comments are key factors in helping us plan future editions.

Thank you,
Edward J. Krueger
Editor

CONTRIBUTORS

Mark C. Bach has worked for the Fort McDowell Police Department, a Yavapai Indian community/Nation, on the edge of the Phoenix metropolitan area, for 13 years and is presently the Patrol Lieutenant. Bach served as Director of the Office of Management and Budget for the Tempe, AZ Police Department before he retired with 20 years of service. He has consulted for the National Crime Prevention Council as a subject-matter expert on crime prevention through environmental design (CPTED) for Federal Weed and Seed sites. Bach earned his Bachelor's degree in Public Administration from the University of Arizona and his Master's degree in Justice Studies from Arizona State University. He is currently an adjunct professor at Fox Valley Technical College. He has reviewed more than 600 grants for the federal and state governments in a variety of fields including domestic violence, juvenile issues, drug abuse, Native American issues, and homeland security.

Norman D. Bates, Esq., President and founder of Liability Consultants, Inc., is a nationally recognized expert in security and the law. For more than 20 years he has been providing security management consulting services to private industry as well as to court-certified expert witness services nationwide for cases regarding inadequate security, negligent hiring or training, and workplace violence. Bates regularly presents seminars on civil liability issues and has authored numerous articles and books on the subject. Actively involved with the drafting of legislation, he authored a bill on criminal stalking in Massachusetts that was passed into law. Formerly, Bates was an assistant professor of Criminal Justice at Northeastern University in Boston and Director of Security and Legal Counsel to the Saunders Hotel Corporation. He received his Juris Doctor from Suffolk University and a Bachelor's of Science degree in Criminal Justice from Northeastern University. He is a member of the Massachusetts Bar, the International Association of Professional Security Consultants (President 2009-2011), the National Crime Victim Bar Association, ASIS International, and Society for Human Resource Management. He is also former member of the American Association for Justice (formerly ATLA) and the Defense Research Institute (DRI).

Ronald H. Beatty, CPP, was formerly Director of Security for North American Aircraft Division of Rockwell in El Segundo, CA, and manager of security for General Dynamics Corporation, at the Electric Boat Division in Groton, CT. Prior to that he was Director of Security for the Lycoiming Division of Avco Corporation in Stratford, CT. Beatty is a past president and chairman of the board of the American Society for Industrial Security (ASIS). He is also past chairman of the Aerospace Industries Association Security Committee and past chairman of the National Industrial Security Program Industry Task Force. Beatty is an alumnus of the State Department's Overseas Security Advisory Council and has served on the board of directors of the Research Security Administrators.

Robert Chapman is a Supervisory Social Science Analyst with the U.S. Department of Justice, Office of Community Oriented Policing Services (COPS Office). He is responsible for conducting, managing, and overseeing program development activities, applied research and evaluation, and program and policy analyses on a variety of community policing and law enforcement topics. He has worked at the COPS Office in the Program and Policy Support and Evaluation Division since 1996. Prior to joining COPS, Chapman assisted in a variety of public safety research projects and managed the legislative affairs for the Police Executive Research Forum (PERF), a membership organization of progressive police executives from the largest city, county, and state law enforcement agencies. Chapman received his Master's degree from Johns Hopkins University School of Government and has also conducted graduate studies in Demography and Public Policy at Georgetown University.

Michael R. Cummings, CPP, has been in the field of security since 1973. He spent the first 12 years of his career in the retail sector. He has been a security practitioner in health care since 1985. He is currently Director of Loss Prevention Services for Aurora Health Care in Milwaukee Wisconsin. He has served in this position since 1987. Cummings has served many years in various capacities on the local ASIS board in Milwaukee.

He served as a member of the Standing Committee for Health Care Security between 1989 and 1999, and he was elected to the board of directors in 2003 and currently serves as its chairman. A senior member of the International Association for Healthcare Safety and Security (IAHSS), Cummings holds a BA from Marquette University and has completed course work for a MA in Security Management and Human Resources Development from Webster University. He received CPP designation in 1986 and has recertified for this designation every three years since.

Cummings has served on several boards including the Southeast Wisconsin Homeland Security Partnership, Inc. (SWHSP), the Editorial Advisory Board for *Corporate Security* and the board for the Avenues West Association. Mike has lectured on a variety of security topics at local and national levels. He has published in *Security Management*, *The Journal of Healthcare Protection Management,* and *The Security Business Practice Reference.*

Francis James D'Addario, CPP, CFE, served as Director of Loss Prevention for Hardee's Food Systems Inc., in Rocky Mount, NC. D'Addario's previous security management positions include Jerrico Inc., The Southland Corporation, Sheraton ITT, and Sears. A violence-prevention and victim's rights advocate, D'Addario has also served as Vice President for The McGruff House National Network board of directors. He has advised Eastern Kentucky University, the International Assets Protection Institute, and the University of North Carolina on loss prevention. He served as chair for the American Society for Industrial Security's Standing Committees on Crime Prevention and Food Service Security. D'Addario authored *Loss Prevention Through Crime Analysis*, and *The Manager's Survival Guide*, in addition to numerous articles in trade journals. He developed LOSSVISION, a copyrighted, investigations and assets recovery software package. He also coproduced "Two Minute Warning," a violence-prevention film and has made numerous radio and television appearances.

Nicholas Dillon is currently Director of Education & Risk Management Services at Aegis Corporation in Brookfield, WI, where he manages the continuing education function of the organization and provides ongoing research, leadership, coaching, and mentor training to the risk management and safety staff. For more than a decade, Dillon has taught courses in occupational safety and health, served as consultant and instructor for the business and industry department of the Waukesha County Technical College, and provided seminars in diversity, ethics, management and leadership, and workplace communications.

Dillon earned a BA degree in Management from Marquette University, a MA degree from the University of Phoenix, and a MS from the University of Wisconsin. He is an authorized OSHA Instructor, a member of the American Society of Training & development, and a board-certified instructor by the State of Wisconsin Technical College System Board.

Michael Jay Fagel, PhD, CEM, has more than three decades of public service, including fire service, emergency medical service, public health, law enforcement and emergency management, as well as corporate safety, security, and threat risk management. Currently, he is an instructor at the University of Chicago in the Threat Response Management program. He also teaches at Northwestern University in Chicago, Benedictine University, Eastern Kentucky University, and is guest lecturer at Northern Illinois University. In addition, Fagel is a Homeland Security Analyst at Argonne National Laboratories in Illinois, engaged in the protection of critical infrastructure.

During his 10-year tenure with FEMA, Fagel responded to critical incidents, including the Oklahoma City bombing and the World Trade Center attack on September 11, 2001. He spent two years in the Middle East working on a national response plan and was a delegate to the European Conference on Emergency Management in Budapest in 2007.

Fagel has authored more than 100 articles on safety and disaster planning. He has published two textbooks on the topics of food safety and emergency management and has served as editor for numerous trade books and articles. His professional membership associations include the Northern Illinois Critical Incident Stress Debriefing team and the International Association of Fire Chiefs Committees on Safety and Health, to name a few.

Kenneth G. Fauth specializes in providing physical security reviews, threat and vulnerability assessments (TVA), and business contingency and emergency response planning for various companies throughout the United States and Asia. Fauth was employed as Director of Security and Loss Prevention for Spiegel, Inc. in Chicago for more than 16 years. During that period he oversaw the protection planning and operation for various distribution and warehouse facilities, office complexes, and travel requirements at domestic and global locations.

As a university professor, Fauth developed curriculum and taught classes in security, asset protection, fire and disaster planning, retail theft prevention, investigations, crime prevention, security management, and loss avoidance at Western Illinois University and Northern Michigan University for more than a decade. Commissioned as 2nd Lieutenant in the U.S. Army, Fauth served in Vietnam and in various capacities at the Pentagon, including security officer, intelligence officer, and with the military police. He retired with the rank of Lieutenant Colonel from the U.S. Army Reserve.

Fauth earned his doctorate degree from Indiana University, a MS degree in security from Michigan State University, and a BS degree from Central Michigan University. He is one of the first groups of individuals to have received his certified protection professional (CPP) designation from ASIS International. He has served on the board of directors for ASIS International.

Eugene F. Ferraro, CPP, CFE, PCI, specializes in the investigation and prevention of workplace violence, employee dishonesty, fraud, harassment, discrimination, substance abuse, and misconduct in the workplace. With more than 27 years as a corporate investigator, he has conducted thousands of public and private investigations throughout the Unites States and abroad and has testified as an expert witness in numerous criminal and civil matters. Ferraro is currently CEO of Business Controls, Inc., in Littleton, Colorado and is a member of several professional organizations including ASIS International, the Association of Certified Fraud Examiners, the Professional Private Investigators Association of Colorado, and the National Council of Investigative and Security Services, to name a few. He earned a BS degree from the Florida Institute of Technology and completed Naval Flight School and Naval Justice School.

Ferraro has authored several books and has published numerous articles for an assortment of trade publications serving the security industry. He is also a frequent book critic and reviewer for the journal *Security Management.*

Douglas E. Himberger is Senior Vice President and Director for Security, Energy, and Environment at the National Opinion Research Center (NORC) at the University of Chicago. He has more than 35 years experience in integrating security, survivability, and resiliency best practices across public, private, and civil sectors. He leads NORC in addressing issues of interest to the public in security, energy, and environment, developing and directing studies involving complex data collection, analysis, and reporting in social and physical science arenas. He has extensive expertise fostering collaboration between industry, academia, and government to build strong information sharing networks and advanced technology solutions.

Himberger created a global risks initiative taskforce for analyzing complex all-hazard threats (e.g., pandemics, natural/man-made disasters) and provided multidisciplinary solutions for government and industry. He serves on the boards, or as a volunteer for numerous organizations, including the Safe America Foundation (SAF) and the Emergency Management Professional Organization for Women's Enrichment (EMPOWER). He writes and speaks extensively on security topics for professional journals and events.

Lauri A. Kingsbury is a police officer for the Detroit Police Department. She is also coordinator for the Detroit Police Department's Basic and Advanced Crime Prevention courses, which are attended by law enforcement personnel and private security professionals. She teaches crime analysis, security survey format and technique, and crime prevention through environmental design. Her police background includes work in conspiracy investigation, financial analysis of assets from narcotic trafficking, and traffic accident investigation and reconstruction. Her work has been primarily in the areas of risk analysis and loss prevention for businesses. She has written numerous commercial security surveys and is responsible for program research and development in the Crime Prevention Section of the Detroit Police Department. Kingsbury serves as a consultant in the field of crime prevention through environmental design to Detroit's Site Plan Review Committee and several neighborhood associations. She has been a guest lecturer at Kingsford College in Sarnia, Ontario, Canada; Central Michigan University; and Macomb Community College. Her publications include *A Basic Crime Prevention Manual*, the textbook for the basic crime prevention course, and the manual for the Citizen's/Business Owner's Police Academy at the Detroit Crime Prevention School.

Edward J. Krueger is the Director for the Criminal Justice Center for Innovation at Fox Valley Technical College. Previously, he served as a police officer and the director of a regional police academy. He has a graduate degree in education and more than 30 years of military, criminal justice, occupational training, and educational experience. Nationally known for his work in the security and criminal justice fields, Krueger developed and managed numerous education and training programs focusing on criminal justice management, security management, and crime reduction/crime prevention management. He is the recipient of numerous awards, including the Wisconsin Technical College System educator of the year. As director of hotel security, Krueger was responsible for securing special events for NFL teams, celebrities, and political dignitaries, and he served as assistant director of security for the 2004 presidential debates. Recently, Krueger provided his expertise to the U.S. State Department and the Peruvian government regarding the security of unearthed historical communities.

Richard S. Post, PhD, CPP, was Managing Director of an international consulting firm in Hong Kong, the previous Director of Human Resources and Corporate Security for Primerica (now Citibank), and Director of Security and Political Risk for the BF Goodrich Company. He chaired the department of Criminal Justice Studies at the University of Wisconsin and has served with the U.S. Army and as a career clandestine services officer with the Central Intelligence Agency. Post attended Michigan State University where he received a BS and a MS degree. He completed the Executive Development program at Cornell University and earned a PhD from the University of Wisconsin-Madison.

Post served as a Chapter Chairman for the American Society for Industrial Security (ASIS). He was a member of the Professional Certification Board, and a National Vice President and Director of ASIS. Post has authored 12 security textbooks and numerous articles. He also completed two business books for the Economist: *Strategic Design for Asia Pacific Business* and *Cost Effective Operations for Asia Pacific Business*. Recently, Post co-authored (with his wife, Penelope) *Global Brand Integrity Management* published by McGraw-Hill, and he continues to provide consulting services through Post & Post LLC and GlobalBrandIntegrity.com.

Norman MacArthur Spain, JD, CPP, is a former faculty member of Ohio University and Pennsylvania State University and was a Fellow in Loss Prevention and Safety at the Indian Institute of Security and Safety Management in New Delhi, India. He is a former security manager and has consulted as an expert witness on security for cases in several states. Spain specializes in security-related litigation, and he has written several articles in this area and lectured on this topic at professional security seminars in more than 30 states.

Part 1

Security Overview

1

Historical Perspectives

Edward J. Krueger

CHAPTER 1

OBJECTIVE

The objective of this chapter is to provide a historical perspective of security—its beginnings and evolution into a profession. This information will provide readers the background needed to accomplish the following:

- develop an understanding of the security field
- determine how security management principles can complement the level of environmental and occupational safety and health management in organizations today

As these areas coalesce, the importance of each continues to increase.

INTRODUCTION

The significance of security to human survival is found in its evolution from primitive self-protective techniques, devices, and strategies, to that of modern sophisticated systems technology. From physical-access barriers created by such rudimentary perimeters as rock fences to electronic access-control systems, there has always been one rationale for security; that is, the need for survival of a species. Arguably, all living organic matter (animal, plant, or a simpler form of organism) developed some need for security. However, humans had the intellect and reasoning necessary for the development of sophisticated security measures for self-protection. Such measures included the use of barriers, height, natural surveillance, and distance.

Such early initiatives as well as those made today suggest a general concept of security. As defined by Post and Kingsbury, security "provides those means, active or passive, which serve to protect and preserve an environment which allows for the conduct of activities within the organization of society without disruption" (Post and Kingsbury 1977). The definition recognizes that a stable and orderly environment is a result of properly applied security; and it suggests that many measures (active and passive) can be necessary to achieve that goal.

The need for security in the primitive stages of human development was for protection of persons (individual, family, or society) and property (land, residence, personal, or business). Eventually, communications security (protection of community, military, or business intelligence) would become a third need. However, with the growth of technology, a fourth category of "cyber-security" must be considered. This need increasingly becomes evident when considering the impact of the September 11, 2001 terrorist attack on the United States. And, compounding this issue has been the continuous "hacking" attempts on government and private business information systems. Emphasizing this point is the fact that the National Cyber Security Division was created within the U.S. Department of Homeland Security. Its responsibility is to manage the requirements for cyber-security in the United States as well as the response system and the risk program. Today, these specific functions of security are grouped into the following four categories:

- physical security (facilities, locks, access control, perimeter security, etc.)
- personnel security (employees, visitors, vendors, and other property users)
- information security (human resources records, customer records, proprietary information, research, etc.)
- cyber-security (information technology, computers, Internet/systems infrastructure, communications systems, etc.)

However, even with all the attention given to security since the September 11 attack, many agencies and businesses have a fragmented security structure focusing on as few as just one of the four categories. Historically, this is not unusual as an organization's security needs are often defined by a security breach and the response to that breach only addresses that specific incident. It is important for the organization's manager of security to develop a program that incorporates the four principal categories of security.

ANCIENT HISTORY

To appreciate the need for modern security, it is important to appreciate its importance to the stability of societies. Members of primitive society found it necessary to defend against animals or other humans, who intended to steal basic needs (food, clothing, shelter) or who threatened life. Threats to safety and security were always present. As a result, societies developed methods to ensure their safety and security.

Using local terrain and climate, humans sought security by using natural caves or openings on cliff faces as shelters. These residences provided a natural view of the surroundings and a form of singular access control. Such facility planning made it easier to protect against rival tribes or approaching animals. Later, other types of homes were built on stilts above the surface of lakes. This isolation served as a natural barrier that would not be crossed by most animals. It also provided the residents with an open view of the surrounding landscape. This early form of access control emphasized the importance of allowing only authorized individuals on owner-controlled property.

For protection of personal possessions and valuables, the ancient Egyptians were one of the first civilizations to develop a locking device that became the forerunner of the pin tumbler lock (Roper 1983). Archaeological discoveries, found on walls of pyramids, displayed many drawings that illustrated the internal workings of locking devices. Such early locks and keys were fashioned from wood. These offered limited physical protection; although an intrusion into a secure area could be determined because of damage to the wood.

As societies developed, the need for rules or laws became evident. Although many of the early societies did have community standards, it was not until about 2100 BC that the earliest form of written law appears. These early sanctions identified expectations that the community had of the conduct displayed by its members to each other, as well as to the general society. These early laws were developed from customs and modified by King Hammurabi of Babylon. They represented an early behavior guide that was to be selectively incorporated into future laws.

Early Greeks and Romans were known for using guards to protect their rulers from harm. This protection eventually was extended to roads and highways for the safety of travelers. As the Roman Empire continued to grow, the need for security became increasingly important. A military unit was formed by Emperor Augustus to protect the city. Members of this unit, which was called a *cohort*, formed the Praetorian Guard. This organized unit represented one of the earliest forms of modern-day policing.

At the time of the fall of the Roman Empire, the importance of community control was evident. Societal customs became laws, early semblances of policing could be identified, and citizen involvement in community decisions was expected.

SECURITY DEVELOPMENTS IN ENGLAND

THE MIDDLE AGES

During the Middle Ages (500 AD to 1500 AD), security and safety was provided for the less fortunate and poor in return for labor. Land was divided by the king and assigned to noblemen to manage—often on behalf of the king. Those laboring the land would receive protection and food for their efforts. Eventually, laboring families banded together for security. Groups of 10 families were called a *tithing* and were responsible for maintaining community law and order.

Eventually, because the existing system of justice was weakened by bribes and a lack of respect for traditional rights, the feudal system was challenged by British barons. The barons demanded that King John of England sign a document, acknowledged as an important source of modern legal procedure. In 1215 AD, King John signed the Magna Carta, which forced him to be bound by law. This major event positioned law over arbitrary decision making and provided a system of due process to all people.

Another major event in community security occurred with the issuance of the Statute of Winchester in 1285 by King Edward I. A system of "watch and ward" was created. This required each householder to take a turn standing guard at entrances to the city, in addition to patrolling the streets during darkness. The statute also required that citizens give a "hue and cry" when crimes were witnessed. It was expected that all citizens would actively seek the capture of the offender.

At this time, several security events began to unfold. Placing guards at gate entrances to walled cities was an early example of today's gated communities. The use of citizen guards and the raising of a hue and cry are early forms of neighborhood watch. Although there are several other historical security events that came about, the Statute of Winchester marks an era of citizen responsibility for community policing.

The Middle Ages was a time when physical security measures were of concern for nobility. Elaborate castles were built on high land and often surrounded by moats. Access was limited to one entrance and controlled by a drawbridge and gate entrance. Narrow vertical slits provided light into castle rooms and protected residents from arrows or rocks, while allowing for a better vantage for protectors of the castle to take aim at marauders.

Castles were built for height and visibility. Attackers could be observed great distances from the castle. For those who reached the base of the castle, hot tar or rocks were dropped upon them. Other buildings and residences also used height and visibility for security. Many residences did not have ground-floor doors or windows. To enter some residences, it was necessary to use a ladder to reach a second-floor entrance. At night, or during times of attack, the ladder would be raised, thereby eliminating easy access to the home.

ECONOMIC CHANGE

Throughout the Middle Ages, societies and economics were primarily agrarian. As communities increased in size, demand was created for small shops and factories. The catalyst for economic growth, however, was the Industrial Revolution, which opened a market for jobs. The timing of this coincided with famine in rural areas, which caused thousands of people to move to town with hopes of employment. The job market became saturated and, with little money for food or shelter, many people turned to crime. This was a disastrous event with which communities were not prepared to cope.

With the growth of small business came a change in the distribution of wealth. A middle class of merchants emerged and began to take issue with mandatory community watch services. They insisted on having the opportunity to pay others to take their places. Unfortunately, this method of purchasing a "security service" was not successful and did not provide protection for homes and businesses. As a result, dissatisfied merchants began to employ individuals whose specific responsibilities were to protect their business property. This was an early form of proprietary business security or "merchant police."

Crime rates continued to increase to the extent that private citizens were carrying weapons for protection. Reform was about to occur when, in 1748, Henry Fielding was appointed magistrate for Middlesex and Westminster, England. Fielding's concept of combating crime was to prevent it by early intervention through a form of policing action. This concept—the basis for modern crime prevention—suggested the following five policing strategies.

1. Policing was a municipal function.
2. There was a need for well-paid men.
3. Some form of mobile patrol was needed to protect the highways.
4. Separate "runners" were needed to move swiftly to the scene of a crime.
5. A separate police court would be instituted.

With the assistance of his brother John, Fielding organized what was to become known as the "Bow Street Runners," an organized effort to investigate crime and apprehend criminals. In spite of the Fielding efforts, crime continued to flourish. The next significant event occurred in 1822, with the appointment of Sir Robert Peel as Home Secretary. Peel's contributions to modern-day policing and security were so significant that he has been referred to as the father of modern law enforcement.

Peel's interest was in creating a strong, unified police force through the employment of full-time, uniformed personnel. He advocated preemployment screening and military principles. His concept of policing also included interaction with the community through distribution of crime information and "beat walkers." Peel's first headquarters was located on a London street called Scotland Yard and his officers eventually became known as "bobbies," a name that still is used today.

SECURITY DEVELOPMENTS IN THE UNITED STATES

As England was experiencing its challenges with increasing crime rates, so was the United States. As communities became more urbanized, citizens' expectations of crime control became a daily concern. During the mid-1850s, security and policing in the United States began its formal development. In 1844, New York City was one of the first communities to organize a police department. Soon to follow were the cities of Chicago in 1850, Cincinnati in 1852, Philadelphia in 1855, and Detroit in 1865. Although the cities developed a formal structure for policing, there remained a demand for dedicated security and investigative services from businesses.

In the late 1880s, Allan Pinkerton was one of the first to form a private security agency. Pinkerton used the investigative skills he developed as a Chicago police detective to provide investigative services to railroad companies. Those services included the protection for the railroad, as well as the investigation of criminal activities that occurred on railroad property.

Pinkerton provided a contractual security service for clients in need of protective services not offered by the police. Soon to follow were other firms including Brinks Armored Car

Service (1889), Burns Investigative Agency (1908), Wells Fargo and Company (1852), Holmes Protection, Inc. (1858), and American District Telegraph (1874). The era of modern security contracting service had begun.

It is important to consider why private security developed in a time when community police departments were organized.

- Security services not limited by jurisdiction could provide dedicated service to a client; whereas the responsibility of police mainly focused on community issues.
- Early policing did not have the financial incentives to apprehend criminals as did private security personnel working for a client.
- With the growth of policing, many departments were not prepared to provide the training necessary for officers to investigate crimes and apprehend criminals.

Such concerns of business are evident today with the continued growth of private security services.

PRIVATE SECURITY TODAY

Through the 1960s, business security efforts in the United States were a combination of physical security and guard service systems. Gradually, two world wars and the Korean and Vietnam conflicts began to influence private business. Government and military facilities were designed to receive maximum protection from saboteurs. Employees were informed of their responsibilities not to discuss their work with others. Military techniques, many of which were continuously tested by special force units, were adopted by public and private agencies. The United States Army Field Manual (FM 19-30) on physical security quickly became a security standard for the physical security practitioner.

In the early 1970s, a significant monograph, Architectural Design for Crime Prevention, examined the relationship between crime, vandalism, and building decline to the perception or fear of crime. A conclusion of this study was that the physical construction of the residential environment can influence how occupants actively participate in providing for their own security. By incorporating "defensible space" design features, which include natural means (such as sight surveillance) for ensuring safety, residents will actively observe and report crime or disorderly behavior to the police—thereby taking charge of their residential areas (Newman 1973).

This report followed an extensive interview by the President's Commission on Law Enforcement and the Administration of Justice, which reported that "fear of crime has led over 50% of citizens to radically change their life-styles." That behavior change included: relocation to safer neighborhoods, avoidance of strangers, and avoidance of nighttime activity. Essentially, the residents gave up their personal control of the neighborhood. Many communities attempted to look for solutions, without great success, by adding more police activity and security service (President's Commission on Law Enforcement and Administration of Justice 1968).

Historically, legislation is often the vehicle used to force employers to provide a safe place for workers. With the passage of the Occupational Safety and Health Act of 1970, a national standard for safety was created through federal legislation. Although the Act contains no specific security regulation, it does, under the General Duty Clause, require employers to provide a safe and healthful workplace (OSHAct 1970).

An extension of the General Duty Clause to include security violations occurred in 1993 when a $5,000 penalty was proposed against an Illinois hospital because the OSHA inspector alleged the hospital failed to protect people on the premises from violence. In 1996 this was followed by a series of guidelines to reduce violent incidents at businesses during evening hours. These events seem to follow the historical precedent that, in the absence of proactive measures by businesses, mandatory regulations will be enacted.

In 1975 one of the first major studies focusing on security in the United States was undertaken. An objective of the report, *Private Security Standards and Goals*, was to "attempt to codify standards dealing with all of the divergent persons and things associated with private security." The report attempted to analyze proprietary security, contract security, and other security services such as alarm systems and investigative services. The task force identified more than 80 professional standards and goals for the security industry (Task Force on Private Security 1977).

Another major series of studies came with the Hallcrest Reports, which examined the relationship among law enforcement, private security, and community interests. Hallcrest II identified an improving relationship between private security and public law enforcement and noted that both sectors participated in numerous cooperative programs. This partnership not only serves to provide opportunities for combining technology and resources but also enhances the overall protection services each industry provides to its constituents (Cunningham, Strauchs, and Van Meter 1990, 1991).

Because the United States government outsourced contracts to business and industry, private security applications became important. Nuclear arms, space exploration, and economic development became concerns of the "super power countries" as each tried to maintain a competitive advantage over another.

This "cold war" atmosphere continued until the late 1980s when Russian glasnost appeared and the Berlin wall disappeared. While countries appeared to improve relationships with each other, economic espionage by foreign governments became a concern for domestic business. Computer and communications technology became prime targets for theft. In addition, white-collar crime added to the woes of private business. Payoffs, bribery, extortion, and electronic eavesdropping demonstrated the extent to which companies would use illegal means to gather competitive intelligence.

Even with these challenges, most businesses were not prepared for legal litigation associated with workplace violence. The growth of premises liability lawsuits from the 1970s to the 1990s was phenomenal. The catalyst for this growth was a 1974 premises liability lawsuit involving a claim of inadequate physical security. This case, which received extensive media attention, involved a well-known entertainer. The assailant was able to gain entry to the victim's room through a sliding glass door that was determined by the jury to have been inadequately secure. The victim filed suit against the Howard Johnson Motor Lodge and was awarded $2.7 million.

In September of 1993 the U.S. Department of Health and Human Services for Occupational Safety and Health (NIOSH) published an "alert" notice regarding workplace homicide. This document, *Preventing Violence in the Workplace*, stated that from 1980 to 1989 homicide was the third leading cause of death from injury in the workplace, accounting for 12% of all such deaths and the leading cause of death from occupational injury among women (NIOSH 1993).

The purposes of disseminating this alert were to:

- identify high-risk occupations and workplaces
- inform employers and workers about risk
- encourage employers and workers to evaluate risk factors in the workplace and implement protective measures

- encourage researchers to gather more detailed information about occupational homicide and to develop and evaluate protective measures (NIOSH 1993)

This awareness infers a responsibility businesses have to implement protective measures and inform employees about workplace crime risks.

In a 1996 study, funded by the National Institute of Justice, researchers found that

> victims of crime are seeking compensation from the owners and managers of properties on which crime takes place with increasing frequency. In these court cases, commonly known as premises liability cases, juries are being told that crime was the result of a perpetrator's ability to take advantage of a lack of security in a certain building or property (Gordon and Brill 1996).

The authors of the study emphasized the following causes for this increase: "crime victims, in part due to the liberating nature of the victims rights movement, are now more likely to seek restitution for what happened to them"; and many courts "have become increasingly willing to let juries hear theories about the relationship between how properties are managed and designed and criminal behavior" (Gordon and Brill 1996).

The implication is, of course, that facility managers, security directors, and business owners need to be cognizant of the human and physical changes that may occur not only on "owner controlled" property but on neighboring property as well. Throughout history, it has been documented that crime will flourish in areas where property is left to deteriorate and where disorderly behavior is uncontrolled. Many communities unceremoniously provide visual evidence of once thriving business zones turned into vacated buildings with graffiti, broken windows, and garbage filled entrances and parking lots.

However, the incident that has had the most profound impact on the Unites States was the terrorist attack of September 11, 2001. Not only did this incident awaken the United States to its vulnerability to attack by foreign dissidents, but it also put businesses on notice that business operations outside of the United States may also be vulnerable to criminal activity. So dramatic was this event that the Department of Homeland Security was created by the Homeland Security Act of 2002. This action brought 22 federal agencies into one department. It was the largest such restructure of the federal government since the Department of Defense was created in 1947. With the creation of the Department of Homeland Security came the need for defining "homeland security." In its 2007 National Strategy for Homeland Security, the department defines homeland security as: "a concentrated national effort to prevent terrorist attacks within the United Sates, reduce America's vulnerability to terrorism, and minimize the damage and recover from attacks that do occur (U.S. Department of Homeland Security 2007).

Even more definitive is the department's mission statement and security strategic plan for fiscal years 2008–2013, which states in part: "We will lead the unified national effort to secure America. We will prevent and deter terrorist attacks and protect against and respond to threats and hazards to the Nation. We will secure our national borders while welcoming lawful immigrants, visitors, and trade (U.S. Department of Homeland Security 2007, 3). The philosophy of "protection" is one that should be emulated by businesses and agencies.

THE FUTURE OF SECURITY

Statistics can provide important information regarding crime trends and the public's attitude toward crime. However, without proper analysis, they can be misleading. For example, a community experiences 100 violent crimes in a previous year and experiences a reduction by 10 in the present year. News reports will emphasize a 10% reduction in violent crime and imply that the community is safer. Some will interpret this to mean a downsizing in police and security personnel is in order. Businesses with an eye on the profit line may actually reduce their commitment to security.

While a decrease in crime should be applauded, it does not mean that crime is eliminated. It is important to understand and appreciate that, in the example used, the community or business still had 90 violent crimes. Any one violent crime may be disastrous for a business (negative publicity, financial loss, employee morale). Facility managers, security and safety personnel, and business owners need to develop, implement, and continuously monitor a zero crime tolerance policy. A commitment to the safety and security of workers and other property users must be an environmental norm.

As gated communities continue to be developed, so will gated businesses. The quality of community life and demographics will be strongly reflected in the work force. Communities will always be challenged by criminal activity.

It is projected that "if trends continue as they have over the past 10 years, juvenile arrests for violent crime will more than double by the year 2010" (Snyder, Sickmund, and Poe-Yamagata 1996). Eventually, many of these violent juveniles, whose records will not be available to future employers, will transition into the job market and be hired by unsuspecting employers.

The future of security does profile that of law enforcement. Both are necessary to maintain social order and to provide protection services to their respective constituents. However, security will continue to outpace law enforcement in personnel and expenditures. The ability to provide client customized prevention services and charge a user fee allows security firms to expand with demand. Law enforcement, restricted by available community tax revenue, is responsible for multiple clients (citizens). The enforcement of community laws and ordinances differ greatly than the enforcement of policies for a client. There is little opportunity for police to provide client-directed, rather than community-directed, protection services. Therefore, security will continue to grow because of its ability to provide direct user-friendly services. Clients will, however, demand that the protection service be provided by qualified, educated, and trained personnel, not the traditional fire watch guard. With the appreciation that the United States is not immune from either domestic or foreign terrorism, the sophistication of security services and service provider knowledge will be expected and demanded.

A 1991 National Institute of Justice study determined that private security outspends "public law enforcement by 73 percent and employs 2½ times the work force." The respective figures are $52 billion and 1.5 million employees for private security and $30 billion and 600,000 employees for the law enforcement sector. Today that figure has escalated because of need and the fact that many government and private businesses are outsourcing services.

SUMMARY

As the professional security field began its slow development in the 1950s, there were few organizations or academic programs dedicated to security. Today, there are numerous opportunities for associate, baccalaureate, or graduate degrees in security or loss prevention. Several security organizations are available for professional networking. One of the most prominent, the American Society for Industrial Security International has more than 37,000 members and more than 200 chapters worldwide. More than 20,000 security professionals attend its annual seminars and exhibits. Such growth in the security field has not gone unnoticed. It has drawn attention to the need for public/private partnerships. Instrumental in the effort, with assistance from such organizations as the American Society for Industrial Security International and the International Association of Chiefs of Police, is the U.S. Department of Justice's Office of Community Oriented Policing Services (COPS). This topic is discussed in depth in Chapter 25, A Public–Private Partnership on Security.

In addition to professional advances in security, the technological innovations have been noteworthy. Systems now incorporate sophisticated intrusion detection sensors, laser technology, biometric access controls, communications, and cyber-security technology—all necessary ingredients for a security program faced with issues such as workplace violence, computer crime, employee theft, drug abuse, and possible terrorism threats, whether they be foreign or domestic in nature. Individuals willing to make a commitment to the security field will find it to be a challenging but rewarding and honorable career.

To appreciate security and loss prevention, a review of history is warranted. Virtually no society was without some form of security. The same can be said today. The same can be said for future societies.

REFERENCES

Cunningham, William C., John J. Strauchs, and Clifford W. Van Meter. 1991. *Private Security: Patterns and Trends,* Washington, DC: National Institute of Justice Research in Brief.

——. 1990. *The Hallcrest Report II: Private Security Trends 1970-2000*, Boston, MA: Butterworth-Hinnemann.

Germann, A.C., Frank D. Day, and Robert Gallati. 1971. *Introduction to Law Enforcement and Criminal Justice*, Springfield, IL: Charles C. Thomas.

Gordon, Corey L., and William Brill. 1996. *The Expanding Role of Crime Prevention Through Environmental Design in Premises Liability*, Washington, DC: National Institute of Justice.

National Institute for Occupational Safety and Health. 1993. *Alert: Request for Assistance in Preventing Homicide in the Workplace*, Cincinnati, OH: U.S. Department of Health and Social Services, 93-109.

Newman, Oscar. 1973. *Architectural Design for Crime Prevention*, Washington, DC: National Institute of Law Enforcement and Criminal Justice.

OSHAct. 1970. *General Duty Cause.* Section 5, Duties, accessed on January 2010. http://osha.gov/pls/oshaweb/owadisp.show_document?p_table=OSHACT&p_id=3359.

Post, R.S., and A.A. Kingsbury. 1977. *Security Administration: An Introduction.* Springfield, IL: Charles C. Thomas.

The President's Commission on Law Enforcement and Administration of Justice. 1968. *The Challenge of Crime in a Free Society*, Washington, DC: National Institute of Law Enforcement and Criminal Justice.

Roper, Carl A. 1983. *The Complete Handbook of Locks and Locksmithing*, Blue Ridge Summit, PA: TAB Books.

Synder, Howard N., Melissa Sickmund, and Eileen Poe-Yamagata. 1996. *Juvenile Offenders and Victims: 1996 Update on Violence*, Washington, DC: Office of Juvenile Justice and Delinquency Prevention.

Task Force on Private Security. 1977. *Private Security Standards and Goals*, Cincinnati, OH: Anderson Printing Co.

U.S. Department of Homeland Security. 2007. *National Strategy for Homeland Security*, Washington: DC: U.S. Department of Homeland Security.

U.S. Department of Homeland Security. *One Team, One Mission, Securing Our Homeland Strategic Plan: Fiscal Years 2008-2013*, Washington: DC: U.S. Department of Homeland Security.

BIBLIOGRAPHY

Germann, A.C., Frank D. Day, and Robert Gallati. 1971. *Introduction to Law Enforcement and Criminal Justice*, Springfield, IL: Charles C. Thomas.

McGoey, Chris E. 1990. *Security Adequate…or Not?*, Oakland, CA: Aegis Books.

Roper, Carl A. 1983. *The Complete Handbook of Locks and Locksmithing*, Blue Ridge Summit, PA: TAB Books.

Smithsonian Institute. 1993. *Timelines of the Ancient World*. New York, NY: Dorling Kindersley.

2

Security Issues and Trends

Mike Cummings, HEI, AHC

CHAPTER 2

OBJECTIVE

The objective of this chapter is to explore where security has come and, more importantly, the emerging trends in security professions, especially since the terrorist attack of September 11, 2001.

INTRODUCTION

Since the first edition of this book was published in 1997, there has been a significant change in the security profession and in the duties and responsibilities of those responsible for security functions. Today, those responsible for security functions must be capable of negotiating through many disciplines and subdisciplines related to security. Being a solid generalist exclusive to the practice of security is challenging enough, but being one as an adjunct to other, equally complex disciplines, such as safety, can be challenging. The complexity of the security discipline has shifted dramatically in the post–9/11 era.

The previous chapter looked at the long and storied history of security, from an introduction to Allan Pinkerton (a pioneer of the private security agency) to a greater reliance by communities on private security because of fiscal constraints within the public sector (see Chapter 1, Historical Perspectives). This chapter will look at a snapshot of what the expectations are for the security industry today, explore the emerging challenges and complexities, and provide some solutions to issues that must be addressed to keep pace.

There is sufficient documentation to support the notion that private security will continue to fill the gaps in tasks and responsibilities that in the past fell under the purview of public law enforcement. Increased criminal activity within the workplace—whether directed toward the organization or individuals within it—will ensure full employment for security practitioners for years to come. Depending on how well a security professional understands the needs and expectations of an organization can mean a heightened respect for the security profession as well as for the organization.

NEW CHALLENGES

Since the terrorist attack on September 11, 2001, a number of new challenges have emerged or taken on new prominence relating to what *security* looks like. The fact that the federal government created a new federal agency, the Department of Homeland Security (DHS), speaks volumes about the impact of that event. The "ripple effect" of that event has led to greater and ongoing security awareness in day-to-day operations. Several areas in particular have been impacted, including:

- regulatory and legislative developments
- guidelines and standards
- emphasis on credentialing
- understanding the business of security
- networking and partnerships
- technology

REGULATORY AND LEGISLATIVE DEVELOPMENTS

Naturally, the topic of security has received ongoing focus, scrutiny, and attention in the wake of 9/11. Beyond the growing role and responsibility of the private security industry in the United States has been an increase in regulatory and legislative developments at the local, state, and federal levels.

A movement to bolster the licensing of law enforcement officers at the state and local levels has led to legislative mandates for increased training requirements in some jurisdictions. Similarly, numerous regulatory agencies in various specialty areas have increased requirements for security programs. Regulations for increased security in banking, health care, port security, nuclear security, and border security, to name a few, have all seen new regulations and standards. It is important that security practitioners and educators stay current and informed about the rapidly changing legislative developments in this area. One way to do this is for security professionals to develop systems, individually or collectively, or through professional associations, to monitor and become involved with regulations regarding security. Professional associations, such as the American Society for Industrial Security (ASIS) International, can be accessed online at www.asisonline.org.

GUIDELINES AND STANDARDS

In addition to increased licensing requirements for law enforcement professionals, there has also been a proliferation of guidelines and standards within and for the security profession. In the past, there was the general belief that it was in everyone's best interest to allow security operations to function with minimal standards. In recent years there has been a pendulum swing in the opposite direction. A number of new standard-writing bodies have emerged, specifically for the security industry, such as ASIS International. Other agencies, such as the National Fire Protection Association (NFPA), that have written standards in other areas for years, have migrated to security standards writing.

The challenge for those charged with managing or directing security operations is to determine which standards or guidelines to follow. The security professional must have a clear understanding of how the standard-writing process works to make informed decisions about what standards apply. It is also helpful for the security professional to understand what it means, for example, for a standard to be certified by the American National Standards Institute (ANSI).

CREDENTIALING

How does one differentiate himself in the workplace as a true security expert or professional? One way is to seek accreditation in the security field. Like standards, accreditation programs have flourished in the past decade. People new to the security industry, and those who have practiced for many years, are well served to determine which credentials are important to their personal growth as well as which credentials demonstrate an expertise that is meaningful to an employer.

Not all security positions are created equal and not all require the same level of expertise for success. With so many options available, it is important that the time, effort, and money needed to achieve certification is not squandered on meaningless certifications. For someone acquiring security responsibilities as an adjunct to other duties, a good starting place is to match the certification being considered with the job description and a frank discussion with a security superior to ensure that the certification matches closely with the skills and attributes needed for success.

For example, if the security role has limited expectations for performing investigations related to the commission of fraud, but greater responsibility for performing research, recommending, and implementing security access systems, one might not get the best use of a certified fraud examiner (CFE) certification or a professional certified investigator (PCI) certification. Both are excellent certifications, but in this scenario, a physical security professional (PSP) certification is likely the designation to seek.

One should bear in mind that the certification not only sends a message to a potential employer about level of proficiency, it also provides the individual with a measure of knowledge about the subject matter. Practitioners need to be prepared for changes in certification requirements.

UNDERSTANDING THE BUSINESS OF SECURITY

Understanding the business of security is no different than understanding the business in which we practice safety. Security professionals must be able to bring value to the organizations in which they work. They must understand that they are not in business to have a security department, but rather to provide a product or service, much like health care or banking services. Security professionals must work to apply security concepts and best practices within the business environment in such a way that is financially beneficial to a company or organization. Often this implies that money is more important than security; quite the contrary. Security professionals must be able to demonstrate the value of security services, technology, practices, or procedures in terms that can be understood by business, such as *return on investment* (ROI).

The timeline shown in Figure 2–1 is a simple, yet powerful display of how private security has evolved in the United States over the past half century. Although this graph is more representative of how the profession has changed at the higher level of security positions, it is also applicable to all levels of security and illustrative of where the profession is heading.

Figure 2–1. Evolution of Security in the United States. *Source: M. Blades, "Successful Organizational Leadership," Security Magazine, July 2009, pp. 22–28.*

NETWORKING AND PARTNERSHIPS

Doing more with less has been the rallying cry for some time in the security profession. The recent downturn in the economy has forced many businesses, including security, to work smarter. One effective way to work smarter is to create partnerships and realize that we do not have to do it all alone. This can be simple or it can require "outside-the-box" thinking to have the greatest impact.

Partnerships can come from teaming with security departments from other organizations or within an organization, joining professional organizations, or networking with public law enforcement in your community for external assistance. Partnering within an organization benefits everyone because when everyone has a stake in security, the outcome is overall improved security. In this case, coalition building and senior leadership support is critical. Looking at who benefits from a requested change in policy or implementation of new technology, can be an incentive to succeed. Selling the human resources manager on the benefits of improved criminal background screening as a way to mitigate violence in the workplace could help move such an initiative quickly. Externally, partnerships with like-minded security professionals, or more formal arrangements through organizations that allow for blogs, policy sharing, and educational networking, are obvious in their benefit to a security program.

TECHNOLOGY

It is not the intent to discuss any specific security technology or applications in this chapter. However, any discussion related to trends in security practice would be incomplete without mentioning the positives and negatives that technology presents. Whether it is access control equipment, closed-circuit television (CCTV) systems, infant protection programs, or point-of-sale (POS) systems that interface with security operations, never before have security professionals had access to as much technical equipment to help them do their jobs.

The volume of options and the complexity of the equipment, however, can be a curse for the ill-informed security practitioner. All security systems require careful consideration before purchase and implementation. Applicability of a system to the security problem at hand, the cost, the expected life span, and the compatibility with other systems must be considered before purchase. The company culture and policies must also be considered when choosing technology for a security program. Keep in mind that technology represents the potential for both great success and great failure. The difference in outcome can be determined by the well-informed security practitioner.

SUMMARY

In light of the elements discussed above, one can see that the security profession is heading in a new direction in terms of what it takes to be a successful practitioner. For anyone inheriting or assuming security as an adjunct to other duties and responsibilities, it is important to consider emerging guidelines and standards, new technology, and which credentials to pursue. The application of strategies and tactics to achieve success will continually grow more difficult and costly. It is basic and fundamental to embrace these changes as part of an individual and organizational approach to security.

Part 2

Security and the Law

PART 2

3

Concepts of Security Law

Norman M. Spain, JD, CPP

CHAPTER 3

OBJECTIVE

The primary legal risks that security professionals can encounter when developing and managing a security program vary, depending upon the laws of the state, type of business or institution, and security services provided. The objective of this chapter is to explore some of the primary legal risks involved for safety professionals responsible for the security function.

INTRODUCTION

The overall goal of private security often is defined as the protection of people, property, and information of a company or institution. To achieve this goal, security departments provide:

1. physical security for protection of people and property
2. information security for protection of proprietary data
3. investigative security services

Each of these functions involves certain legal considerations and risks, and although a security manager may be involved with more than one function, the most likely function is overseeing the physical security of a facility.

PHYSICAL SECURITY LAW

Physical security law can be divided into the four following functional topics:

- protection of people
- protection of property
- use of force
- labor strikes

PROTECTION OF PEOPLE

The primary goal of many security programs is the protection of people from acts of physical violence. Providing a reasonably safe environment from crime for employees and nonemployees makes good business sense. An unsafe workplace can make employee recruitment difficult; result in high employee turnover and low productivity; and drive away nonemployees, such as customers, clients, and students. A failure to provide reasonable security can also under certain circumstances result in corporate or institutional liability.

The victims and perpetrators of violence are either employees or nonemployees. This business relationship can determine the liability of a company; therefore, a discussion on the legal duty of businesses to protect people from acts of violence is divided into the following topics: nonemployees as victims and perpetrators, employees as victims, and employees as perpetrators.

Acts of Violence against Nonemployees. Typically, a business owes no duty to protect people on its property from the criminal acts of third parties. However, if a business has a history of prior activity, prior acts of violence, or other similar criminal activity, it may create a scenario for which a court will impose upon the business a reasonable duty to protect nonemployees on site (Restatement [Second] of Torts § 344 [1977]. *Madden v. C & K Barbecue Carryout, Inc., and*

Decker v. Gramex Corporation, 758 S W 2d 59 [Mo banc 1988]). As a general rule, the more recent in time, the more frequent, and the more serious the prior acts of crime on the premise and surrounding areas, the more likely a court is to find a business has a duty to take reasonable precautions to protect against foreseeable criminal assaults committed by third parties on the premise. In a few jurisdictions this duty to protect may extend to adjacent property that the business may not own or control but uses regularly in the course of business; for example, a common entrance to a facility (*Banks v. Hyatt Corp.*, 722 F.2d 214 [5th Cir. 1984]) or adjacent parking areas (*Southland Corp. v. Superior Court*, 250 *Cal Rptr* 57 [Cal App 2 Dist 1988], and *Holiday Inns, Inc. v. Shelburne*, 576 So 2d 322 [Fla App 4 Dist 1991]).

Moreover, if the victim of a violent crime had a relationship with the business wherein the victim was dependent on the business to provide reasonable security, the court can impose such a duty on the business even if there is no history of prior similar acts. For example, a sedated patient who was raped was permitted by the Alabama Supreme Court to sue a hospital for negligent security even though the hospital had experienced no prior similar crimes (*Young v. Huntsville Hospital*, 595 So 2d 1386 [Ala 1992]).

Acts of Violence against Employees. When an employee is assaulted on the job, the first issue to determine is whether the employee is entitled to workers' compensation benefits. A compensable injury is one that arises out of and during the course of employment (see, e.g., N C Gen Stat § 97-2[6]).

Acts of violence in the workplace can be divided into the following three categories: occupational, neutral, and personal. The states agree that employees injured as a result of occupational risks (such as an officer injured while frisking a shoplifter) are entitled to benefits (*Domanoski v. Borough of Fanwood v. The Great Atlantic & Pacific Tea Company*, 568 A 2d 123 [N J Super A D 1989]). Further, employees injured at work as a result of personal risks (e.g., an employee shot at work by a distraught suitor) are not entitled to benefits (*Kendrick v. Peel*, 795 S W 2d 365 [Ark App 1990]).

State laws differ as to whether so-called *neutral risks*, an employee in the wrong place at the wrong time, are covered by workers' compensation. There appears to be a trend by the courts, using a positional risk analysis test, to award benefits if the employee's position exposed him to a risk of violence to which he would not have otherwise been exposed; for example, a store clerk shot by a man on a random shooting spree (*K-Mart Corp. v. Novak*, 521 N E 2d 1346 [Ind App 1 Dist 1988]). Further, nonsecurity employees who risk their own personal safety to help fellow employees (e.g., an off-duty employee returning to assist a supervisor being assaulted in a company parking lot) might also be covered by workers' compensation (*Brind v. Leasing Corp. v. Workman's Compensation Appeal Bd [Dougherty]*, 584 A 2d 1102 [Pa Commw Ct 1990]). But nonsecurity personnel who chase robbers fleeing the scene may be doing so at their own peril. For example, a delivery driver shot in pursuit of robbers who had fled without any company property was denied benefits (*Alder v. Mid-South Beverages, Inc.*, 783 S W 2d 541 [Tenn 1990]).

Employees who are assaulted on their way to and from work may or may not be covered by workers' compensation. For example, the Minnesota Supreme Court has twice found that employees assaulted in or abducted from company parking lots were covered by workers' compensation (*Bear v. Honeywell Inc.*, 468 N W 2d 546 [Minn 1991] and *Foley v. Honeywell, Inc.*, 488 N W 2d 268 [Minn 1992]). This same court has also held that an employee who was shot on a public sidewalk away from company property while walking back to work after taking a break was not eligible for workers' compensation benefits (*Gibberd by Gibberd v. Control Data Corp.*, 424 N W 2d 776 [Minn 1988]). In contrast, a Louisiana appellate court held that an elderly contract security officer who was assaulted and seriously injured while walking from a public bus stop to his post in a dangerous part of the city was entitled to benefits on the grounds that his work exposed him to the risk (*Stewart v. Louisiana Plant Services, Inc.*, 611 So 2d 682 [La App 4 Cir 1992]).

If an employee is covered by workers' compensation, the second major question is whether the employee is barred under the exclusive or restrictive remedy provision of the statute from suing the employer for negligent security and other related actions. The trend by the courts is to answer this issue in the affirmative. For example, the Supreme Court of Colorado ruled that a secretary assaulted by a company janitor while she was walking from her office to the employee cafeteria was entitled to workers' compensation and barred from suing her employer for negligent security and negligent hiring (In Re Question Submitted by the U.S. Court of Appeals for the Tenth Circuit, *Tobert v. Martin Marietta Corp.*, 759 p 2d 17 [Colo 1988]). Likewise, the Ohio Supreme Court held that the family of a convenience store clerk who was murdered during a robbery was barred by the state workers' compensation act from suing the employer for intentionally failing to take reasonable security precautions (failing to train the employee to respond in a robbery, failing to install cameras, and failing to take other security precautions) (*Mitchell v. Lawson Milk Company*, 532 N E 2d 753 [Ohio 1988]).

If an injured worker is not covered by workers' compensation, the employee may be able to sue and obtain greater damages against the employer for negligent security and other related torts. However, this is not always the case. In the Louisiana case involving the nurse who was stabbed in the hospital elevator, the trial court held the hospital was not immune against her negligent security lawsuit under the workers' compensation law. It awarded her a $125,000 judgment with an approximate $6,000 set aside for the money she received under workers' compensation. The appeals court reversed and ruled her negligent security claim was barred by the workers' compensation act (*Mundy v. Dep't of Health and Human Resources*, 609 So 2d 909 [La App 4 Cir 1992]). After the supreme court ruled she was not covered by workers' compensation, it reinstated her negligent security claim (not judgment) (*Mundy*, 593 So 2d 346 [La 1992]). Subsequently, she lost the case for failing to introduce sufficient evidence to prove the assault was foreseeable and to establish the hospital breached its duty to provide reasonable security (*Mundy*, 609 So 2d 909 [La App 4 Cir 1992], aff'd, 620 So 2d 811 [La 1993]).

Acts of Violence by Employees. When an employee is the perpetrator of an act of violence, in addition to raising a claim of negligent security, the victim can also argue that the employer was negligent in hiring the employee into a position of trust, was negligent in its supervision and retention of the employee, or a combination of these factors.

The duty of an employer to hire with care depends on the level of trust that a position entails. The more others depend on an employee to perform a job in a reasonable and safe manner, the greater the trust and the greater the duty of the employer to hire with care.

Traditionally, innkeepers and common carriers, such as taxi and bus companies, have had a high duty to hire with care because the public is highly dependent on these employees to perform their duties in a reasonable and safe manner (*Salinas v. Fort Worth Cab and Baggage Co.*, 725 S W 2d 701 [Tex 1987]). In recent years this duty to hire with care has been extended by the courts to other businesses, including but not limited to home sales companies (*McLean v. Kirby Company*, 490 N W 2d 229 [N D 1992]), home delivery personnel (*Tallahassee Furniture Co.*

v. Harrison, 583 So 2d 744 [Fla App 1 Dist 1991]), health care employees (*Fisher v. Beverly Enterprises*, U.S. District Court, E D Ark, No J-C-83-201 29 *ATLA L Rep* 419 [Nov 1986]), and contract security officers (*Kanne v. Burns International Security Service, Inc. Cal.*, Los Angeles Superior Court, No SWC 61883 28 *ATLA L. Rep* 78 [March 1985]).

If an employee demonstrates a propensity toward violence, an employer fails to take corrective action, and the employee subsequently injures someone, the injured party can bring suit against the company for its negligent supervision, retention of the employee, or both. In one case, a company rehired an employee after he was released from prison, where he had been serving a sentence for murdering another employee at the same company. After being rehired and after repeated problems and transfers to different facilities, the employee was terminated. A few days later he murdered an employee at her home for having refused his advances when they worked together. Because of the public policy need for rehabilitation of ex-felons, a Minnesota appellate court held the family of the deceased employee could not sue the company for negligent hiring. But it also ruled that the company could be sued for negligent supervision, given the perpetrator's behavior at work after he was rehired (*Yunker v. Honeywell, Inc.*, 496 N W 2d 419 [Minn App 1993]).

As a general rule, employers cannot be held liable for subsequent acts of violence committed by ex-employees. However in recent years, two arguments have been raised by plaintiffs in an attempt to carve out exceptions to this rule:

1. The company was negligent for failing to warn subsequent employers of the employee's known dangerous propensities.
2. The company misrepresented the employment status of the former employee.

Although relatively new and not widely tested or accepted, managers responsible for the security function should be aware of these emerging risks.

In a *failure to warn* case, the victim will claim the former employer knew of the dangerous propensities of the ex-employee and had an affirmative duty to warn the subsequent employer. Had this been done, the victim will further submit, the assault would not have occurred because the assailant would never have been hired and would never have had the opportunity to attack the victim. For example, an appellate court in Washington ruled that a former employer could be sued for failing to warn that an ex-employee hired as a youth and adult drug counselor was a known pedophile (*Does 1-9 v. CompCare, Inc.*, 763 p 2d 1237 [Wash App 1988]). But in an opinion more reflective of the prevailing view, an appellate court in Michigan rejected a failure-to-warn claim in a case involving the murder of a security officer by a contract maintenance employee who previously had been terminated for 24 disciplinary actions, including drug use and violence. The court noted to impose such a duty might result in open employment records, it might permit disgruntled employers to harass competent former employees, and it might increase the risk of defamation lawsuits against employers. Nonetheless, the court did state, "we may have reached a point where people should make this type of information known, we restate our belief that this is a substantive change in our laws, the type of change best left to our Legislature" (*Moore v. St. Joseph Nursing Home, Inc.*, 459 N W 2d 100, 103 [Mich App 1990]).

Misrepresentation is the second emerging issue. In a pending case, a trial court ruled that a former employer can be sued on the grounds that it misrepresented the reason for an employee's termination. The employee, who had threatened female co-workers, thought he was an alien and was a known devil worshipper. He was discharged when he brought a concealed weapon to work. The former employer, in a negotiated discharge, wrote a letter stating the termination was not related to the employee's job performance. The employee was hired by another company that subsequently discharged him. In retaliation, the employee returned to the offices of the second employer and shot several employees. The victims and their families were permitted by the court to bring a lawsuit against the former employer on the ground that it misrepresented the status of the ex-employee (*Jerner v. Allstate Ins. Co., Employment Coordinator*, Sept. 7, 1995, at 1 & 4 [Fla Circuit Court, No 93-09472, Division F]).

Managers need to recognize that the more others depend on an employee for their welfare and safety, the greater the duty of the employer to conduct a reasonable background investigation (see section on investigative law). The duty to hire with care often includes entry-level employees in basic skill positions (hotel maids, apartment maintenance workers, home delivery personnel, nurse's aids, etc). Managers also have an obligation to supervise employees so that they perform their duties in a reasonable and safe manner. Employees who engage in intentionally dangerous or reckless behavior should be counseled, retrained, and disciplined, up to and including termination when appropriate.

Progressive companies now are training managers to recognize behavior patterns of potentially violent employees and providing them guidance on how to counsel and discipline these employees. Further, such companies identify the corporate support that is available to managers when dealing with such employees (the company may provide additional security or send a representative from a remote office to handle the discharge). Finally, when discharging potentially violent employees, managers need to recognize and avoid the emerging risk of providing these employees with letters of reference that misrepresent the employees' job performance.

PROTECTION OF PROPERTY

Excluding the transportation, storage, and hospitality industries, lawsuits against businesses for theft and vandalism of property are not often reported. Transportation and storage companies do have contractual obligations to return or deliver the property in the same condition in which it was entrusted. Failure to do so can expose the corporation to traditional negligence liability, liability under the Uniform Commercial Code, or both. However, these risks are usually standard or otherwise known and specified by the parties, contractually limited, and/or covered by insurance.

The hospitality industry also is vulnerable to lawsuits for lost, stolen, and damaged property. These cases usually involve alleged theft of property left in guest rooms and vehicles. But because of the potential for fraudulent guest claims, the most states have statutes that restrict the maximum liability of innkeepers (see, e.g., *Ark Code Ann* § 20-26-302 [1975], and *Cal Civil Code* §§ 1859 and 1860 [1985]). These statutes are usually upheld by the courts (see, e.g., *Nova Stylings, Inc. v. Red Roof Inns, Inc.*, 747 p 2d 107 [An 1987], and *Duvall v. The Ritz Carlton Hotel Co.*, 946 F 2d 418 [5th Cir 1991]); however, they may not protect innkeepers in every circumstance. For example, the Arkansas Supreme Court held that a hotel can be liable for the full value of jewelry stolen from a guest's room, $110,000 versus the few hundred dollars set by statute, if the hotel did not post proper notices of its limited liability as mandated by the innkeeper's statute (*H.D. Grimes v. M.H.M., Inc.*, 776 S W 2d 336 [Ark 1989]). Similarly, the Nevada Supreme Court ruled that the statute limiting innkeeper liability for stolen property from guest rooms was not applicable to the theft of a van and its contents from a hotel valet parking lot (*Tienda v. Holiday Casino, Inc.*, 853 p 2d 106 [Nev 1993]). In contrast, the Supreme Court of Kansas, in an advisory opinion, affirmed a hotel was liable for only $250 under its state statute for jewelry valued at $650,000. The jewelry was taken from the hotel office but the guest did not, as required by law, declare the value of and itemize the items he left in the care of the hotel (*Nova Stylings, Inc. v. Red Roof Inns, Inc.*, 747 p 2d 107 [Kan 1987]).

Museums and galleries also have liability exposure, by severity not frequency, for acts of vandalism and theft. If exhibits on loan are vandalized or stolen, the host museum or gallery can be subject to lawsuit for failure to provide reasonable security and breach of contract to protect and return the property to its owner (*Colnaghi, U.S.A., Ltd. v. Jewelers Protection Services, Ltd.*, 595 N Y S 2d 381 [Ct App 1993]). These risks are traditionally identifiable and insurable.

Hospitals also need to develop patient property procedures to reduce the risk of patient ill-will and lawsuits over lost or stolen property. A procedure designed to encourage patients to leave their valuables at home, coupled with a process to collect and safeguard it in the hospital safe if patients bring valuables with them, will reduce the risk of lawsuit. If patients insist on keeping their valuables with them, they should be requested to sign a waiver that will minimize the exposure of the hospital.

Lawsuits against other types of businesses for stolen or damaged property are reported on occasion. For example, a New York court held a department store liable for the theft of a customer's purse, which held $531 in cash, since the store accepted only cash and was having a problem with theft of cash from its registers (*Lacelle v. Hills Department Store*, 553 N Y S 2d 1014 [City Ct 1988]). But another New York court ruled that a bank was not liable for the theft of cash from the purse of a customer in its lobby, even though the customer had just made a cash withdrawal (*Urgo v. Jamaica Savings Bank*, 546 N Y S 306 [N Y City Civ Ct 1989]). Similarly, an Ohio court held a bank was not responsible for alleged theft of jewelry from a safe deposit box, absent proof the box had been entered or the bank violated its access procedures (*Moynihan v. Fifth Third Bank of NW Ohio*, NA, No L-94-097, Court of Appeals of Ohio, [Lucas County, Oct 28, 1994]). But a New Jersey court held that an employer can be liable for damage to an employee's car based on prior incidents and the fact that employees were required to park outside the security fence where company vehicles were parked (*Berrios v. United Parcel Service*, 627 A 2d 701 [N J Super L 1992]). Even in light of this case, when employees claim their cars were vandalized or broken into while on company property, employers should determine the authenticity of these claims as it is very easy for employees to have their cars damaged off site and seek payment from their employers.

In general, excluding industries that transport, store, or exhibit property, the frequency and severity of the risk of lawsuit for theft or vandalism to property appears to be modest and manageable for most companies.

USE OF FORCE

Security managers responsible for physical security need to understand when security officers can and cannot make arrests, when they can and cannot use force, and how much force they can use. The illegal detention of another or the unauthorized or excessive use of force can subject the security officer and corporation to liability for false arrest and false imprisonment, assault and battery, defamation of character, malicious prosecution, and infliction of emotional distress.

Felonies, Misdemeanors, and Violations. Criminal conduct can be divided into three categories: *felonies, misdemeanors*, and *violations*. *Felonies* include serious offenses wherein the penalty for incarceration ranges from one year to life imprisonment, or death. *Misdemeanors* are those acts wherein the penalty for incarceration is less than one year. *Violations*, also known as *summary offenses and infractions,* are wrongful acts that are less serious than misdemeanors. Violations can be proscribed by state and local governments and are often not considered crimes (speeding or violating a city curfew). The typical penalty for a violation is usually a fine, but it can include a jail sentence usually not to exceed 90 days (see, e.g., 18 *Pa Cons Stat* § 106 [1990]). Safety managers need to understand these categories in their respective states because the severity of the crime determines the authority of private security officers to make arrests.

Arrest Authority of Security Personnel. Unless state statute provides differently, a private security officer has the same authority as a private citizen to make an arrest. This authority differs by state. As a general rule, a private security officer can make an arrest when the suspect has in fact committed a felony; a felony has in fact been committed and the security officer has reasonable grounds to believe the suspect committed it; the suspect is committing a felony in the presence of the security officer and the officer immediately makes the arrest; or the suspect is committing a *breach of peace* (*Restatement (Second) of the Law of Torts*, § 119 [1966]). A breach of peace is public conduct that is likely to provoke an immediate disturbance (*Restatement (Second) of the Law of Torts* at § 116). An example of this is a party threatening to assault another or using vulgar or derogatory language to provoke another.

Unless authorized by statute to arrest or detain for misdemeanors, such as a retail theft statute (see, e.g., 18 *Pa Cons Stat* § 3929 [1990]), security officers generally have very limited or no authority to make arrests for misdemeanors. Further, they generally have no authority to issue citations for violations. Accordingly, it is critical for security officers to be trained in their authority under state law to make arrests and to be given guidance by corporate policy as to when they should exercise this authority. For example, the security officers should know if they have the authority to arrest or detain someone for creating a disturbance or for defiant trespass, or if the company wants the officers to observe, document, and report this conduct.

Use of Force by Security Personnel. The permissible use of force under law also varies by state. Unless otherwise provided for in state law, security officers can use force in the same circumstances as private citizens. As a general rule, security officers can only use deadly physical force to protect themselves or others from imminent serious bodily injury or death (18 *Pa Cons Stat* at §§ 505 and 506. [1990]). They also can use reasonable force to protect themselves and others from physical injury (18 *Pa Cons Stat* at §§ 505 and 506). Security officers also can use reasonable physical force when effecting a lawful arrest (18 *Pa Cons Stat* at § 508).

Whether physical force can be used in defense of property, to prevent a wrongful trespass or the removal of one's property, is a more sensitive issue. In some states, reasonable force can be used to protect property from wrongful conversion if the taking of the property is occurring at that time and the party has been given ample notice and opportunity to leave the property undisturbed (18 *Pa Cons Stat* at § 507). Similarly, private security officers can, in some states, use reasonable force to remove trespassers if the trespasser has been advised of the authority of the officer to demand the trespasser leave, has been given ample notice and opportunity to depart, and still refuses to leave (18 *Pa Cons Stat* at § 507).

Unauthorized and Excessive Use of Force. The unauthorized or excessive use of force by security officers can result in corporate and individual officer liability. Lawsuits of this nature usually involve civil tort claims for assault and battery, false arrest and false imprisonment, defamation of character, malicious prosecution, and infliction of emotional distress. Each of these actions is further discussed:

- *Assault and Battery*. A civil assault is when one party (security officer) creates in the mind of another a reasonable apprehension of an imminent and forthcoming battery. A battery is the unconsented and offensive touching of one party by another. In some states, assault and battery are treated as a single tort. If a security officer is sued for assault and battery, he can raise an affirmative defense that the use of force was lawful and not excessive under the circumstances (the officer was defending himself, herself, or another); the officer was affecting a lawful arrest or detention; or the officer was authorized to use force to prevent a conversion or trespass of property. If the officer acted unlawfully or excessively in the use of this force then he can be held liable for an assault and battery.
- *False Arrest and False Imprisonment*. False arrest is an unlawful arrest without probable cause. When one is unreasonably denied voluntary freedom of movement, the detained party can bring a tort action for false imprisonment. A false arrest is almost always grounds for a false imprisonment claim (when one is arrested, he or she is almost always denied voluntary freedom of movement). However, an act of false imprisonment is not always a false arrest (a suspect detained for an unreasonable period of time for questioning by a security officer can be falsely imprisoned but not arrested). These actions are often filed in conjunction with assault and battery claims (i.e., the officer's use of force was unlawful and the injured party was unlawfully arrested, falsely imprisoned, or both).
- *Defamation of Character*. A defamatory statement is a false statement of fact made by one party to a disinterested party that injures the community reputation of a third party. Libel refers to defamatory statements that are written, whereas, slander refers to defamation by the spoken word, conduct, or both. For example, if a security officer physically detains someone without cause while others are in the area, the detained party can claim the officer's conduct defamed her reputation in the community by implying she was engaged in criminal or wrongful conduct.
- *Malicious prosecution*. If a suspect is arrested by a security officer and the charges are dismissed, or there is a judgment in favor of the accused in the criminal case, the accused can file a civil suit for malicious prosecution. In order to prevail, the accused must prove that the officer initiated or continued a criminal proceeding without probable cause and for a malicious purpose against the accused. There must also be a decision in favor of the accused in the criminal proceeding. The mere fact the accused was found not guilty at trial is insufficient by itself to prove malicious prosecution. The accused must show there was no probable cause for criminal charges and that the officer acted maliciously (i.e., the officer arrested the accused to make an example of him. (See the section in this chapter on investigative law for further discussion on defamation and malicious prosecution.)
- *Infliction of Emotional Distress*. Victims of unlawful or excessive use of force often claim they have suffered *intentional* or *negligent infliction of emotional distress*, or both. *Intentional infliction of emotional distress* refers to outrageous conduct that is highly probable to result in the injured person suffering severe psychiatric harm (e.g., beating a suspect after he has been subdued). *Negligent infliction of emotional distress* refers to wanton or reckless conduct that is highly likely to result in the victim suffering severe psychiatric harm with physical manifestations (e.g., management knew the officers were using excessive force and never took appropriate action to prevent future incidents).

Managing the Risk of Use of Force. Given the differences in state law regarding the use of force, security officers must be trained in their authority under their respective state law to use force in defense of self and others and in defense of property. Further, they should be guided by corporate policy as to when the company will authorize its officers to apply force. For example, one major corporation has restricted the use of force by its officers to self-defense and defense of others, even though its state law permits the use of force to defend property.

Security officers also should be trained how to properly use force. For example, if officers are provided chemical weapons, such as a pepper or gas spray, handcuffs, or nightsticks, the officers should be trained in the proper use of those items. Similarly, if officers are provided firearms, they need to be properly trained and certified. If weapons are involved, a rough rule is security officers should receive the training that is recommended by the manufacturers, training that is equal to that of law enforcement officers who use those weapons in that jurisdiction, or both. Officers should also be periodically retrained and recertified.

LABOR STRIKES AND ACTS OF VIOLENCE AND VANDALISM

Under the National Labor Relations Act, employees have the right to collectively organize and to negotiate with a company to improve their conditions of employment. If unsuccessful in their good faith attempts to negotiate a contract, the employees can vote to strike. Although employees have a right to strike and to picket a corporation, they do not have the legal right to vandalize property, or to physically threaten or harm working employees or corporate suppliers and customers. Such acts are unfair labor practices and are criminal. If a strike escalates to vandalism or violence by union supporters, and is documented by observers or cameras, a corporation can use this information to obtain a court injunction to restrict the union and its members and to prosecute them. If a corporation has reason to expect a strike will turn violent, it should consider contracting for additional security with an experienced firm. The use of a seasoned security firm will minimize the liability risk and provide the company with experienced officers equipped and trained to document and handle potential destructive acts.

INFORMATION SECURITY LAW

In addition to protecting people and property, the security area can be responsible for protecting proprietary information and data systems. This section provides an overview of the basic categories of public and business information, explains the legal duty of businesses to protect their trade secrets, and discusses state and federal laws that protect corporate trade secrets and data systems. The chapters in Part 5 of this book offer more information on the management of proprietary information.

PUBLIC AND BUSINESS INFORMATION

Information used regularly in the course of business, excluding government classified information, can be divided into the following categories: public domain, trademarks, copyrights, patents, corporate proprietary, and trade secrets. The right of use and the protection of law differ with each category, as does the obligation of ownership.

- Public domain. Unless information is otherwise designated it falls within the *public domain.* Public information can be useful or even essential to a corporation, but the corporation has no legal claim of ownership or exclusive use. Information in the public domain can be used at any time by anyone without authorization or payment of any fee.

- Trademarks. *Trademarks* are names, symbols, devices, or combinations thereof that identify a company's products. Although well-known to the public (IBM® and Coca-Cola®), trademarks are not part of the public domain and cannot be used except by authorization of the trademark holder. Trademarks are issued for 20 years and can be renewed by the trademark holder (15 *USCA* § 1051 et seq). It is the responsibility of the trademark holder to detect abuse, be it intentional or unintentional, and to seek appropriate legal sanctions.
- Copyrights. Information that can be *copyrighted* includes original writings and artistic works, such as books, newspapers, articles, songs, photographs, movies, and software. Technical manuals and software are often copyrighted to deter competitors from misusing and profiting from the information. Parties that infringe upon a copyright can be sued for lost profits, acquired profits, and legal costs. An injunction to destroy all unauthorized copies and to prevent further infringement can also be obtained. When a copyright expires (50 years after the death of the last surviving author or 75 years from the date of issue if a work for hire), the information becomes part of the public domain (17 *USCA* § 101 et seq).
- Patents. *Patents* can be obtained for new and useful products, processes, machines, composite materials, or combinations thereof that are not obvious, novel, and useful. A patent becomes public knowledge when it is filed, but during the life of the patent this information cannot be used without authorization. When a patent expires (20 years after issue), the information becomes part of the public domain (35 *USCA* § 101 et seq).
- Proprietary information. Information that a corporation desires to keep confidential is known as *proprietary information*. This information is not necessarily protected by law. Customer lists, for example, are often considered proprietary, but the taking or disclosure of this information cannot be a violation of any law if the list was readily obtainable. For example, an auto-parts supplier who delivered goods to garages around a city would have a difficult time claiming its customer list was protected under law if it was memorized by one if its drivers who went to work for a competitor or opened a competing business. A company might be better served by having employees sign reasonable, noncompete agreements.
- Trade Secrets. A *trade secret* is technical or scientific information that is not commonly known and gives a company a competitive advantage in the marketplace. It often includes leading-edge technology and research, such as a new method for manufacturing computer chips. But this is not always the case. The formula for Coca-Cola®, for example, is a trade secret. Trade secrets are a form of proprietary information and are protected by law.

A few states also recognize so-called *business trade secrets*, information that is not commonly known and gives a business a competitive edge in the marketplace. If a corporation had a confidential business plan to buy a large tract of land for a new facility and an investment realtor bribed an employee for this information, resulting in the corporation having to pay an inflated price, the aggrieved corporation in these states can file a civil suit for theft of a business trade secret.

A trade secret can be kept indefinitely or for a lesser period of time depending on the needs of the corporation. The formula for Coca-Cola® is a trade secret that probably will be kept indefinitely. On the other hand, a computer company developing new software or hardware may wish to keep this information a trade secret during the developmental stages and then copyright the program and/or patent the equipment prior to releasing it on the market.

A corporation has no legal claim against competitors for use of public information or proprietary information that is not otherwise protected; however, it can bring a civil suit and sometimes seek criminal sanctions against other parties and businesses that violate its trademarks, copyrights, patents, and trade secrets.

CIVIL LAW—THE DUTY TO PROTECT TRADE SECRETS

Before an aggrieved company can prevail in a civil suit against a former employee, competitors, or others for misappropriation of a trade secret, the company is required by the court to show that it took reasonable precautions to protect the information. If a company is careless in its handling of its trade secrets, it can lose the legal right to claim the information as a trade secret and to seek redress against those who misappropriate it.

A reasonable security program includes the following basic components: identification, notification, and restricted access.

Identification. In order to claim a trade secret existed, a corporation must be able to reasonably *identify* what information was secret (i.e., research project, portion of an otherwise standard manufacturing process, unique formula, or novel piece of equipment). A business must be careful not to claim too much information as a trade secret, as an overly broad claim of otherwise common knowledge can defeat the legal right to claim any of the information is protected (*Motorola v. Fairchild Camera & Instrument*, 366 F Supp 1173 [1973]).

Notification. A corporation must take reasonable steps to *notify* persons having access to the information that it is a trade secret. Depending on circumstances, notification can occur in many ways:

- Job applicants can be asked to sign confidentiality and noncompete agreements upon employment.
- Employees can be asked to sign confidentiality and noncompete agreements at exit interviews when leaving a company.
- Visitors, suppliers, and potential buyers of a company can be asked to sign confidentiality and noncompete agreements.
- Notices can be placed on visitor sign-in logs.
- Signs can be placed in working areas where a process is considered confidential.
- Documents and lab notebooks can be marked.

The key is for the company to take reasonable steps to ensure that everyone working with or having access to the information knows it is confidential.

Restricted Access. The third component of a trade secret security program is the *restriction of access* based on a need-to-know basis only. If it is a manufacturing process, suppliers should only be shown this area if they need to know and after having signed a nondisclosure agreement. Employees should not be allowed access, unless they work in that section and have signed confidentiality and noncompete agreements. If it is formulas or drawings, these documents and their storage containers should be labeled confidential, properly secured at all times, and access restricted. If a company permits open access to information, the courts will not permit it to claim the information was a trade secret.

It is not necessary for a corporation to take extraordinary precautions to protect its trade secrets. For example, it was not necessary for a chemical company building a plant that involved a new process to erect a cover over the construction site to block pictures being taken from a plane (*E.I. du Pont de Nemours & Co., Inc. v. Christopher*, 431 F 2d 1012 [1970]). If a company has taken reasonable steps to protect a trade secret and it is misappropriated, the company can seek a civil remedy in the courts. This remedy can include damages and an injunction to prohibit the other party from further use of this information.

CRIMINAL LAW—CORPORATE TRADE SECRETS

Many states have enacted criminal statutes that prohibit the theft of trade secrets. But these statutes differ as to the types of trade secrets protected and the penalties for misappropriation. If the misappropriation involves interstate theft, it may also be possible to prosecute under federal law (Spain, NM. Prosecution options for trade secret thefts. *Security Management*, Sept. 1983, pp. 59-68.)

State Trade-Secret Statutes. The criminal codes of many states specifically prohibit the theft of trade secrets. The level of protection varies, depending on the following factors:

- *Scientific, technical, and business secrets.* Most states protect scientific and technical trade secrets. Only a few protect business trade secrets. In many states, if a party misappropriates a business trade secret, the injured business would not be able to criminally prosecute the wrongdoer.
- *Tangible and intangible secrets.* Some states protect only tangible secrets, whereas other states protect both tangible and intangible secrets. For example, if an employee stole a notebook or a computer disk with a secret formula, this would be a tangible theft for which the offender might be prosecuted. If the employee memorized the formula and stole it, this would be an intangible theft and the company may not be able to prosecute the offender in some states.
- *Criminal sanctions.* Penalties under the various criminal statutes also vary. Trade secret theft in some states, regardless of the value of the information, is not always a felony. In these states, the criminal sanction may not be appropriate for the value of the loss.

Federal Criminal Statutes. In addition to the state criminal statutes that protect trade secrets, there are a few federal criminal statutes that prohibit certain methods of theft and that may be applicable to trade secret cases. A few of the more relevant statutes include the following.

- *Mail Fraud Act.* Under this act it is illegal for any party to use the U.S. mail or private interstate carrier as a part of any scheme to defraud another (18 *USCA* § 1341). Thus, if a competitor mailed a check to an employee for theft of a trade secret or the employee mailed the information to the competitor, under either scenario both might be prosecuted under this statute.
- *Wire Fraud Act.* This act makes it illegal for any party to use electronic communications interstate to defraud another (18 *USCA* § 1343). If a dishonest employee faxed a trade secret interstate to a competitor or the competitor used the phone to conspire with the employee, both parties under either fact pattern might be prosecuted under this act. Further, if an out-of-state intruder used electronic means to invade and remove a trade secret from a corporate database, the intruder might be prosecuted under this statute (*U.S. v. Seidlitz*, 589 F 2d 152 [4CA 1978], cert den, 441 U.S. 922).
- *Right of Privacy Act.* Under this law it is illegal to intercept a wire communication, and it is unlawful to use electronic or other means to overhear business-related conversations on the premises of or that pertain to the operation of a business. Further, it is illegal to knowingly use or disclose this information (18 *USCA* § 2511).

- *National Stolen Property Act.* This act makes it illegal for a party to transport in interstate or foreign commerce stolen goods in excess of $5,000 in value, or to knowingly receive or sell stolen goods in excess of $5,000 in value that have been moved in interstate or foreign commerce (18 *USC* §§ 2314 and 2315). Although this act does not specifically protect trade secrets, the federal courts have permitted prosecution under it for theft of tangible and intangible scientific and technical trade secrets.
- *Conspiracy Act.* If two or more parties agree to break a federal law and take an overt step in furtherance of that agreement (offering a bribe to an employee to steal a trade secret), they can be prosecuted under the Federal Conspiracy Act (18 *USCA* § 371).

The intent of this section has been to make safety managers aware of the more prominent federal and state criminal statutes that might be invoked in the event of the misappropriation of a trade secret. If a security manager believes a misappropriation has occurred and the corporation wishes to prosecute, the manager should contact either the state or federal prosecutor's office to discuss all known facts and to assess the feasibility of prosecution.

CRIMINAL LAW—CORPORATE DATA SYSTEMS

Computer systems are vital to the operations of many companies, but these systems are also vulnerable to internal and external abuse. Employees and nonemployees without proper authorization can remove data or sabotage a data system in numerous ways, such as worms, viruses, and other attacks designed to temporarily or permanently disable a system.

There are two federal statutes that indirectly and directly prohibit attacks on data systems. If access to a system is gained via interstate telecommunication systems and data was removed, the offending party can be prosecuted under the Federal Wire Fraud Act. An intruder might also be prosecuted under the Federal Computer Security Act, which makes it illegal to use interstate telecommunication without authorization to intentionally or recklessly launch a program, information, code, or command to a computer to damage the system or data, or to deny access to or use of the computer (18 *USCA* § 1030).

Computer data systems also can be protected under state criminal statutes (see, e.g., 18 *Pa Cons Stat* § 3933 [1990]). This is an option that may need to be considered in the event of an attack on a company's computer system that did not involve interstate communications. If a disgruntled employee planted a computer virus from her computer at work, this might not be a federal offense involving interstate activity, but it still may be prosecuted under state law.

Safety managers responsible for information security need to become familiar with the civil and criminal laws that govern the use of trade secrets and regulate the abuse of the data systems in their respective states.

INVESTIGATIVE SECURITY LAW

The third major function of security is to conduct proactive and reactive investigations. A proactive investigation is one that is conducted to avoid future problems, such as employee background checks and due diligence checks on suppliers, acquisition targets, credit applicants, and others. A reactive investigation is one conducted in response to suspicious circumstances. For example, a tip that a purchasing agent is seeking kickbacks, a complaint that a manager is engaged in sexual harassment, or a report that shows a store employee has an unusually high number of refunds, are reasons for a reactive investigation. All investigations involve decisions and conduct with potential legal consequences, many of which are discussed in this section.

DUTY TO INVESTIGATE

The decision to conduct an investigation may or may not reside with the corporation. A corporation can assume that it has a duty to conduct a good-faith investigation into suspicious circumstances that involve illegal activity in violation of the rights of their employees or others (claims of discrimination or harassment, a tip the company is engaged in price-fixing or illegally dumping toxic materials). If the suspicious activity involves only employee misconduct, such as being late to work or abuse of sick leave, the decision to investigate resides with the corporation. Similarly, the decision to investigate illegal activity wherein the company is the victim and the loss is not material (e.g., theft of a VCR from an office) also resides with the company.

Good-Faith Standard of Investigative Care. An investigation is a sensitive matter and must be conducted with reasonable care. If challenged in court as to scope or method of investigation, a corporation should be prepared to show that it conducted a reasonable and good-faith inquiry into a matter in which it had a valid interest.

EMPLOYEE BACKGROUND INVESTIGATIONS

Employee background investigations are routine in today's business world. Companies can reduce their liability associated with these investigations by having applicants execute waivers permitting the company to contact former employers, references, educational institutions, and certification boards. These waivers should also contain hold-harmless clauses that state parties releasing information to the employer will be held harmless for the information released and any decisions and actions based on that information (*Smith v. Holley*, 827 S W 2d 433 [Tex App San Antonio 1992]).

Employers need to understand what information they can obtain. Depending on the position for which the applicant is applying, typical background checks can include prior employment, military, education, certification, driving records, credit history, and reference checks. Depending on the state law, employers can also check criminal convictions, but not arrest records, related to the job. Preemployment drug tests with proper waivers can also be obtained. Preemployment inquiries, direct or indirect, about age, race, nationality, marriage status, children, disability, medical condition, and other potentially discriminatory topics should not be made.

EMPLOYEE DUTY TO COOPERATE IN AN INVESTIGATION

Under common law, employees have a general duty to cooperate in reasonable inquiries by employers. Nonetheless, it is advisable for a company to have a policy that states employees have a duty to report misconduct and to cooperate in corporate investigations. With this type of policy in place, it is easier for a company to justify taking disciplinary action if it uncovers employee misconduct that was known to others and not reported. However, such a policy is not absolute. For example, union contracts can prohibit union members from testifying against each other in disciplinary proceedings. Similarly, if reasonably afraid of physical reprisal, an employee might be justified in not reporting misconduct or cooperating in a company investigation.

MIRANDA RIGHTS IN INTERVIEWS AND INTERROGATIONS

If a police officer conducts a custodial interrogation of a suspect in a criminal investigation, the officer must first advise the suspect of the *Miranda Rights* (the right to remain silent, the right to terminate the meeting at anytime, the right to counsel, the right to have counsel appointed if the suspect cannot afford an attorney, and a notice that any statements made by the suspect can be used in a court of law). Off-duty police officers working security and security officers who hold special commissions should also "mirandize" suspects when conducting custodial criminal interviews.

Unless deputized or otherwise acting under the authority of the police or prosecutor, private security officers are not generally required to "mirandize" people suspected of criminal activity. Further, most corporate investigations are internal and noncriminal in nature, such as discrimination and employee misconduct investigations, and there is no obligation to "mirandize" in such circumstances.

DISCIPLINARY INTERVIEWS OF UNION MEMBERS

The National Labor Relations Board (NLRB) has ruled that union members are entitled to union representation at any meeting where employee discipline is a likely outcome (*NLRB v. J. Weingarten, Inc.*, 420 U.S. 251 [1975]).The union representative can help clarify the facts, advise the employee of his rights under the union contract, and serve as an observer. The NLRB has vacillated on whether nonunion employees are entitled to the right of representation (Hartsfield WE. *Investigating Employee Conduct.* Deerfield, IL: Callahan & Company, 1988, § 4:11). As a practical recommendation, in a unionized corporation, nonunion employees should be granted the same rights as union members—the right to have an observer of their choice present as a witness.

FALSE IMPRISONMENT DURING INTERROGATIONS AND INTERVIEWS

One legal risk that investigators face when conducting interrogations and disciplinary interviews is the risk of false imprisonment, unlawfully denying an individual voluntary freedom of movement. The potential for false imprisonment can arise when an employee is called or taken to a room to be interrogated or interviewed for disciplinary purposes. The employee can claim he was detained against his will, especially if the company later prosecutes or takes adverse employment action. For example, the employee can claim he was trapped in the room by the investigator who sat closest to the door, or felt trapped by two or three other security or management personnel. The employee might further claim that the investigator would not let him go to the restroom, kept him long after he wanted to go, verbally threatened him by telling him to sit down, or physically pushed him back into his chair when he attempted to leave.

When conducting interviews and interrogations, investigators should document the start and end times of the meeting, the room location, and the persons present. Further, the employee should be permitted unobstructed access to the door, unless the investigator has reason to be concerned for her safety; for example, if the employee is known to be violent. Experts disagree on how long an interview or interrogation should last, but if it does take a period of time, the interviewee should be permitted reasonable use of the restroom and this use should be documented, as well as any refreshments obtained or provided. If not required, experts also differ on whether there should be a witness present for an interrogation. Because of the potential for claims of sexual misconduct, if the interviewee and interviewer are of opposite sex, a witness of the same sex as the suspect should be present, when practical *(Hoke v. May Department Stores*, 891 p 2d 686 [Or App 1995]).

CONFESSIONS

If a confession is obtained and not recorded, the employer should request that the individual either write or sign the confession. Any changes to the confession should be initialed by the suspect. At least one investigator has stated that he prefers to have some changes made to a confession to show the suspect did read and understand it. The confession should contain a statement it was made freely, without promise or duress, and it should be witnessed. Note that even a signed confession can be excluded or disregarded at trial if the court concludes the confession was coerced. As a result, many companies videotape confessions as proof that the confession was voluntarily given without coercion or false promise.

DEFAMATORY INFORMATION

Sensitive and derogatory information is often the product of investigations. This information should be shared only on a need-to-know basis. Careless spreading of derogatory information can result in a lawsuit for defamation of character.

Recall that a *defamatory statement* is a false statement of fact that is made by one party to a third disinterested party and that hurts the community or business reputation of a second party. A *disinterested party* is one who has no need to know. *Libel* refers to a written statement that is defamatory. *Slander* refers to a spoken comment that is defamatory. Slander also includes defamation by conduct.

There are two affirmative defenses to lawsuits for defamation. First, *truth* is a defense. If a company told a potential employer that an ex-employee was fired for theft and the employee sued for defamation, the company can prevail if it can prove the employee did indeed steal from the company. However, if the company could not prove theft it might lose a judgment for defamation of character.

Privileged communication is the second defense. There are two types of privilege. Absolute privileged communications usually occur in judicial, administrative, and legislative proceedings. For example, if a manager accused an employee of theft at the criminal trial and the employee was found "not guilty," the manager may not be sued for defamation, as statements made at trial are absolutely privileged.

A *conditional or qualified privileged communication* is one that is made in good faith to an interested party, that is, someone who had a need to know. For example, if a safety manager, in good faith, tells a plant manager and personnel manager that an employee is suspected of drug use, then that communication is a conditionally privileged remark because these managers would need to know.

Defamation lawsuits are a major legal risk for security personnel. Therefore, it is advisable to exercise reasonable care with sensitive information and to share it within the corporation with only those persons who need to know in performance of their duties. Comments outside the corporation should be made with extreme caution. Reports to the police are normally protected if full and truthful disclosure of all facts is made. Similarly, complete disclosure to administrative agencies, such as workers' compensation and unemployment boards, are also normally protected. Comments to other employers, if made in good faith, also can be protected. Even so, when making references about former employees, many companies prefer to give only the dates of hire and separation and last position held. Managers need to act with reasonable caution when sharing potentially defamatory information.

INVASION OF PRIVACY

Another risk associated with corporate investigations is that of *invasion of privacy*. This is an intrusion by one party into the reasonable expectation of privacy of another, which is highly offensive to a reasonable person. The intrusion can be a physical intrusion or an intrusion into one's private life (*Restatement (Second) of the Law of Torts*, 652B [1977]). For example, it can be reasonable for a company to put a hidden camera above a cash register to determine if an employee is stealing. But it is a far more sensitive proposition and one that should receive more serious consideration before that same company put a camera in an employee restroom to detect employee drug use. The following questions need to be answered:

- What specific reliable information is available to justify the camera?
- Is there another less intrusive alternative for investigating the matter?
- Who will make the ultimate decision to install the camera?
- How will the tapes be controlled, reviewed, and used?
- How long will the investigation proceed?
- Under what circumstances will the camera be removed?

An employee's expectation of privacy is far greater in a restroom than on the work floor. Similarly, a corporate investigator can photograph an employee who claims to be on disability, working in her yard if the investigator is in an area open to the public, but the investigator would not want to physically trespass on the employee's property to take the same photo.

Employees can also enjoy a degree of privacy in their lockers, toolboxes, lunchboxes, offices, and so forth. Therefore, it is advisable that companies have an inspection policy, reserving the right to inspect all areas, packages, and vehicles entering or remaining on site. This policy can further advise employees of contraband that is prohibited, such as drugs and weapons or even unauthorized company property; and it can stipulate the right of the company to discipline for violation of the policy, including discipline for failure to cooperate in an inspection (*Simpson v. Unemployment Compensation Bd. of Review*, 450 A 2d 305 ([a 1982]), cert denied, 104 U.S. 88 [1983]).

Society places a high value on individual rights. Not only is this protected under common law, an individual's rights are also protected in the constitutions and statutes of some states. Therefore, a company must exercise reasonable care over the scope and method of inquiry when conducting an investigation, to ensure that it does not exceed a person's reasonable expectation of privacy.

Decision to Terminate—Wrongful Discharge. A security investigation can result in the documentation of evidence on which a company decides to terminate an employee. Most employees are *at-will*, which means they can quit at any time for any or no reason, and the employer can fire them at any time with or without a reason.

But, there are exceptions to the employee at-will doctrine.

- First, union members are contract employees and can only be fired under the terms stipulated in the union contract (the company must show cause or good cause to discipline the employee).
- Second, employees cannot be fired in violation of public policy (an employee fired for reporting illegal company activity, such as dumping of toxic waste, might claim the discharge violated public policy). If an employee filed a workers' compensation claim and was discharged, he might sue on the grounds the discharge violated public policy because the company was

trying to discourage other workers from filing for workers' compensation.

- Third, in some states employees cannot be fired in violation of the covenant of good faith and fair dealings. This is a common law doctrine that holds that employers and employees have a duty to deal reasonably with each other. Violation of this conduct requires outrageous conduct.
- Finally, employees cannot be fired for discriminatory reasons.

The decision to terminate based on investigative results should be made with caution and sometimes factors beyond the investigation also must be considered prior to the decision to terminate. For example, if two employees of different sex, race, and age were involved in a fight and only one was fired, the fired employee might claim the discharge was discriminatory. The decision to terminate an employee must be made with due reason or else the company can find itself unable to defend a suit for wrongful discharge, even though the individual was an at-will employee.

Decision to Prosecute—Malicious Prosecution. A safety manager faced with the decision to prosecute someone needs to understand the risk of malicious prosecution or abuse of process, as it is known in some jurisdictions. Malicious prosecution occurs when a party initiates or continues a criminal proceeding against another without probable cause and for malicious purposes, and there is a decision in the criminal case in favor of the accused.

The first issue is *initiation* of a criminal proceeding. If a security officer arrests someone, this is clearly an initiation of a criminal proceeding. Similarly, if a manager knowingly makes a false or incomplete report to the police, withholding vital information so the police will arrest the suspect, this too can be the initiation of a criminal proceeding. If the manager subsequently uncovers information in favor of the accused and fails to share it with the police, this withholding of information might be a *continuation* of the criminal proceeding.

When filing police reports or signing criminal complaints, managers must be certain of their motives. Employees should not be criminally charged to make examples of them or to force them to confess. Conduct of this nature can indicate that the company acted without probable cause and with malicious intent. Further, if a corporation has filed criminal charges and subsequently wishes to withdraw the charges, this should be done in cooperation with the prosecutor who can move to have the charges dismissed without a finding on the merit.

If a company elects to proceed criminally against an individual, it should be certain to act in good faith and to make all facts and information available to the authorities, to minimize the risk of a lawsuit for malicious prosecution, should the charges subsequently be dropped or dismissed, or the individual be found not guilty.

Labor Investigations. Under the National Labor Relations Act, employees have the right to unionize. When conducting investigations into employee misconduct of union members, companies should not use the investigative mechanism as a means to harass or discourage employee unionization or to disrupt the operation of a union. Investigations, however, can be used to document misconduct, unfair labor practices, illegal acts, and violence by union members.

Provision of investigative services is an important security function. Responsible safety managers should see to it that such investigations are conducted in good faith and in a reasonable manner, to avoid exposing the company to unnecessary liability. Derogatory information that is uncovered as a result of an investigation needs to be shared on a strict need-to-know basis to avoid liability for defamation. Adverse employee decisions, such as termination and prosecution, should be made in good faith and be based on available facts and not as a means to coerce or intimidate the accused. Investigations must be handled and managed professionally.

SUMMARY

The objective of security is the protection of people, property, and information. To achieve this goal, security programs often provide physical security, information security, and investigative services. Each of these services entails specific legal risks that differ by state and must be understood and managed by the executive responsible for the security function.

The goal of this chapter has been to identify some of the legal risks associated with each of the security functions and to highlight some general ways these risks can be managed. However, this chapter is not written as legal advice. It is the obligation of the managers responsible for security to understand the laws of their respective states, to ensure that that their security officers are properly trained to minimize the risk of liability, and when appropriate, to seek the advice of competent legal counsel.

BIBLIOGRAPHY

BOOKS

Hartsfield, W. *Investigating Employee Misconduct*. Deerfield IL: Callaghan, 1988.

Inbau, F.E., Farber B.J., and Arnold, D.W. *Protective Security Law*, 2nd ed. Boston: Butterworth-Heinemann, 1996.

Keegan, F.R., and Fogleman, J.L. eds. *Avoiding Liability in Premise Security*. Atlanta: Strafford, 1987.

Maxwell, D.A. *Private Security Law: Case Studies*. Boston: Butterworth-Heinemann, 1993.

Nementh, C.P. *Private Security and the Law*. Cincinnati: Anderson, 1989.

Risser, R. *Stay Out of Court: The Manager's Guide to Preventing Employee Lawsuits*. Englewood Cliffs, NJ: Prentice Hall, 1993.

NEWSLETTERS

Law Reporter. Washington DC: Association of Trial Lawyers of America.

Private Security Case Law Reporter. Atlanta: Stafford.

Security and Special Police Legal Update. Chicago: Americans for Effective Law Enforcement.

Security Management Bulletin. Waterford CT: Bureau of Business Practice.

4

Foreseeability of Crime and Adequacy of Security

Norman D. Bates, JD

CHAPTER 4

OBJECTIVE

The objective of this chapter is to explain basic legal principles and practical considerations to those responsible for facility security and those called upon to develop crime-prevention programs. Most security professionals are concerned about potential civil liability but frequently misunderstand the logic behind the case law. Sometimes the case law is not logical, and that fact must be accepted.

INTRODUCTION

Knowing that this chapter is likely to be read by members of the legal profession, as well as those in the security profession, it is important to point out that the case law in negligent security litigation has generally lagged behind the practical considerations of everyday life. The legal system needs to be more cognizant of these everyday considerations when evaluating the risks of crime and the adequacy of a security program.

From the perspectives of both legal theory and applied security practice, determining the foreseeability of crime and the adequacy of security measures requires an understanding of basic principles of law and an appreciation for the need to keep current in business practices.

While case law typically looks at issues of foreseeability and adequacy in somewhat narrow terms, able security practitioners will take a broader approach to evaluating and planning the security needs of their organizations. This does not mean to suggest that the law is at odds with the everyday practices of business. The legal analysis involved in the evaluation of what constitutes "legally foreseeable crime" and what constitutes adequate security measures is rather narrowly focused. While the security practitioner should understand the more limited approach taken by the legal system, that practitioner must go beyond the minimum required by law to be an effective professional.

If security professionals limit their determination of foreseeability or crime risk to the definition stated by a single jurisdiction, they will likely ignore numerous other important factors that can be critical to the overall success of their security programs. Conversely, the lawyer or court that ignores those factors relevant to the security practitioner to satisfy only the case law can do a disservice to the intended beneficiaries of a well-planned security program. What is "foreseeability" and how does it fit into the scheme of an inadequate security case?

BASIC PRINCIPLES

Inadequate security liability is the application of principles in the civil tort of negligence to a scenario where, in most cases, an individual is the victim of a violent crime. Also referred to as *premises security liability* and/or the *liability for the criminal acts of third persons*, this form of liability arises when a business or other private institution is sought to be held civilly liable for monetary damages when a customer, patron, guest, or other category of person is assaulted on the property of that business or institution. Government agencies are commonly immune from these claims.

A typical premises security claim occurs when a resident of multiunit residential housing sues the property owner, landlord, or both and other relevant parties for the failure to protect that resident against the violent acts of an intruder. For example, perhaps a woman is raped by an assailant who was able to enter her apartment illegally via a sliding glass door with a defective lock. It would be argued that the broken locking mechanism is the cause of the intruder's access to the apartment.

A claim for inadequate security is a claim that the property owner was negligent because that owner failed to provide a reasonable level of security, given a particular set of circumstances. The "circumstances" will be explored in greater detail in this text, but essentially include the level of risk in existence at that property at the time of the assault. What constitutes "risk" will also be discussed in the foreseeability section of this chapter.

The fact that the intruder actually committed the rape is not overlooked. If that intruder has been caught, he too could be liable civilly (and criminally, of course) but under different legal principles. A claim against the assailant, if brought, would be in battery, which is not a negligent tort action.

The claim against the property owner is made pursuant to the laws of negligence for the particular jurisdiction. The vast majority of negligence-based lawsuits are brought under an individual state's laws and rules pertaining to such actions. As such, businesses and other private institutions that operate in more than one state will need to know each state's rules if negligent security liability is to be avoided.

The *tort of negligence*, as with all other tort claims, contains a number of elements or components that the plaintiff must prove in court to successfully make his or her case. The essential elements to the tort of negligence are as follows:

- duty
- breach of duty
- injury and causation

DUTY

Duty is the legal responsibility that a property owner has for maintaining reasonably safe premises. The extent of the duty, or how much is required of the property owner, is directly tied to the level of risk or foreseeability of crime. Absent a legal duty or responsibility imposed by operation of a statute, ordinance, or code, the degree of responsibility to provide reasonably safe premises is proportionate to the level of risk at any given time.

The legal duty of a property owner to provide reasonable levels of security can be imposed in several different ways. A duty may arise by operation of the following:

1. via statute, ordinance, or code
2. through appellate case law
3. under the terms of a contract
4. where a special relationship between the plaintiff and defendant exists (e.g., student and university)
5. where the defendant has voluntarily undertaken the responsibility

A *duty created by statute or code* is one in which the property owner is required by legislative action or regulations of a state or federal agency to provide certain measures. For example, in South Carolina the state law requires hotel and motel rooms to be equipped with certain specified locking devices.

A *duty arising from case law* tends to be more general in nature, in that one or more appellate opinions of a state will indicate that certain types of property owners (e.g., hotels, office buildings, residential landlords) have a duty to provide reasonable levels of security. Inversely, a small number of states have indicated that no duty exists. Case law that imposes a duty often fails to provide clear specifics about what actions are required, in contrast to a building code or regulation, for example.

When a contract exists between the parties (e.g., landlord-tenant) that includes terms that the landlord will provide specified security measures, the *duty exists by operation of that agreement* and can be enforced in *contract law* as well as in *tort law*. Occasionally, a special relationship can be found to exist between the plaintiff and defendant, giving rise to a *duty of care*. In some cases, a college student could have a special relationship with the college due to the former's dependency on the latter.

The most common circumstances in which a duty will be found is when the defendant has voluntarily undertaken the responsibility for certain security measures. For example, a security staff of a hospital attempts to provide additional protection to a patient who is the victim of spousal abuse. The security staff fails to adequately protect the patient-victim from a subsequent attack by the offending spouse. A *duty voluntarily assumed* is likely to be found. The rule of law that is a *duty voluntarily undertaken* must be performed with equal diligence as a *duty imposed by law*.

BREACH OF DUTY

Breach of duty is the easiest element in the *tort of negligence* to understand. *Breach* simply means failure. It is the failure of a defendant in a civil action to perform the duty imposed by law or the duty assumed voluntarily. For example, a hotel can be required by law to provide dead-bolt locks on all guest rooms. Inherent in the law is the presumption that the locks will be in good working order. The innkeeper who fails to keep these locks working properly has breached the *duty of care*.

In another example, a hospital's administration has decided to employ a security staff to provide protection to the hospital facility. The security department, however, is poorly trained in how to handle the problems of the institution and is understaffed. The deficiencies of poor training and insufficient staffing also "breach" the hospital's duty of care.

INJURY

For the security practitioner, the element of "injury" is also fairly straightforward and is not likely to be a matter of direct concern when planning a security program. This is not meant to imply that security professionals do not care about injuries. On the contrary, a security program exists to prevent injuries and other losses.

Injury, in the legal context, ranges from the more obvious physical injuries one may suffer as a crime victim, to the less obvious mental trauma also suffered as a result of victimization. *Injury in the legal context of negligence* also encompasses losses sustained from the damage to property in its various forms.

CAUSATION

Causation or *proximate cause* is clearly the most difficult concept for security practitioners, lawyers, juries, and others to grasp. In simple terms, for a defendant to be found negligent, there must be a connection or link between the failure of the defendant to meet the duty of care and the ultimate harm suffered by the plaintiff. The law of negligence requires that the failure or deficiency (breach) be a cause for the injuries. Determining the legal cause for the injury is a jury function because it is a question of fact that can be arrived at only once all the evidence is heard at trial. However, there can be more than one cause of an injury. This is where the law of negligence can be confusing.

Examples of which failures are causally related to an injury and which are not should help the reader. Those deficiencies or breaches that are legally causative of an injury should be a matter of common sense. Unfortunately, common sense is not always that common. Some examples follow.

Example 1. A man is assaulted and robbed in a parking lot. Investigation reveals that the parking lot was poorly lit. In fact, in the area of the assault, there was no measurable light (according to a light meter). A local municipal ordinance requires all parking lots to have illumination of at least 1.0 footcandles.

The deficiency in the parking lot lighting would constitute a breach in the duty of care imposed upon the defendant property owner by the local ordinance. However, if this assault occurred during daylight hours, artificial lighting meant for nighttime use would be irrelevant or not causally related to the plaintiff's case.

Should the incident have occurred at night, the lack of lighting could be causally related to the assault. Ultimately, the jury would have to decide whether the assailant was able to commit the crime at least in part due to the poor lighting conditions.

Example 2. A woman is raped in her apartment by an intruder who was able to gain access to her unit via the ground-floor window. The lock on the window was broken and the victim had previously complained to the management company, seeking its repair. The defendant management company would most likely have a duty to provide reasonably operating locking devices on unit windows as either a matter of law (such as building codes) or through evidence of the community standard. The failure to repair the lock in a timely manner would constitute a breach of duty. Is the broken lock causally related to the plaintiff's injuries?

If the intruder was able to enter the windows because the plaintiff could not lock it, then the answer most likely is yes. But, what if the plaintiff left the window open for ventilation? Would the broken lock have mattered to the intruder?

Some defendants would argue that the broken lock is irrelevant or not causally related to the assault. The analysis does not stop here though. The plaintiff's response to the defendant's argument would probably be that other secondary locks (i.e., ventilation locks) should have been provided to accommodate the resident's need for fresh air without fear of assault.

Confusing? Not necessarily. The reader must ask the question: Had the deficiencies complained of by the plaintiff been corrected, would that have prevented the assault, using a "more likely than not" standard? Under the law of negligence, generally, a plaintiff does not have to prove that correcting a deficiency would have absolutely prevented the crime.

FORESEEABILITY—THE RISK OF CRIME

The rules on what constitutes *foreseeability* vary from state to state, and as such, readers would benefit from an understanding of them. From a purely preventive mode, practitioners should not wait until a crime has occurred before acting appropriately to prevent it. Ultimately, whether a crime is considered legally foreseeable will depend on the court and jury in each case.

Property owners are not expected to predict with absolute certainty when someone may be victimized, by whom, or exactly how the crime will occur. However, owners are expected to consider the various factors that constitute risk at their properties according to the law of negligence in their jurisdiction. This risk factor is referred to as *foreseeability*. Generally, as a matter of law, a property owner cannot be held responsible for those events that could not have been foreseen.

In contrast, the more foreseeable or probable the event is, the higher the risk of crime and the greater the duty imposed upon the owner. Just what legally constitutes a foreseeable criminal event varies from state to state. The states take two basic approaches to this concept with some minor variations. Further, some states have yet to adopt a position due to a lack of sufficient case law on the subject in their jurisdictions.

The two rules followed by most states are the *prior similar crime rule* and the *totality of circumstances rule*. The *prior similar crime rule* is the older, more conservative approach taken by several state appellate courts. This rule requires, as a matter of law, that there be some evidence of prior crimes that are similar in nature to the one complained of in the plaintiff's case.

For example, under this rule a plaintiff who was robbed and physically assaulted in the parking lot of a shopping center must provide evidence of a prior robbery in that parking lot before the case even reaches a jury. If there were no prior robberies, the defendant may be able to get the case dismissed well before trial.

The major problems with the *prior similar crime rule* are the lack of clear direction on what constitutes "similar," how far back in time the prior crime must have occurred to be relevant in the instant case, and the inherent unfair result for the plaintiff who just happens to be the first victim of such a crime. In a highly mobile society, it is hard to imagine a crime that could not occur anywhere at any time. It is important to note that no court requires evidence of absolutely identical crimes—this is not the standard anywhere.

The legal effect of the prior similar crime rule is to take a black-and-white position on the issue. The risk of crime is not black and white. The appropriate inquiry should be to determine just how much risk is present. There is always some degree of risk present, although it can be quite low. In that case, a low level of risk simply requires much less of a response or lower levels of security compared to an environment of higher risk where greater security measures may be warranted.

The more contemporary approach to analyzing foreseeability is the *totality of circumstances rule*. This rule became well known after the *Isaacs v. Huntington Memorial Hospital,* 695 P2d 653 (1985) case. The California Supreme Court ruled that property owners should consider other factors beyond prior similar crime to determine the level of risk. In *Isaacs*, the plaintiff was assaulted and robbed in the parking lot of a hospital. Lacking evidence of prior similar crimes, the case was subject to dismissal under the older rule. The plaintiff successfully convinced the appeals court to adopt the broader *totality of circumstances rule*. Under the newer rule, evidence was allowed to show the existence of prior dissimilar crime, crime in the neighborhood, and other relevant circumstances, such as the busy and often tumultuous emergency room at the hospital. The court in *Isaac* reasoned that the old rule on foreseeability "had the effect of discouraging landowners from taking adequate measures to protect premises which they know are dangerous" *(Isaac, supra,* p. 658).

In 1993, the California Supreme Court revisited its earlier ruling in *Isaacs* in *Ann M. v. Pacific Plaza Shopping Center et al.,* 25 Cal Rptr 2nd 137 (Cal 1993). The plaintiff, Ann M., was raped while working in a retail store in the defendant's mall. The plaintiff was unable to show the existence of prior violent crimes at the mall. The Supreme Court of California subsequently ruled that before a landowner could be required to provide security personnel as a matter of law, the court would require a high degree of foreseeability and in its opinion, such a high degree of risk could only be evidenced by a showing of a prior similar act.

While at first it appeared that there was a return to the prior crime rule, a closer reading of the *Ann M.* opinion indicates clearly that given the substantial economic burden created by having to employ security personnel, the California court was merely attempting to balance the interests of the plaintiff against the costs or burden imposed upon the defendant.

Regardless of the rules applied by any particular jurisdiction, *the basic principle is to provide a level of security that logically responds to the level of risk present.* A practitioner in the security industry will benefit from current information about risk levels. Further, given the fluid state of the case law and its unpredictable changes, practitioners also would do well to keep abreast of legal developments in their states.

By using the "totality of circumstances" approach to eradicating the level of risk, security practitioners will be better able to assess the true nature of the risk of crime at their properties than they would if they restricted their analysis to just prior crime at the same site.

RISK FACTORS

Practitioners can consider a number of factors or sources of information that may affect the risk of crime at a property. The following list is representative of the more common sources of crime risk information; however, the reader should be aware that additional sources may be developed at any time.

PRIOR CRIME AT THE SITE

Both property crime and violent crime should be considered when evaluating risk. Both security and police practitioners commonly accept that an environment conducive to criminal activity in the form of property crimes (e.g., theft, vandalism, burglary, etc.) creates a higher risk of more serious violent crime. The rationale for this belief has two parts: (1) Some property crimes could either be a "threshold" offense for the violent crime (e.g., burglar enters a hotel room and subsequently rapes the guest); or (2) the property crime could escalate to a more serious violent crime (e.g., a thief caught attempting to steal a car subsequently assaults the person who caught him).

CRIME HISTORY OF A NEIGHBORHOOD

Defining what constitutes the "neighborhood" is difficult because of differences in geographic terrain, economic status, or other demographic factors. The local police agency may be able to indicate how it defines the "neighborhood" or geographic area where the subject property is located. Once the geographic boundaries are defined, crime data for that area can be obtained and analyzed.

CRIME IN SIMILAR BUSINESS ESTABLISHMENTS

It is well known that certain types of businesses are prone to the same kinds of criminal activity:

- Taverns have problems with assaultive behavior and disorderly conduct.
- Parking lots and garages frequently have car theft problems.
- Convenience stores are subject to robberies.
- Apartments and guest rooms have the problem of burglary to address.

This area of risk analysis requires security practitioners to keep current on crime trends affecting their types of businesses. By reading reported cases and trade publications, and through interaction with their peers in professional associations, they will learn about developing problems and potential solutions.

One of the cases that illustrates this concept is a 1984 decision from Nevada in *Early v. NLV Casino Corp,* 678 P 2d 683 (Nev 1984). Beverly Early, the plaintiff, was a patron at the defendant's hotel casino. While at the casino, Early left her family to use the restroom facilities. When she was in the restroom, a male assailant attacked and robbed her. At the trial, the evidence showed that although the hotel casino had never experienced a crime of the type suffered by Early, there had been numerous instances of this crime occurring at other hotel casinos in town. This information was shared among the hotels through the hotel security directors' association, but the security chief for the defendant testified that due to the small size (in number) of his staff, he was unable to attend the monthly association meetings.

The Nevada court reasoned that this information was readily available and that given the relatively slight burden or effort required of the defendant hotel casino, casino staff should have made themselves aware of the problem and taken appropriate steps to address it. From a practical perspective, had the casino been aware of the crime trend, it could have conducted regular inspections of the restroom facilities. Even if the crime still occurred, the hotel casino would at least have been in a defensible position.

PRIOR COMPLAINTS

Prior complaints made to a property owner about problematic tenants, unsafe conditions, suspicious activity, and so on can be good indicators of the propensity for a security problem or crime to occur subsequently. For example, a resident in an apartment community complains to the property manager about high levels of foot traffic in and out of an adjacent apartment at all hours of the day and night. Also, the resident adds that these "visitors" appear to stay for only a short period of time. There may be a legitimate reason for the activity at the apartment; however, a property owner would be wise to investigate the circumstances, seeking out the aid of the police to help determine whether there is drug activity taking place.

In this hypothetical situation, if the landlord does not take action, and an incident occurs in which another resident or visitor to the complex is injured as a result of drug-related activity (e.g., being struck by a stray bullet in a crossfire situation), then the property owner could be held liable for failing to take any action to investigate the problem. In contrast, should the landlord in this scenario contact the police and make a reasonable effort to discover what is going on, then a subsequent injury will be far less likely to result in liability.

KNOWLEDGE OF CRIME AND ACKNOWLEDGMENT OF RISK

Frequently in negligent security lawsuits, defendants will claim that they were unaware of the existence of certain criminal activity or other significant problems that could establish the foreseeability component of the plaintiff's case. However, when it can be shown that the defendant had actual knowledge of a particular risk, that evidence can have an impact on the outcome of the case.

For example, when in the course of discovery (i.e., plaintiff has requested specified documents from the defendant) the defendant turns over to the plaintiff certain documents, such as security incident reports, memos, correspondence, or other papers indicating that the defendant was aware of the risk and in some manner acknowledged it, the plaintiff may have satisfied the legal requirements for a foreseeable event more easily than if the defendant claims no knowledge of prior crime.

Acknowledgment of the risk can be more subtle than just awareness of a history of prior crimes, but could include some evidence of the defendant's awareness of the potential for problems in designing its organization's security program. The following two examples illustrate this point.

Example 1. In the case of *Mullins v. Pine Manor College,* 389 Mass 47 (1983), 449 NE 2nd 331 (Mass 1983), the Massachusetts Supreme Judicial Court ruled that the college recognized the risk of rape on the campus of an all-women's school because the college provided rape prevention programs during new student orientations. The court indicated that the rape of the plaintiff was not only foreseeable generally, but was actually foreseen (acknowledged) by the college because the college had addressed the risk in their orientation program.

Example 2. In a second Massachusetts case, *Roe v. The Marriott Hotel* (unreported, trial-level case), the plaintiff who was raped in the defendant's parking lot sought to introduce evidence of the foreseeability of the crime through the hotel's own security manual. A portion of the manual outlined the responsibility of the security staff for external patrol and most significantly, the reasons for such patrol. Among the reasons cited for patrol were the prevention of crime such as auto theft and "rape." While the hotel's documented history of crimes in the parking lot was limited to property crimes such as theft and vandalism, the hotel's patrol guide acknowledged the risk of rape.

PLAINTIFF'S THEORIES

Most inadequate security cases fall into these three categories:

1. defective equipment and devices
2. crime deterrence theory
3. operational failures

Defective equipment and devices are fairly self-evident. They would include deficient lighting, inoperative alarms or cameras, and broken locking devices.

Deterrence theory cases are those in which the plaintiff asserts that the assailant would have been deterred from committing the crime had certain security measures been in place at the time of the assault. A common example of such a claim is when the plaintiff is assaulted in a parking lot in an area where security has not been patrolling regularly. The *deterrence theory* argument in this case would state that had the defendant's security staff been patrolling as it should, the assailant would more likely than not have been deterred from committing the crime. Often, the plaintiff will seek to introduce evidence of the assailant's behavior before, during, or after the assault that points to his or her deterrability. Efforts to plan the crime in advance, conceal one's identity (e.g., wearing a mask, using a condom in a rape to avoid DNA detection), or evade capture—see *Sharpe v. Peter Pan Bus Lines,* 401 Mass 788 (1988), 519 NE 2nd 1341, or a discussion by the Massachusetts Supreme Judicial Court on efforts to avoid capture as evidence of deterrability—are all evidence of the assailant's deterrability to be considered by the jury.

The third category of plaintiff theories is *operational failure*. This is a fairly broad category and can include several different deficiencies in an organization's security program, alleged simultaneously. In fact, the reader should note that it is common for a plaintiff to claim several deficiencies in more than one previously discussed theory. Operational failures include such issues as the following:

- failure to train
- failure to supervise
- inadequate preemployment screening (negligent hiring)
- lack of specific policies and procedures or violation of one's own policies
- inadequate staffing levels

Two examples of a violation of one's own policies and procedures are the following:

Example 1. The defendant organization has a policy and procedure that requires regular accounting for master keys and distribution of such keys on a controlled basis. If the evidence in the plaintiff's case indicates that an intruder gained access to the plaintiff's apartment using an unauthorized master key, the failure to comply with one's own key control procedures becomes the central argument in the plaintiff's case.

Example 2. A security officer who fails to properly handle a situation that results in injuries to the plaintiff may be found not to have been properly trained. This type of problem arises in retail environments where the security officer is required to detain and/or arrest a shoplifter, but either too much force is used or an improper search of the person is conducted.

ADEQUACY OF A SECURITY PROGRAM

The adequacy of security is determined in one of two scenarios, either in the *context of a negligent security lawsuit* or in an *overall nonlitigation evaluation for planning purposes*. This distinction can be further explained by describing the analysis of security in a lawsuit. This is when a given set of facts and an identified risk level at a certain time are compared to specific measures, actions, or inactions and evaluated. In a lawsuit, usually only a portion of the defendant's security program is being examined relative to the crime. For example, if the assault occurred outdoors during daylight hours, exterior lighting would not be evaluated.

In a nonlitigation security assessment, when a property owner wants to obtain information about the quality of security at a facility, the overall security program is evaluated. While there can be numerous parallels to the case-analysis approach, *security program evaluations* look at broader issues and are not limited to a single criminal event or set of circumstances. Security program evaluations, in their broader approach, will consider four basic components with numerous subparts. The four basic components include the following:

- crime-risk analysis
- security equipment
- staffing and related issues
- management practices

CRIME-RISK ANALYSIS

A crime-risk analysis is conducted to consider many possible sources for criminal activity, including prior violent and property crime at the site; crime in the surrounding community; crimes common to that type of business or institution; prior complaints by guests, tenants, or customers; and other sources that can be developed during the evaluation process.

Sources of data to conduct the crime-risk analysis are numerous. At the site level, security incident reports, daily logs, complaints, previous claims of inadequate security, and periodic (monthly, annual) summary reports all are helpful to the practitioner in identifying both previously experienced problems as well as potential threats.

Beyond the records of the individual site, the practitioner will want to obtain and consider data from the local police department for the site (not all crimes get reported to site management) and the area considered to be the neighborhood around the site. Computer printouts of reported crimes and calls for service can be obtained from the records division of the local police department.

While there can be some confusion about the availability of such records to the public, these records are considered to be in the public domain. As such, the police agencies must provide this information on request. The protocol in obtaining these printouts will vary from jurisdiction to jurisdiction, and most police departments will charge a fee (it should be nominal). Anyone seeking this data who meets resistance from the local police authority can make the data request pursuant to the Federal Freedom of Information Act.

A security-risk analysis using a standard multistep approach works well at this stage.

The Risk-Analysis Formula. One common approach to analyzing the crime risks and security needs of a facility is the standard risk-analysis formula. This formula has the following seven basic steps:

1. *Identify the assets of the institution.* Assets are divided into two general categories: *people* and *property*. *People* includes tenants, guests, customers, visitors, or anyone else who is lawfully on the property. *Property* includes the tangible assets, such as cash and other valuables, that can be desirable to criminals. The reputation of the business, or "goodwill," should also be included in this category.
2. *Identify the types of risks or threats to those assets. Risks* or *threats* are those crimes likely to occur at a site either due to a history of such events at that property or a history of occurrences at other similar facilities.
3. *Establish probability of loss and frequency of exposure. Frequency of exposure* relates to the regularity of the exposure. For example, if the threat is the assault of patrons at a shopping mall, the frequency would be each day that the mall is open and during the hours of operation.

   Probability of loss is a concept based upon considerations of such issues as prior crime, trends in criminal activity, (e.g., carjacking is currently a "popular" crime), and the general decline or improvement in the neighborhood. This is a consideration of the likelihood of crime occurring at a particular property.
4. *Determine the impact (financial and otherwise) of the loss.* The *cost* of a loss will, to some extent, affect a decision on how many resources are allocated to prevent the occurrence. Some losses, such as an assault on a patron or the negative impact on a business's reputation, are extremely difficult to calculate. Both direct and indirect costs should be evaluated.
5. *Identify solutions to the risks and threats.* What security measures were available at the time of the incident that could have prevented or deterred the criminal conduct?

6. *Determine the feasibility of application.* Will the implementation of the various security solutions seriously impede the operation of the business or facility? For example, a retail store may solve the problem of shoplifting by closing its doors. Without shoppers, it is out of business and thus, this would not be a feasible solution.
7. *Conduct a cost-benefit analysis.* Is it financially feasible to implement the level and/or type of security that would have been required to prevent the crime? Is the cost too high given the level of risk and the impact potential of the loss?

The following hypothetical case provides an illustration of the risk-analysis approach in action.

Hypothetical Case: Parking Garage Assault/Robbery. The Facts: Two elderly women are walking to their car on the fourth floor of a garage in a downtown location on a Saturday afternoon. Shortly after they arrive at the vehicle, a tall male approaches the driver as she opens the door. He grabs her handbag and, in the struggle, she falls to the ground and breaks her hip. The other woman comes to her aid and is slightly injured in the process. An action is pursued on behalf of the women.

Was the level of security in the garage at the time of the incident adequate? Applying the risk-analysis approach to this problem, the following issues and questions are raised.

1. *What are the assets of the garage?* Two categories will usually arise—people and property. In this situation, the "people" are employees of the garage and patrons or other lawful visitors. We already know that the "asset" is the patron, so now the potential threats and risks must be identified.
2. *What are the risks?* In this environment, the obvious risks to a patron include robbery, assault, and theft of property. To the employees, similar risks exist due to the presence of cash on the site.
3. *What is the probability of loss and the frequency of exposure?* The frequency of exposure is simple. Every time a patron enters the garage, there is a theoretical exposure. In this hypothetical case, the garage is open to the public seven days a week from 7:00 a.m. to 11:00 p.m. The exposure to the particular threat, therefore, is during business hours only.

   The probability of loss is more difficult. Probability is based on the same concepts as foreseeability and as such, one factor (there are many) would be the existence of prior crime at this facility. In this example, the garage has a history of numerous thefts and damage to parked vehicles; this history shows a presence of criminal activity inside the facility. There have also been a few robberies and assaults during a three-year period before the incident. Given the history of crime and location of the garage in an urban setting, it is more probable than not that an incident such as this could occur.
4. *What is the impact of such a loss (injury)?* In this instance, the damages incurred by the patrons, the cost of negative publicity, and litigation all make this type of loss substantial, and all should be weighed when considering possible solutions to prevent the risk from occurring.
5. *What potential solutions exist that could have been used to prevent this crime?* Simply controlling access to the facility and patrolling the interior spaces where patrons are exposed are the most logical solutions.
6. *Is it feasible to use access controls and patrols?* Do the solutions substantially impair the

normal operation of the facility? If not, then they may be feasible. In contrast, the doors could be closed and crime eliminated, but then the garage would be out of business, making this "solution" not feasible.

7. *What is the cost versus benefit (burden of the defendant)?* In this hypothetical case, the garage had only one security guard on duty at the time of the assault. During regular business hours Monday through Friday, there were two guards on duty. The question is: *Would it have been feasible and cost-effective to have a second guard on duty when the assault occurred?*

Since there is no interference with the operation of the facility, the remaining issue is one of cost. Based upon the cash flow of the garage, the business could well afford the minor expense of a second guard, and thus the burden was slight to the defendant.

NEGLIGENT HIRING LAW

Negligent hiring is a claim which asserts that the defendant business failed to exercise reasonable care in the selection and hiring of an employee through proper preemployment screening to avoid hiring an individual with an unsuitable background, and where that employee subsequently injures the plaintiff.

Traditionally, employers have been held responsible for the actions of their employees under the legal doctrine of *respondeat superior,* which operates on the principle that employers are responsible for the acts of their employees when tasks are performed on behalf of the employer's interests. If, through the carelessness or negligence of the employee, a member of the public is injured, the employer is held financially and, occasionally, criminally responsible vicariously (e.g., a moving van driver causes an accident and injures others).

However, this doctrine has its limitations. If the employee committed a harmful act outside the scope of his employment and not in the interests of the employer, the company would not likely be held responsible. If the harmful conduct clearly falls outside the scope of his employment, then the doctrine of *respondeat superior* does not apply, and no recovery can usually be made against the employer under that legal theory.

Negligent hiring liability, however, is the failure of an employer to exercise reasonable care in selecting an applicant in light of the risk created by the position to be filled. This means employers must screen individuals adequately *before* they are hired.

Employers are not always exposed to liability just because they failed to check an applicant's background. Liability usually results only when an inadequate screening effort is logically connected to the wrongful conduct. If a reasonably conducted investigation would have revealed facts indicating the applicant was undesirable, the failure to obtain the information may be considered negligence. Conversely, if a reasonable background check did not reveal any negative information about the applicant, then liability would not likely result.

The legal cases emphasize that the amount of screening to be conducted on an applicant should be proportionate to the degree of risk presented by the position to be filled. Hence, the greater the risk, the greater the effort to investigate a potential employee's background.

The case law is also helpful in identifying what constitutes risk. It suggests the risk posed by a particular position depends on *access*. If employees are able to subject others to harm as a result of having *unsupervised access* to them, then risk exists. *Access* is the focal point. To determine risk factors in employment, the focus must be on the position to be filled, not the applicant, the job title, or the wage rate.

When conducting background checks, employers should consider an employee's unsupervised *access* to vulnerable people or dangerous instrumentalities such as:

- children
- elderly persons
- persons who are disabled (mentally or physically)
- patients
- private homes
- master keys
- narcotics (licit or illicit)
- explosives
- dangerous chemicals
- weapons/ammunition

Security employees who exercise arrest power, use force, or carry weapons should also be included, although generally employees who misuse their authority or force subject their companies to *respondeat superior* liability as well. The concept of negligent hiring also encompasses incompetent conduct that causes harm, as illustrated by the overzealous security officer, bouncer, or doorman.

In conclusion, documentation and consistency are the keys to avoiding negligent hiring liability. Whatever options an employer chooses to use for screening, *all* information gathered from legitimate sources should be well documented. Since most litigation is not resolved for years after the employee has been hired, the written word becomes invaluable. Consistency in preemployment screening practices is also important because of the evidentiary effect a company's practices can have at trial. For example, if a decision was made to obtain criminal history data on all positions of risk, be sure to conduct such a search for each and every applicant who becomes an employee.

Employers tend to become lax, especially when they have not experienced any problems. Once a plan is adopted, it represents a self-imposed standard. The failure to follow one's own plan becomes evidence of negligence for a jury to consider.

EQUIPMENT, STAFFING, AND MANAGEMENT

Security equipment, staffing, and management practices are the three major areas making up most security programs. To determine the adequacy of any one of these areas or its components, one must first understand the role of laws and standards.

Role of Statutes, Ordinances, and Regulations. Throughout the United States, there are numerous examples of *statutes, ordinances, and codes* that require specific security measures. In some municipalities, minimum measurable lighting levels are required; in others, certain types of locking devices may be required. Nationwide, for example, OSHA has established guidelines for workplace violence in health care institutions.

One of the first steps the security practitioner must undertake is to inquire whether his or her security program is subject to any statutes, codes, ordinances, or federal regulations. If so, those components specified in the statutes or regulations must meet the stated criteria of the applicable law.

Unfortunately, some laws are written with a degree of ambiguity. An ordinance may require "reasonably secured windows and doors" without guidance on what "reasonably" constitutes. Interpretation may be required by counsel, an experienced consultant, or both. In addition to codes, ordinances, and the like, most states have some amount of appellate case law, which will also state the nature of the legal responsibility (i.e., duty) of the property owner with respect to security measures. Again, all too often the case law does not define with particularity what is required of the property owner.

The important point of consideration is that failing to meet the minimum requirements of the law is very strong evidence of negligence in a lawsuit and, additionally, may subject the institution or business to civil fines.

THE ROLE OF STANDARDS AND GUIDELINES

Historically, the private security industry has been poorly regulated. Frequently, such regulation has only taken the form of limited state statutes that set forth licensing requirements—and on rare occasions, minimum training standards—for contract security agencies or so-called guard companies. Proprietary security staff—individuals who are the direct employees of, for example, a hotel, shopping center, or office building—traditionally have not been regulated by states or municipalities.

Since the early 1970s, when the Connie Francis rape case against a motel in New York received widespread publicity, there has been a multitude of civil litigation alleging inadequate security against privately owned businesses. With many verdicts of more than $1 million and increased public awareness of this alternative remedy for victims of crime, business owners have become motivated to improve the quality of their security services to guests, tenants, visitors, and employees. Unfortunately, given the dearth of standards guiding property owners on how much or what type of security to provide, many failed to take the appropriate steps to properly analyze the risks of crime associated with their businesses. As a consequence, these businesses have failed to provide adequate protection for the public despite their legal duty to do so.

After 30 years of claims against property owners for poor security, a public outcry for nationwide security standards requiring some minimal measures to prevent crime would seem inevitable. In fact, during that 30-year period, only a handful of technical standards were developed by standard-setting organizations. However, these standards typically have been limited to technical items such as locks, fencing, safe construction, or lighting levels. There were no standards or guidelines for the management of security services or the use of security devices in any given application. This means that the landlord of an urban apartment building or the general manager of a downtown hotel would not be able to refer to a written standard regarding what type of locks should be installed on sliding glass doors. The liability of the motel in the Connie Francis case was predicated on the poor-quality locks that were provided for the singer. She was raped in her room by an unknown intruder who gained access via a defective locking device on a sliding glass door.

As recently as the early 1990s, there was still opposition by three major industries to the development of any type of security standard or guideline. The apartment, hotel, and shopping-center industries, through their respective trade groups, fought an effort by the American Society for Testing and Materials (ASTM) to develop minimum guidelines for security measures in all types of privately owned businesses open to the public. A three-year effort to develop the guidelines dissolved with threats to the nonprofit ASTM that it was working outside its charter. Although it is doubtful that there was any charter violation, the organization could not afford the cost of litigation and consequently disbanded the committee.

In late 2000 and early 2001, the National Fire Protection Association (NFPA), another standard-setting organization, made public its intentions to start the process of writing national security standards.

Standards or Guidelines? The difference between a standard and a guideline is to some degree a matter of semantics, and yet, there are distinctions. A standard usually refers to an adopted standard of practice for the construction, design, use, or application of a product or service. For example, there are national standards for the manufacturing of certain types of locking devices. An adopted standard usually goes through a time-consuming consensus-setting process where all interested parties have input on the content. Words such as "shall" are frequently used. Standards can be and often are adopted by municipalities in codes or ordinances, such as a building code.

Guidelines are generally less restrictive than standards, using language such as "it is recommended" or "courses of action may include." By definition, guidelines are meant to provide guidance to the end user—the private business owner or manager—who needs help in identifying options that may be available for a certain type of application.

The legal implications of a standard versus a guideline are somewhat blurry. While a standard is developed over a longer period of time and goes through a more rigorous process, the effect in the courtroom of invoking standards or guidelines is not likely to be very different. For the plaintiff who is introducing a guideline, the objective is to show a jury that there was a business practice that, arguably, the defendant company should have followed in this case. The alleged failure to adhere to that practice or guideline becomes evidence of negligence in most jurisdictions.

The Rationale for Standards. At least two views have emerged on whether standards or guidelines that attempt to regulate the security of private organizations should be adopted. The more conservative view is that no standards or guidelines can be written to fit all circumstances. The "one-size-does-not-fit-all" argument has been made numerous times, including during the early 1990s ASTM effort. However, this argument is misleading. It fails to recognize that many efforts can be undertaken by any size organization to improve the quality of its security program.

The more progressive view on standards development is that they are necessary to ensure a higher level of professionalism within the security industry and to render a more consistent approach to the provision of security measures in any private-sector application. Security standards or guidelines can be written to apply in any given setting or circumstances, a fact which is well illustrated by the *General Security Risk Assessment Guideline* written by the American Society for Industrial Security International Guidelines Commission and approved on November 13, 2002.

General Security Risk Assessment Guideline. The *General Security Risk Assessment Guideline* was written to develop a standardized approach to conducting security risk assessments. Regardless of the application or the business or organization type, there is a long-recognized, logical method of analyzing security risks and identifying the options that are available to manage security-related problems. The *General Security Risk Assessment Guideline* outlines this method. (The guideline is available free online at www.asisonline.org.)

The guideline is described as being "applicable in any environment where people and/or assets are at risk for a security-related incident or event that may result in human death, injury, or loss of an asset." The phrase "a security-related incident or event" is not limited to criminal activity. It also includes natural disasters, war, and other activities that could result in a loss of life or property.

The guideline is a "seven step process that creates a methodology for security professionals by which security risks at a specific location can be identified and communicated, along with appropriate solutions." (It also includes definitions, a flow chart, appendices, and a bibliography.) The guideline's seven-step framework for conducting a security risk assessment is broken down as follows:

1. Understand the organization and identify the people and assets at risk. The first objective for a security practitioner in the risk-assessment process is to understand the nature of the organization being evaluated, including its peculiarities, business purpose, methods of operating, and corporate goals. In addition, the nature of the assets and the type of people at risk are essential pieces of information in a proper risk assessment. The guideline's appendices include two sections: a qualitative approach to risk assessment and a quantitative approach. In the first appendix—which addresses the qualitative approach that will be described further in this section—numerous examples illustrate such issues as what constitutes an "asset" or the type of "people" that the practitioner should consider when making the assessment.
2. Specify loss risk events/vulnerabilities. The guideline defines risks or threats as "those incidents likely to occur at a site, either due to a history of such events or circumstances in the local environment. They can also be based on the intrinsic value of assets housed or present at a facility or event." For clarification of this definition, the reader can again refer to the appendices. For example, the concept of "loss risk" events includes prior crimes at the site or in the immediate vicinity and crimes that may be common to that type of industry (e.g., robberies in convenience stores or burglaries in apartment communities). Loss risk events are not just crime or security-related problems. They also include non-criminal events such as human-made or natural disasters such as storms, power outages, and labor disputes.
3. Establish the probability of loss risk events and frequency of events. In establishing the probability of loss, one should consider such factors as prior incidents, trends, warnings, and threats. The probability is not based on mathematical certainty, but simply a consideration of the likelihood that an event will occur, based on historical data, events at similar establishments, and so forth. For instance, it is well known within the industry that convenience stores are targets for armed robbery. This is primarily because they are cash businesses, often are open 24 hours a day, frequently have only one clerk, and commonly are located at major intersections where there are more escape routes for the criminal. The security practitioner would take this "inherent risk" into account when assessing the probability of future robberies in similar establishments and would provide the appropriate recommendations.
4. Determine the impact of the events. The impact of an event refers to financial, psychological, and other related costs incurred by an organization. "Other related costs" may not be so obvious, such as negative media coverage, poor consumer perception, the inability to obtain insurance coverage (especially after a catastrophic event), or poor employee morale, which affects worker productivity.
5. Develop options to mitigate risks. It is understood and accepted within the security industry that one cannot eliminate all risks or prevent all losses. Frequently, however,

there may be several options or security solutions that can be applied to the same set of factors. Examples of security solutions include staffing, security equipment (e.g., card access systems, closed-circuit television cameras, alarms, lighting, and locks), the transfer of financial risk of loss through insurance coverage, indemnification agreements with security service providers, and a number of creative approaches to address a problem. Security solutions often involve a compromise arising out of the long-standing conflict between security and "convenience." Convenience is the argument that "we have always been doing it that way and it wouldn't be convenient to change the way we operate." The example of forcing employees to use a single entrance to a facility to enhance access control illustrates the problem.

6. Study the feasibility of implementation of options. The questions are whether the security measures available are feasible for an organization and whether the measures would substantially interfere with the organization's operation. If they do substantially interfere, the security measures may not be practical. As an absurd example, if a retail store had severe shoplifting problems, one possible "solution" would be to simply lock the doors of the store. In doing so, the shoplifters would be prevented from stealing the merchandise. Of course, legitimate shoppers would also be prevented from purchasing the merchandise and the store would go out of business. The "solution" here would obviously substantially interfere with the operation.
7. Perform a cost-benefit analysis. Security measures should be proportional to the risks involved. The impact of a loss that involves the death or injury of people can be substantial in a variety of ways—from the obvious emotional costs to the economic harm caused by the loss of key employees. On the other hand, some property losses are more bearable than others, and as such, the security practitioner would be expected to compare the cost of the various options against the cost of the loss. While many people would insist that no cost is too great to save a human life, most would also concede that it makes no sense to spend $100,000 on security equipment to prevent the loss of property worth $1,000.

SUMMARY

Those in charge of security require an explanation of both basic legal principles and practical considerations when called upon to develop security programs that are both reasonable and defendable. While a security program can meet the minimum legal standards, security professionals must be more progressive in developing their programs in order to have a successful operational program. A security program is not static and should continually be evaluated and modified to meet the changing needs of the organization and legal requirements.

BIBLIOGRAPHY

The process of evaluating the risk of crime at a specific location or geographical area is widely recognized and has been adopted nationwide by private industry, public law enforcement, municipalities, and other governmental agencies. The following published sources reference the process used to perform a crime-risk analysis. Note that this bibliography is provided solely as a resource tool for those interested in researching the issue of crime-risk analysis. It is not an endorsement of the publications or content therein.

American Society for Testing & Materials—Draft Standard Guide for Determining Design Criteria for the Physical Protection of a Facility. Philadelphia: American Society for Testing & Materials, August 1996.

Apo, A. "Security for Apartment Buildings." *Crime Prevention Report*. no. 93.10. New York: American Insurance Services Group, October 1991.

Apo, A. "Security for Parking Facilities." *Crime Prevention Report*, no. 93.40. New York: American Insurance Services Group, July 1992.

Bates, N. D. *Premises Liability for the Criminal Acts of Third Parties*. Boston: Massachusetts Bar Association Continuing Legal Education, April 1996.

Bates, N. D. "Security Experts' Preparation of Plaintiff and Defense Cases." In *Negligent Security Cases*. Boston: Massachusetts Bar Association Continuing Legal Education, 1993.

Bellmio, P. "Crime Analysis and Crime Prevention." In *Handbook of Loss Prevention and Crime Prevention*, 2nd ed., edited by L. J. Fennelly. Newton, MA: Butterworth-Heinemann, 1989.

Biery, K. D. Jr., and J. L. Schaub. *The Ultimate Security Survey*. Newton, MA: Butterworth-Heinemann, 1994.

Broder, J. F. *Risk Analysis and the Security Survey*. Newton, MA: Butterworth-Heinemann, 1984.

Bursik, R. J. Jr. *Neighborhoods and Crime*. New York: Lexington Books, 1993.

Chuda, T. J. *Basic Crime Prevention Curriculum*. Columbus, OH: International Society of Crime Prevention Practitioners, 1990.

Crowe, T. D. *Crime Prevention through Environmental Design*. Newton, MA: Butterworth-Heinemann, 1991.

D'Addario, F. J. *Loss Prevention through Crime Analysis*. Newton MA: Butterworth-Heinemann, 1989.

Department of the Navy, Naval Facilities Engineering Command. *Physical Security Design Manual 13.1*. Washington, DC: GPO, 1983.

Fennelly, L. J. *Effective Physical Security: Design, Equipment and Operations*. Newton, MA: Butterworth-Heinemann, 1992.

Fennelly, L. J. *Handbook of Loss Prevention and Crime Prevention*. 3rd ed. Boston, MA: Butterworth-Heinemann, 1996.

Girard, C. M. Planning, Management and Evaluation. In *Handbook of Loss Prevention and Crime Prevention*, 2nd ed., edited by L. J. Fennelly. Newton, MA: Butterworth-Heinemann, 1989.

Green, G. (revised by R. J. Fisher). *Introduction to Security*. 4th ed. Newton, MA: Butterworth-Heinemann, 1987.

"Guide Being Developed for Determining Design Criteria for Physical Protection of a Facility." *ASTM Standardization News* 24, no. 8 (August 1996).

Harold, V. "Site Survey and Risk Assessment." *Effective Physical Security: Design, Equipment and Operations*. Newton, MA: Butterworth-Heinemann, 1992.

Imwinkelreid, E. J. "Admissibility of Nonscientific Expert Testimony." *Trial* 32, no. 10 (October 1996).

Lyons, S. L. *Security of Premises: A Manual for Managers*. London, UK: Butterworth-Heinemann, 1988.

McCartney, L., ed. *Kids and Violence*. National Governors Association. Washington, DC: 1994.

McGoey, C. E. *Security Adequate . . . or Not?* Oakland, CA: Aegis Books, 1990.

Morrison, Mahoney, and Miller. *Defending the Negligent Security Claim*. Boston: 1992.

National Crime Prevention Council. *350 Tested Strategies to Prevent Crime*. Washington, DC: National Crime Prevention Council, 1995.

National Crime Prevention Institute. "Designing Crime Risk Management Systems." In *Handbook of Loss Prevention and Crime Prevention,* 3rd ed., edited by L. J. Fennelly. Newton, MA: Butterworth-Heinemann, 1996.

National Crime Prevention Institute. *Understanding Crime Prevention*. Newton, MA: Butterworth-Heinemann, 1986.

Purpura, P. P. *Security and Loss Prevention: An Introduction*. 2nd ed. Newton, MA: Butterworth-Heinemann, 1991.

Sennewald, C. A. *Effective Security Management*. 2nd ed. Newton, MA: Butterworth-Heinemann, 1985.

Sherman, L. W. "Violent Stranger Crime at a Large Hotel: A Case Study in Risk Assessment Methods." *Security Journal* 1, no. 1 (1989).

U.S. Army Corps of Engineers. *Security Engineering Manual*. Missouri River Division/Omaha District, January 1990.

U.S. Department of Justice. *Vulnerability Assessment of Federal Facilities*. Washington, DC: GPO, 1995.

U.S. Department of Justice, Federal Bureau of Investigation. *Crime in the United States, 1995, Uniform Crime Reports*. Washington, DC: GPO, printed annually.

Waldhuber, R. M. "Assessing Security Needs." In *Security and Safety: Issues and Ideas for Shopping Center Professionals*. New York: International Council of Shopping Centers, 1989.

Walsh, T. J., and R. J. Healy, eds. *Protection of Assets Manual*. Vol. 1, Chap. 2. Santa Monica, CA: Merritt, 1996.

5

Aspects of Professional Liability

Norman M. Spain, JD, CPP

CHAPTER 5

OBJECTIVES

Civil procedure refers to the rights and obligations of parties to a civil trial. *Liability* means a debt owed or legal obligation. The objectives of this chapter are as follows:

- examine the procedures of a civil trial
- highlight the procedures of particular interests to safety professionals responsible for the security function
- describe how courts determine liability
- understand how damages are allocated under a variety of circumstances
- explore pretrial, trial, liability, and post-trial issues

OVERVIEW

Security-related lawsuits are most often filed in state courts, but under some circumstances they can be filed in federal court. The litigation proceedings of the state and federal courts are governed by their respective rules of civil procedure. These rules vary in some ways; however, the rules are essentially the same regardless of jurisdiction. All plaintiffs must file timely complaints that specify the nature of the wrong and the damages sought; and both sides are permitted to engage in discovery, which includes the exchange of interrogatories and requests for documents and the taking of depositions. Both sides can file pretrial motions, such as a motion for summary judgment, and at trial both sides are permitted to present witnesses for their cases, to cross-examine witnesses against their cases, and to appeal an adverse judgment.

Liability can include compensatory damages to make whole the victim and, if the defendant's conduct was malicious or willful and wanton, punitive damages to further punish the defendant. An appellate court can affirm or reverse the decision in total or in part, and/or remand in total or in part the case to the lower court for further proceedings. From the filing of the complaint to the final appeal, it normally takes the courts approximately three to five years to decide a case.

FEDERAL AND STATE RULES OF CIVIL PROCEDURE

Each state and the federal government has its rules of civil procedure that govern the judicial proceedings in that jurisdiction. The similarities between the jurisdictions are far greater than the differences. Although the exact names of documents and procedures vary from state to state, our discussion is based on the United States Code and Federal Rules of Civil Procedure. (*The Federal Rules of Civil Procedure*, which govern the federal courts, were adopted by the United States Supreme Court in complement to 28 *USC* § 1 et seq. and provides practitioners a quick overview of the steps and risks in a civil trial, using a hypothetical trial for negligent security as the example case study.)

SELECTION OF A COURT—STATE VERSUS FEDERAL

Most security-related lawsuits are tort actions governed by state law, and the vast majority of people who file suit do so in state court. Generally, a plaintiff can choose to file suit in the state court of the county where the incident took place or the county where the defendant's business is headquartered.

Under federal rules, a plaintiff can file suit in federal court if the defendant and the plaintiff are from different states (28 *USCA* § 1332[a]). This is the case, for instance, if a traveler is assaulted while out of state or a person assaulted on the property of a business that is incorporated out of state and the plaintiff is seeking more than $50,000 in damages (28 *USCA* § 1331). But the federal court will apply state law, most probably the law of the state where the incident occurred. The U.S. Supreme Court opinion in *Erie R.R. v. Tompkins*, 304 U.S. 64 (1938) is the premier case that directs the federal courts to apply state law, rather than develop a separate federal common law, when deciding issues involving areas of law that are essential state issues and the states have a vested interest.

If the case involves a point of law already decided by that state supreme court (e.g., the state high court has held that employees assaulted on the job are entitled to workers' compensation and cannot sue their employers for inadequate security under the exclusive remedy provision of the state workers' compensation act), the federal court will apply that law to the case before it.

On the other hand, if the state supreme court has not decided a legal issue, the federal court can either submit the case to the state supreme court for an advisory opinion (e.g., *In Re Question Submitted by the U.S. Court of Appeals for the Tenth Circuit, Tobert v. Martin Marietta Corp.*, 759 p 2d 17 [Colo. 1988]) or it can decide the case based on how it believes the state high court would decide the issue (e.g., *Cooke v. Allstate Management Corporation, d/b/a The Park Apartments*, 741 *F Supp* 1205 [DSC 1990]). A federal court opinion applying state law is not subsequently binding on the state courts when deciding later cases.

STATUTE OF LIMITATION

Under the *statute of limitation* in every jurisdiction, a plaintiff must initiate the lawsuit within a proscribed period of time, usually within one to two years after the incident for personal injury cases. Failure by the plaintiff to file in a timely fashion can result, upon motion by the defendant, in dismissal of the case regardless of its merit (*Johnson v. Overnite Transportation* Co., No. CV 3-92-180, U.S. District Ct., D. Minn., Third Div., Jan. 13, 1993).

Practitioners must note that plaintiffs can gain a strategic advantage by filing their cases toward the end of the time periods permitted by statute. When plaintiffs wait to file their complaints there is an increased risk for defendants who may unknowingly misplace or destroy essential records. Typically, the security records that are misfiled or destroyed are not directly related to the incident, such as an investigative report, but usually are more indirectly related, such as shift and post assignments, training records, radio logs, vehicle logs, and similar documents that might aid in documenting the activities of the security department.

PRETRIAL

PLEADINGS COMPLAINT

After the appropriate court for filing has been selected, the plaintiff must file with the court and the defendant a *complaint*, which sets forth the authority of the court to hear the case, the allegations of wrong against the defendant, and the request for damages or relief (*FR Civ* p 8).

A typical complaint in a negligent security case contains allegations that the plaintiff was on the property of the defendant at a specific date and time, he was assaulted on the defendant's property, the assault occurred because of the negligent security of the defendant, and as a result, the plaintiff suffered physical and psychological injuries for which he is requesting specific monetary damages.

PLEADINGS—ANSWER AND/OR MOTION

Upon receipt of a complaint, the defendant must within a specified period of time file an *answer* (*FR Civ* p 8), *motion* (*FR Civ* p 12), or both. In the defendant files an answer, the defendant can do one of the following:

1. Admit specific points as true (e.g., the plaintiff was on the property of the defendant).
2. Give a specific denial to a particular allegation or portion thereof (e.g., the plaintiff was not assaulted as a result of the alleged negligence of the defendant).
3. Give a qualified denial on the grounds the defendant lacks sufficient information to render an opinion (e.g., defendant has no knowledge of the plaintiff being on or assaulted on its property).
4. Give a general denial to the allegations if the defendant wishes to contest all the allegations (e.g., the plaintiff was never on or assaulted on property owed by the defendant, as defendant sold the property prior to the incident).
5. Give a qualified general denial of all facts not specifically admitted (e.g., defendant denies all allegations not specifically admitted in this answer).

The defendant, in lieu of or as part of the answer, can file certain motions challenging the complaint (Michigan armed officer case, and *FR* Civ p 12). For example, if the complaint or portion thereof was vague, the defendant might file a motion for a more definitive statement. Or, if the plaintiff failed to file the complaint in a timely manner, the defendant would file a motion to dismiss on the grounds that the plaintiff failed to comply with the statute of limitation. Or, the defendant might file a motion to dismiss on the grounds the plaintiff failed to state a recognized cause of action upon which relief might be granted. For instance, the plaintiff claimed the defendant was negligent for failing to provide armed security officers and the state supreme court had ruled in another case that businesses had no duty to provide armed officers. (In *Williams v. Cunningham Drug Stores, Inc.*, 418 N W 2d 381 [Mich. 1988], the Michigan Supreme Court ruled businesses had no duty to provide armed officers.)

DISCOVERY

After the complaint and the answer have been properly filed, both sides will engage in pretrial *discovery*, wherein each is permitted reasonable access to all relevant, nonprivileged information held by the other in order to identify sources of evidence and to gather factual evidence about the incident. The primary methods of discovery include the following: *interrogatories* (*FR Civ* p 33), *requests for production of documents* (*FR Civ* p 34), *requests for examination* (*FR Civ* p 35), *requests for admission* (*FR Civ* p 36), and *depositions* (*FR Civ* p 30).

Interrogatories are written questions to which the sending party asks the receiving party to respond. For example, the plaintiff may want to know the number of security officers employed, the number of officers on duty at the time of the incident, and the posts or patrol patterns of the officers. The defendant, on the other hand, might wish to know why the plaintiff was on its property, what the plaintiff was doing several hours before and up to the incident, and whether the plaintiff knew the assailant.

Requests for production of documents are demands by one party for the other party to produce relevant documents. Although both parties have equal right to request documents, the defendants usually have significantly more discoverable documents. A plaintiff can request copies of the security manual, post orders, time cards and shift assignments, investigative and incident reports, department training records, memos and correspondence relating to security, department fiscal records, personnel records, and other relevant records. The defendant typically requests medical and insurance records that detail the extent of the plaintiff's injuries.

Request for an examination is typically used by a defendant to obtain an independent medical or psychological evaluation of the plaintiff's injuries. Although a defendant can depose the plaintiff's physician and obtain by a request documents, including the plaintiff's medical records, a request for examination provides the defendant an opportunity to obtain a second opinion on the cause and severity of injuries.

Depositions are sworn statements by the parties, witnesses, experts, and other individuals with relevant information about the case. Depositions are taken in interviews with counsels for both parties present and recorded by a court stenographer. The vast majority of questions are asked by opposing counsel. Counsel for the party being deposed can interrupt and ask that a question be clarified, object to a question as being out of line, or even threaten to end the deposition if opposing counsel is unreasonable or abusive. After opposing counsel has completed the questioning, counsel for the despondent can ask questions, typically to clarify earlier answers.

A deposition can later be admitted at trial to prove facts of the case, to challenge the witnesses' knowledge of the facts, and to challenge the credibility of witnesses. For example, if the plaintiff stated in deposition, he did not know his assailant and the defendant introduced at trial two impartial witnesses who stated they saw the plaintiff talking with his assailant prior to the incident, a jury might reasonably conclude the plaintiff was not a reliable witness. On the other hand, if a security officer stated in deposition it was dark where the incident occurred and subsequently stated at trial the area was well lighted, the plaintiff's counsel would bring forth the officer's deposition to challenge the veracity of his courtroom statement.

Depositions also provide an opportunity for attorneys to evaluate the credentials and the testimony of security practitioners and expert witnesses. Opposing counsel can challenge the expertise of the witness, test his knowledge of the facts, question what opinions he holds about the case, try a few difficult questions to evaluate how the witness reacts under pressure, and evaluate how well the practitioner or expert might relate to a jury.

Because depositions often involve the recollection of facts that occurred years ago, involve detailed and difficult lines of questioning, and involve the giving of sworn statements that can later be admitted at trial to prove facts in dispute or to challenge testimony or credibility, proper preparation of all witnesses to be deposed is a critical step for both the plaintiff and defense counsels. Lack of preparation at the deposition stage, regardless of the experience and knowledge of any witness, can result in critical, irreversible mistakes.

Practitioners need to remember that front-line officers and supervisors who do not possess any direct knowledge about the case can still be deposed by the plaintiff. The objective of the plaintiff is to obtain an inside view of the department to evaluate its professionalism or lack thereof. Therefore, any officer who is to be deposed, just like any witness for the plaintiff or the defendant, must be properly prepared.

Finally, depositions can be introduced at trial if the witness can no longer be located or has died. A deposition, therefore, is also one way to preserve testimony.

Request for admission is a document used by both parties to identify facts of the case that are undisputed and to narrow the focus of the case to the facts that are in dispute and must be proven. Once a request for admission has been filed, within a set period of time, the other side must respond with its *answers and objections*. The responding party can admit to a statement, such as "the plaintiff was on the property of the defendant at the time of the incident." Or, the responding party can object to an assertion, such as "there were no lights in the parking lot the evening of the incident," and explain its answer. Failure by a party to timely respond to a request for admission can be considered as an admission of all the facts claimed by the opposing side, which means the facts can no longer be contested before the court.

OUT-OF-COURT SETTLEMENTS

At any time during litigation—before, during, or after trial—the two sides can reach a settlement. Settlements are often reached after discovery because both sides have had the opportunity to evaluate the strengths and weaknesses of the case. The reasons, often unrelated to the merit of a case, for settlement are many. It can be strictly cost-related, that is, less expensive for a party to settle than to incur any or further litigation costs. Either party can be emotionally drained or it may not wish to risk losing at trial. A prevailing party can even settle a case for less money if the other party gives up the right of appeal. To reduce case backlogs at pretrial conferences, judges also can exert substantial pressure on the parties to settle.

PRETRIAL MOTIONS

During discovery and prior to trial the parties can file different pretrial motions with the court, including motions to *compel discovery* or to grant a *protective order* to shield against excessive discovery demands; or to enter an *involuntary dismissal*, *default judgment*, or *summary judgment*.

If the two parties, for example, cannot agree on what prior security incident reports are discoverable—the plaintiff wants reports from the last five years, but the defendant wants to produce reports from only the last two years—either might file a motion with the court. The plaintiff might move for an *order compelling discovery* on the grounds that the defendant was unreasonably obstructing the plaintiff's efforts to obtain evidence (*FR Civ* p 37). On the other hand, the defendant might file for a *protective order* on the grounds that the plaintiff was abusing the discovery process by making excessive demands for production of documents (*FR Civ* p 26). If a defendant fails to provide requested documents, to respond to interrogatories, or to attend depositions after having been directed to do so by the court, the plaintiff can file a motion for default judgment (*Fed R* Civ p 37). Similarly, if a plaintiff fails to move forward with the case or to cooperate with discovery after being so ordered by the court, the defendant can move for an *involuntary dismissal* of the case (*FR Civ* p 41).

Either party also can move for a *summary judgment* (*FR Civ* p 56). This is a request that a judgment be entered for the party making the motion. For the court to enter a summary judgment, all parties must agree on the essential facts of the case, and the court must review the facts in a light most favorable to the opposing party. If the facts of the case are such that a reasonable person could reach only one decision, the judge, before or during a trial, can grant a summary judgment (*FR Civ* p 56). These judgments expedite justice; however, they can be appealed.

TRIAL

VOIR DIRE

If the parties are unable to settle or to obtain a pretrial judgment, the case will be brought to trial. As common law tort actions, most security related lawsuits are usually tried before a jury; however, the parties can waive this right and try the case before a judge (*FR Civ* p 38).

If it is a jury trial, attorneys for both sides will be allowed to question prospective jurors, either directly or indirectly by submitting questions to the judge. The examination of jurors is called *voir dire*.

Each side is permitted an unlimited number of *cause challenges* to prevent individuals who have knowledge, opinions, or beliefs that adversely affect their ability to render a fair and impartial decision from being sworn in. The decision to excuse or not to excuse the potential juror for *cause* rests with the judge.

Each side is also permitted a limited number of *peremptory challenges*, where a potential juror can be excused without cause. Attorneys often use peremptory challenges based on their intuition and research into the attitudes of jurors.

Contrary to the popular belief that a jury is always made up of 12 people, jury size can vary. The U.S. Supreme Court has affirmed that juries in civil cases can number as few as six people (*Colgrove v. Battin*, 413 US 149 [1973]).

PRESENTATION OF EVIDENCE

After the jury has been sworn in, the plaintiff's counsel will make an *opening statement* that outlines the case against the defendant. Afterwards, defense counsel will give an opening statement that summarizes why the defendant should not be held liable for the incident.

After opening statements, the plaintiff will call its witnesses. After all the plaintiff's witnesses have been examined and cross-examined, the defendant will present its witnesses. Typical witnesses include the plaintiff, defendant, witnesses at the scene, security manager, supervisors and officers, responding police officers, employees, and experts from the disciplines of security, medicine, and economics.

Direct examination refers to the questioning of a witness by the attorney who called the witness to testify. When conducting a direct examination the attorney is not permitted to ask *leading questions*, that is, questions that contain the answer. For example, an attorney should not ask his witness, "*Was the parking lot dark?*" Instead, on direct examination the attorney should ask open questions so the witness can freely answer in her words. The attorney might ask, "*Please describe the lighting in the parking lot at the time of the incident.*"

After a witness has been examined, opposing counsel can immediately conduct a *cross-examination*. On cross-examination, the attorney is permitted to ask leading questions that often require the witness to respond with a "*yes*" or "*no*" answer. If the plaintiff, on direct examination, stated the parking lot was dark, on cross-examination, opposing counsel might ask the following questions: "*Do you know the total number of lights in the lot?*" "*Do you know how many lights were on at the time of the incident?*" "*Do you remember if the sky was clear that evening?*" "*Do you remember if there was full moon that evening?*" "*Were you drinking that night?*" The intent behind cross-examination, since a trial is an adversarial process, is to subject the testimony and character of the witness to careful scrutiny so that a jury can properly evaluate the merit of the testimony.

If the testimony or credibility of a witness was successfully challenged on cross-examination, his attorney can conduct a *redirect examination* to clarify the testimony or resurrect the credibility of the witness. After redirect examination the opposing attorney can conduct a *recross examination* of the witness.

Once all testimony is completed, the defendant and the plaintiff respectively present their *closing arguments* to the jury. This is to summarize the evidence and reasons why the jury should reach a decision in favor of their respective clients.

JURY CHARGE, BURDEN OF PROOF, AND VERDICT

After closing arguments, the judge will summarize the evidence, explain the burden of proof and points of law the jury needs to reach its decision, and charge the jury to weigh the evidence according to the law and to render a fair and impartial verdict.

The plaintiff has the burden of proof. He must prove by a *preponderance of the evidence* that the plaintiff's injuries were caused by the defendant. When the evidence of both sides is considered, the weight of the evidence favors the plaintiff.

Usually a jury is instructed to reach a *general verdict*. Occasionally, in a complex case, a jury can be directed to answer specific *interrogatories* (questions) set forth by the judge and then to set forth its verdict (*FR Civ* p 49). The interrogatories permit the trial judge to be certain the jury did not misapply the law.

After it has completed its deliberation, the jury will return to the courtroom to announce its decision. The judge typically asks the jury foreman to announce the verdict. Or if the jury was provided with a set of interrogatories, the judge can also ask that the interrogatories be answered in open court. Upon request, or at her discretion, the judge can poll each juror to verify the verdict.

LIABILITY

COMPENSATORY AND PUNITIVE DAMAGES

If a court finds a defendant liable, it can award *compensatory* or *actual damages* to financially *make whole* the plaintiff for his actual losses. This typically includes compensation for medical expenses, lost past and future wages, pain and suffering, emotional distress, and lost or damaged property. When translated into dollars, these damages can be quite substantial and difficult to predict. For example, how much might a jury award the victim of a rape for emotional distress if the rapist was infected with HIV?

A court can also award *punitive* or *exemplary damages* if it finds the defendant acted with malice or reckless indifference to others. The dual purposes of punitive damages are to punish the defendant and to deter similar conduct by others (see e.g., *Jardel Co., Inc. v. Hughes*, 523 A2d 518 [Del 1987]). Punitive damages should be reasonably related to the compensatory damages; however, these damages can vary widely. But trial and appellate courts have the authority to reduce excessive damages and have done so on many occasions (see *Jardel Co., Inc. v. Hughes*, 523 A 2d 518 [Del 1987], a quarter-million dollar punitive judgment overturned, and *Craigo v. Circus-Circus Enterprises, Inc.*, 786 p 2d 22 [Nev 1990], punitive damages of $1 million overturned).

MULTIPLE DEFENDANTS

If the plaintiff prevails and there are multiple defendants, the court must determine the liability of the different parties. If the parties are unrelated, the court can assess separate damages against each defendant. If the parties are related, the court must evaluate the relationships, usually by application of the *doctrine of respondeat superior* or independent contractor rule, to determine the collective or individual liability of the parties, or combination of these.

DOCTRINE OF RESPONDEAT SUPERIOR

A corporation is a legal entity that can be sued and held liable in its own name. In addition, a corporation (principal) can be held vicariously liable, under the *doctrine of respondeat superior,* for the acts of its employees (agents) that are committed with the *scope of employment.* An agent is deemed to have acted within the scope of employment if the conduct was under the direction of or for the benefit to the principal (e.g., a tavern held liable for security officers' excessive use of force when removing a disruptive patron (*Market Tavern Inc. v. Bowen*, 610 A 2d 295 [Md App 1992])). But if an agent acted outside the scope of employment, the principal is typically not liable (e.g., a company was not liable for an armed security officer who, as a practical joke, drew a weapon and accidentally shot someone (*Kensington Associates v. West*, 362 S E 2d 900 [Va 1987])). However, if an employee used his fists, which was not a permissible use of force, and the corporation was aware of and failed to take corrective action, the corporation can be found liable on the theory it approved of the conduct (*Market Tavern*, supra note 33).

INDEPENDENT CONTRACTOR RULE

Many corporations hire private security firms. These security companies are independent contractors. An independent contractor performs work per the standards and practices of its profession or trade for a principal. Under the *independent contractor rule*, a corporation is not liable for the acts and omissions of an independent contractor, such as a contract security officer.

However, a corporation can be liable if it reserves or exercises the right of control over the agents of the independent contractor. For example, it reserves the right of control in a contract for guard services agreement or it exercises that right by directing contract officers to perform duties not listed in the contract. Further, it can be liable if it knows of and fails to take reasonable efforts to stop past or current misconduct by an independent contractor. In addition, corporations in some states are liable for the intentional torts, not negligent torts, of contract security officers, as a matter of law (*Peachtree-Cain Co. v. McBee*, 327 S E 2d 188 [Ga 1985]). The principle of law underlying this reasoning is that security is a personal or inherently dangerous duty that cannot be delegated away.

Corporations can manage the risk of liability for the acts and omissions of contract security officers by requiring the contract agency to carry adequate liability insurance and the client be named as an additional insured party. The corporation can also require the contract agency hold the client harmless for the acts and omissions of its officers and to indemnify the client for any damages or losses it may incur as a result of the officer's conduct. A corporation can also reduce its liability exposure by evaluating the contract agency's ability to provide qualified and properly supervised officers, requiring the contract agency do background checks of its officers, requiring the officers be trained, and requiring proper policies and procedures be prepared, followed, and enforced.

INDEPENDENT CONTRACTOR RULE AND POLICE OFFICERS WORKING SECURITY

A corporation may or may not be liable for the conduct of a police officer hired to perform security duties. If the court holds the officer was acting as a police officer at the time and was engaged in a police function, the court can find the officer was not an agent of the client corporation (*Beck v. Paideia School, Inc.*, 381 S E 2d 132 [Ga App 1989]). In the alternative, a court can find that the officer was acting as an agent of the client and hold the corporation liable for the officer's misconduct (*Pardon v. Finkel*, 540 N W 2d 774 [Mich App 1995]). If the officer acted under color of law in violating the plaintiff's rights, a court may even hold the officer and client liable for violation of the plaintiff's civil rights as specified under 22 *USC* §1983 (*Murray v. Wal-Mart, Inc.*, 874 F 2d 555 [8th Cir 1989]).

LIABILITY FOR INJURY TO CONTRACT SECURITY AND POLICE OFFICERS

If a contract security officer is injured due to negligence of the client, the officer might be able to collect workers' compensation benefits from the contract security employer and sue the client corporation for its negligence. For example, a contract officer was permitted to sue a client for requiring him to walk across an icy path, where he slipped, as part of his duty (*Remas v. Duquesne Light Co. v. Gregg Security Services, Inc.*, No. GD 86-12384, Court of Common Pleas, Allegheny County, PA; 136 *Pittsburgh Legal Journal* 93 [1988]).

Police officers and firefighters who are injured in response to calls for assistance cannot sue under the Firefighters' Rule (e.g., a police officer was denied the right to sue when he was injured helping to subdue a disruptive hospital patient in *Seibert Security Services, Inc. v. Superior Court*, 22 Cal Rptr 514 [Cal App 4 Dist 1993]). Such suits would deter the public from seeking assistance. However, if a business increases the risk of injury to the officer by willful and wanton conduct, an injured officer may be able to sue. For example, an injured police officer was permitted to sue a tavern with a long history of fights on the grounds its owner had not taken appropriate security precautions (*Adelsperger v. Riverboat, Inc.*, 573 So 2d 80 [Fla App 2 Dist 1990]). Further, an off-duty officer who is injured while working security as an independent contractor may be able to sue the business under limited circumstances. Instead of bringing a lawsuit, a police officer injured while working off-duty security can file a claim for workers' compensation. These benefits can be charged against the corporation (e.g., cost of workers' compensation benefits for a police officer injured while detaining a suspected shoplifter at a grocery store was split between the city and the company (*Domanoski v. Borough of Fanwood v. The Great Atlantic & Pacific Tea Company*, 568 A 2d 123 [NJ Super A D 1989])).

PERSONAL LIABILITY

An employee can also be held personally liable. If an employee acts outside the scope of employment, the employee can be held personally liable for the conduct, whereas, the principal and employee can be held liable together if the agent acted within the scope of employment. The principal often pays all compensatory damages under such circumstances. If the court awards punitive damages, these damages can be assessed separately against the employer and employee. To minimize the risk of personal liability, employees must be certain to always act within the scope of their employment and to avoid engaging in conduct that could be considered to justify punitive damages.

CONTRACT LIABILITY

Corporations often contract with security alarm companies to provide burglary protection. In reviewing such contracts for services, safety personnel should recognize that most alarm companies do not assume any liability for their acts of negligence or gross negligence, and in the event of any loss, these contracts often contain stipulated damage clauses that restrict the liability of the alarm company to a predetermined amount or percentage of the annual premium for service. Since alarm companies are not viewed as insurance companies, these contract limitations on liability are often enforceable in court regardless of the magnitude of the loss suffered by the client corporation (a $50 limit of liability clause was upheld in a suit involving a loss of approximately $92,000 (*Fretwell v. Protection Alarm Co.*, 764 p 2d 149 [Okla 1988])). However, as a matter of public policy, the courts in some jurisdictions have held that alarm companies cannot limit their liability for their acts of gross negligence (e.g., the owner of fur store that suffered approximately a $300,000 loss was permitted to sue an alarm company for gross negligence on grounds that it failed to install systems per Underwriters Laboratory standards (*David Gutter Furs v. Jewelers Protection Services*, 571 N Y S 2d 702 [A D 1 Dept 1991])). But in another case, the New York Court of Appeals held that the failure of an alarm company to wire a skylight was a significant act of negligence but not a gross act of negligence. Therefore, it held the alarm company was entitled to summary judgment in a suit brought against it resulting from a theft of $10 million in paintings from an art gallery (*Colnaghi, U.S.A., Ltd. v. Jewelers Protective Services, Ltd.*, 595 N Y S 2d 381 [Ct App 1993]). Alarm companies can also be sued if they misrepresent the capabilities of their alarm systems. For example, an employee who was raped was permitted to sue an alarm company when she was trained that the alarm system being used had a panic feature and it did not (*Elizabeth E. v. ADT Security Systems West, Inc.*, 839 p 2d 1308 [Nev 1992]).

TRIAL AND POST-TRIAL MOTIONS AND APPEALS

TRIAL AND POST-TRIAL MOTIONS

During and after a trial either party can file motions with the court for a *directed verdict, judgment notwithstanding the verdict*, and *new trial* (*FR Civ* p 50).

After the plaintiff has presented his case the defense can move for a *directed verdict* on the grounds the plaintiff failed to meet the burden of proof. The plaintiff also, after the defendant has presented his case, can move for a *directed verdict* on the grounds the defendant failed to rebut the prior proof of fault.

After a trial, the losing party can move for the judge to overturn the jury verdict by issuing a *judgment notwithstanding the verdict* on the grounds the verdict was contrary to the evidence (*Elizabeth E v. ADT Security Systems West, Inc.*, 839 p 2d 1308 [Nev 1992]). The losing party might also file a motion for a *new trial* on the grounds the verdict was against the weight of the evidence (*FR Civ* p 50). The difference between the two motions is that the judge substitutes his opinion for that of a jury when issuing a judgment notwithstanding the verdict, whereas, by granting a new trial, the judge merely sets aside the jury verdict and calls a new jury to decide the case.

APPEALS—THE COURTS

If unsuccessful with post-trial motions either party can appeal directly to an intermediate appellate court or supreme court. Generally, states with smaller populations have a two-tier judicial system: the trial courts and a supreme court. States with larger populations and the federal government have a three-tier system: trial courts, intermediate courts of appeal, and supreme court. A note of caution: in some states the highest court is not called the supreme court. For example, in Maryland and New York, the highest court is the state court of appeals.

APPEALABLE ISSUES—QUESTIONS OF FACT VERSUS QUESTIONS OF LAW

In a jury trial, the judge decides *questions of law*, whereas, the jury decides *questions of fact*. Generally, only questions of law are appealable. For example, what evidence is admissible at a trial is a question of law. If a judge ruled before trial that under the prior similar crimes rule only prior crimes of violence on the property of the defendant were admissible to prove the assault on the plaintiff was foreseeable, this is a question of law. Whether there is sufficient number of prior incidents to prove the assault was foreseeable is generally a question of fact for the jury.

If a plaintiff wanted to introduce evidence of all prior crimes, including thefts and other nonviolent incidents, and lost the case, she might appeal the trial judge's ruling. The plaintiff would argue that the appellate court should reverse the trial judge's ruling and order a new trial permitting the plaintiff to use evidence of all crimes on defendant's property to show the assault on the plaintiff was foreseeable.

On occasion, if the facts are such that a reasonable person could reach only one decision—there were no prior crimes—a court, as a means to further the ends of justice, might hold as a matter of law, rather than fact, an assault was unforeseeable.

APPEALS—APPELLATE COURT OPINIONS

A higher court can *affirm* or *reverse* all or part of the lower court's decision, and it can enter a *final judgment* or *reverse and remand* the case to the lower court for a new trial on all or some of the issues. It typically takes three to five years, depending on the jurisdiction, from the time a complaint is filed until a final decision is rendered on an appeal.

The opinion of an appellate court is called the *majority opinion*. It sets forth the decision and reasoning of the majority of justices. A justice(s) can file a *concurring opinion* if he agrees with the decision but for different reason(s) than expressed in the majority opinion. If a justice(s) disagrees with the majority opinion, she can write a *minority* or *dissenting opinion* explaining why. The writing of a concurring or dissenting opinion is not always a fruitless exercise. Either opinion can be adopted by a higher appellate court, a sister appellate court from the same jurisdiction, an appellate court from a different jurisdiction, or since the law is ever changing, it can be adopted by the same court at a future date in a similar case.

SUMMARY

A civil lawsuit begins with the filing of the complaint by the plaintiff in a state or federal court. The complaint must be filed within the statute of limitations and it must set forth the facts upon which the plaintiff seeks relief. Thereafter, both sides can engage in discovery to obtain relevant evidence by exchanging interrogatories, requests for admission, and requests for production of documents, and taking depositions.

Either side can file pretrial motions with the court, including motions to compel discovery, to protect against abuse of discovery, to dismiss for failing to comply with the statute of limitation or failing to state a cause of action, or to grant summary judgment. At trial, both sides will be permitted to evaluate the jury during *voir dire* and to challenge, either for cause or preemptory reasons, prospective jurors who might be incapable of rendering an impartial verdict. After the jury has been sworn in, each side is permitted to make opening statements, to call witnesses for, to cross-examine witnesses against, and to present a closing argument.

If the plaintiff prevails, the court can award compensatory and/or punitive damages. These damages can be assessed directly against a corporation, contractor security provider, or employee, or a combination of these. Typically, a corporation is liable only for the acts of its employees that are committed within the scope of employment. A corporation is not generally liable for the acts of a contract security officer; however, it can be if the corporation reserved or exercised the right of control over the officer, knew of past or current misconduct, and failed to take corrective action, or it can be held liable for intentional acts of the officers in some states as a matter of law.

After a verdict has been announced, the losing party can file motions with the court to issue a judgment notwithstanding the verdict or to grant a new trial. If unsuccessful, the losing party can appeal. An appellate court can affirm or reverse the judgment, or remand the case for further proceedings on all or some of the issues.

BIBLIOGRAPHY

BOOKS

Hartsfield, W. *Investigating Employee Misconduct*. Deerfield IL: Callaghan, 1988.

Inbau, F.E., Farber, B.J., and Arnold, D.W. *Protective Security Law*, 2nd ed. Boston: Butterworth-Heinemann, 1996.

Keegan, F.R., and Fogleman, J.L. eds. *Avoiding Liability in Premise Security*. Atlanta: Strafford, 1987.

Maxwell, D.A. *Private Security Law: Case Studies*. Boston: Butterworth-Heinemann, 1993.

Nementh, C.P. *Private Security and the Law*. Cincinnati: Anderson, 1989.

Risser, R. *Stay Out of Court: The Manager's Guide to Preventing Employee Lawsuits*. Englewood Cliffs, NJ: Prentice Hall, 1993.

NEWSLETTERS

Law Reporter. Washington DC: Association of Trial Lawyers of America.

Private Security Case Law Reporter. Atlanta: Stafford.

Security and Special Police Legal Update. Chicago: Americans for Effective Law Enforcement.

Security Management Bulletin. Waterford CT: Bureau of Business Practice.

Part 3
Organizational Management of Security

PART 3

6

Defining a Security Management Program

Richard S. Post, CPP

CHAPTER 6

OBJECTIVE

The objective of this chapter is to present the common key elements of all effective security management programs.

INTRODUCTION

The past 20 years have seen more changes in how security is viewed by society and corporate management than in any other time in history. In the post–9/11 world, how people think about security and the nature of threats posed to the average person and corporation have changed dramatically. The days of operating a security department as an extension of facility maintenance, building services, or even a facility "police department" are over. Security cost-effectiveness is as important as that of any other area of the organization. While the value the security function provides to the company must be shown, pressure from government and shareholders has forced changes to structure, operations, and focus of security programs. If the Department of Homeland Security designates a company's facilities as part of our nation's critical infrastructure or considers it a terror target because of chemicals stored or used, its risk profile and security requirements will change. Likewise, liability concerns may drive the question of how much security is enough, and a high incidence of crime near facilities may establish an environment where more rather than less security is required.

A range of protective services and options is available to companies, including:

- fences, lighting, locks, doors, access control, guards, ID cards, surveillance systems
- procedures, policies, employee involvement
- protection of information, products, intellectual property, the supply chain
- protection of key facilities
- executive protection, political risk management
- compliance with government-mandated programs or contract terms
- specific government regulations to obtain or keep business (from the Department of Energy, Federal Aviation Administration, Drug Enforcement Agency, etc.)

In addition, security may be provided in a number of ways and through a variety of services, including:

- company (proprietary) employees, company security officers, security management
- contract guard services, contract alarm monitoring services
- off-duty law enforcement personnel
- consultants, contract security managers
- hybrid company and contract operations
- specialty services: executive protection, investigations, physical security audits, security policy development, and training

What is security management? It is providing required and desired services based on the specific needs of each and every company that has such a corporate function. There is a range of organizational approaches, management structures, funding mechanisms, benchmark standards, value propositions, metrics, and degrees of professionalism associated with each program.

Security program expectations and characteristics in a company depend on what business or industry the company is in and its geographical location. There are three considerations for these programs: core expectations, variable requirements, and mandated programs.

- core expectations
    - protect employees
    - protect facilities
    - protect physical assets
- variable requirements
    - protect information, intellectual property
    - protect products
    - protect the supply chain
    - protect information technology systems (IT)
    - protect international executives
- mandated programs
    - federal/state/local government program compliance
    - executive decisions within the company (key employee protection, payroll movements, special event planning, etc.)

Consequently, the focus of any security program development must be the protection of the facility and its employees. The main questions to answer when creating a security program in your organization include the following:

1. How can the security function enhance profitability?
2. What must be done to protect the main interests of the company and facilities?
3. What programs are essential to provide required protection?
4. How much staff is needed to carry out required tasks?
5. How much money is needed to fund required programs?

Clearly defined security goals and objectives are necessary for a cost-effective operation. Simple in concept, these key elements include providing physical security and employee protection, protecting key profit-making activities, and securing information. In program development and operation, both senior management and the staff responsible for security must understand the philosophy of the security program, its reason for being, key goals, and program design criteria. All programs are based on the following hierarchy of needs:

- corporate goals
- products
- facility
- personnel
- environment

In essence, all security management systems attempt to place a series of barriers between what the company wants protected and what the company wants to be protected against. Each type of barrier has value in providing protection against certain types of risks or threats. Proper selection of barriers, combined with other aspects of the system, provides the overall protective framework for the company. Each of these barriers has associated costs and expected levels of effectiveness. Effective security programs will optimize the associated barriers (time delays) economically.

The management techniques and controls required to implement and operate successful systems are also critical. Operational plans need to be developed along with required budgets. Staffing and support requirements and associated standards, as well as emergency response actions, must be established, approved by senior management, and provided to all involved levels of management.

In business today, security program managers must continually refine their approaches to managing losses in their companies. Finding the correct formula for a company requires considerable analysis, creativity, and ingenuity. Remember, regardless of how good your service is, if the customer—company management—won't buy it, you are out of business. Adding other corporate functions such as safety, fire, facilities, and supply-chain controls to the security portfolio may be required to maintain an adequate budget for security program operations. Multitasking and being viewed as a generalist, cross-functional manager is a part of the current security landscape and should be accepted. Conversely, some organizations want only extremely specialized security personnel to handle closely circumscribed protection tasks such as IT firewall security, brand protection, or international security risks.

Local business unit security managers often select resources that appear to add value to their operation. Companies organized in this way hope to build closer relationships with their customers and anticipate their requirements. This is something that security management must do as well. Using proactive techniques and emphasizing support for general management concerns of the business is essential. A more integrated use of security concepts in business activities is essential if management is to create value and support overall corporate goals.

The manager of a security function must ensure that the following fundamental objectives are properly addressed:

1. determine security needs for the company—what is important to the continued well-being of the company, its employees, and shareholders
2. determine the availability of company resources to accomplish the required or desired level of protection for the operations
3. develop an acceptable strategic company response to the threats, risks, and vulnerabilities present
4. obtain company management acceptance of plans and programs
5. obtain the necessary budgetary approval for plans and ensure that required capital expenditures and operating budgets are available
6. organize the approved resources through the application of acceptable corporate philosophy and style
7. develop and implement appropriate control systems to effectively manage the approved plans
8. determine that adequate levels of properly trained and experienced staff are available to provide required services/support
9. implement the required reporting and data management system to provide timely information for adequate communications to all levels of management
10. select key areas of responsibility for direct involvement by corporate staff
11. determine that adequate backup support is present to guarantee continuity of service in the event of crisis situations or unusual workloads
12. develop plans and strategies for active involvement by all levels of management and employees in asset protection activities

13. schedule a timetable so program results are evaluated on a regular and consistent basis following approved management guidelines for this purpose
14. make certain that the improvement process is coordinated with the planning program for both current operations and future budget planning

It is helpful to know how other companies provide for the protection of their assets. While it is not possible to merely copy what another company does, understanding how it has adapted its plans and strategies to its requirements is helpful.

BACKGROUND

The primary purpose of any security management system is the protection of company assets. Both the people and non–people problems of the company need to be completely addressed during the risk analysis and program recommendations. The resulting plan tells how asset protection may be accomplished by cost avoidance, risk reduction, claim avoidance from incidents, and cost-effective operations. The manner in which the program is implemented, procedures are employed, and lost or stolen property is recovered is important for a successful program. However it is accomplished, it must be a total solution for a total problem. The vulnerability analysis, while important, is only as useful as the development of the programs and plans which are the countermeasures used to ensure the protection of company assets.

The security program uses people, software, and hardware to accomplish its goals. The appropriate use of each of these essential elements of the security management program will determine just how effectively program goals will be accomplished. The major reasons losses occur is a failure to recognize risks or vulnerabilities, to change programs to respond to a changing environment, or to use the correct software, hardware, or people to provide protection.

Not all security programs can be cost-justified, but they are nevertheless required for company operations. Whether they can be quantified or not, providing a safe work environment for employees is essential from both a practical and legal perspective. In the era of the downsized company, different techniques are required.

What should management expect from the security function? It is reasonable for general management to expect the equivalent of a business plan to describe how company assets will be protected. This plan must clearly state who is responsible, at all levels of management, for protecting company assets on a day-to-day basis. A regular report on the status of both programs and losses should highlight strengths and weaknesses requiring improvement. Management must be informed about why reported losses are occurring and what can be done to minimize future occurrences.

Furthermore, management should be provided with a clear statement of protective objectives for the company, along with the plans formulated to accomplish them. Since cost is critical, they must be told why proposed expenditures are absolutely necessary. In other words, the security manager must determine how much expenditure is enough.

It is impossible to completely protect the company. Regardless of how much money a company might provide for programs, there are no guarantees that losses will not occur. If the activities or products of a company are sufficiently attractive and a determined criminal attack takes place, it will probably be successful. Even the best security programs will only delay or deter an internal or external criminal attack, not prevent one. What then can a security manager or corporate executive do to lower risk or minimize losses?

First of all, getting the focus of any protection program correct is the key to a sound, cost-effective approach to lowering risks and threats. How these are evaluated depends on the specific issues present in the company at a particular time. It is a truism with the security profession that most losses result from insufficient attention to details of the operation. If key issues are overlooked in planning, or if they are identified but not properly provided for, well- designed programs will not be developed. For example, are strategic corporate goals coordinated with security plans and programs?

The development of a security plan must follow the same corporate criteria as any other function in both development and performance evaluation. If this is done, it will quickly become apparent if specific protective issues are being managed without significant problems developing and if costs are within allowable tolerances. Using such criteria is likely to produce a list of corporate "must-do" activities rather than "should-do" or "nice-to-do" activities, as often appear in security planning documents. Cost-effective security program management presents a system for ensuring that key issues are properly identified and managed (see Figure 6–1) .

All "must-do" items should be given priority, and there should be no surprises for management on security issues. Getting and maintaining support from corporate, business units, employees, law enforcement, and vendors is critical to success. How do you do it?

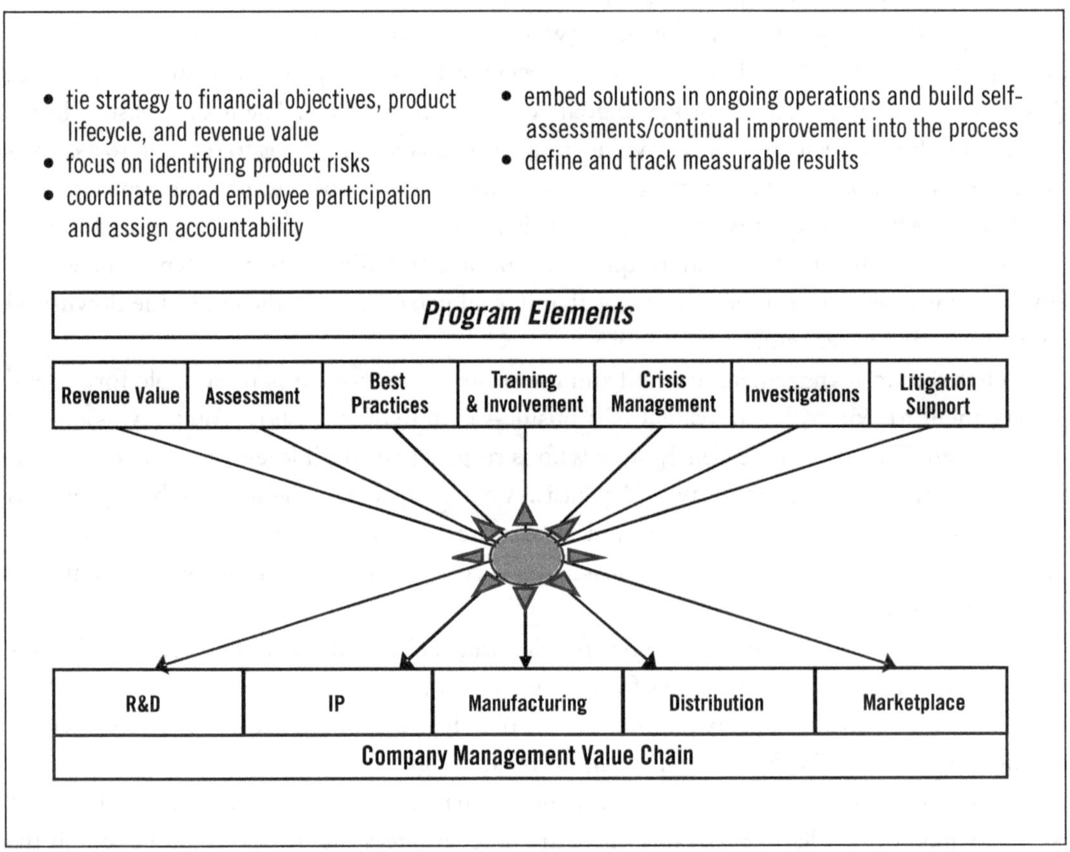

Figure 6–1. Cost-Effective Security Program Management. *Source: Copyright 2009 Post & Post LLC. Used with permission.*

GOAL SETTING

Every corporation has a minimum level of security requirements (must-do activities), a series of additional activities that are a professional's responsibility (should-do activities), as well as some nonessential but desirable activities (nice-to-do activities). Often the problem is deciding which are which! You must, however, not only decide which are which, but be very selective within each category to ensure that only the most essential are actually determined to be the objectives of the security department.

In order to reach this point, some analysis of corporate requirements is required. To do this properly, a database must be either developed or identified to provide the statistical support for decisions. Likewise, an analytical framework must be selected and validated for use within the company. The model for the security program to follow must also be developed.

REQUIREMENTS FOR APPLICATION

The security plan defines responsibilities for the function in the company. The analysis employed during the development of the plan will determine its cost-effectiveness.

How do you get started in managing an effective security management system? What are you selling when you involve senior management in loss-prevention programs?

Security is the most atypical activity in business and industry. Most activities in an organization are concerned with the competition produced by rival organizations and attempt to move the organization toward its goals of profit and operating effectiveness. They are concerned with profit and competition. Security and loss prevention, on the other hand, are concerned with sustenance and safety. (Security attempts to provide them as a primary activity.) If the management staff is unsuccessful in sustaining its activities, security will have nothing to protect. Conversely, if security is not effective, the organization will suffer damage ranging from barely appreciable losses to complete destruction.

Security of the organization is the goal. If the goal has been attained, security is said to be successful. This use of the term is relatively empty in much the same sense that one could say the goal of life is death. The death of a person is unquestionably the terminal condition of life. It, however, says little about the process through which the end is attained. Similarly, to merely state that "security is complete" or "total" is to say nothing. How was that condition brought about? What means have been employed for attaining the desired state? Progress toward a goal such as security must be measured in human terms as well as the degree of organizational protection provided.

The organization that induces its members to work for its advancement or for the development of new and better products or services is most likely to survive and move forward at a rapid rate. Any procedure or practice that makes it difficult for the members to perceive the benefits from full and open participation in organizational survival is harmful and counterproductive to the overall security of the organization. Thus, security should be considered, not as a terminal state to be attained, but as a means or functional activity to assist senior management in the overall attainment of organizational goals.

The underlying motivations for security directors to organize effective security programs include:

- desire for reduced losses
- need to create employee awareness of the consequences of violation of policies

- need to inform employees of their responsibilities for asset protection
- desire to involve employees in loss prevention actively
- desire to modify values, behavior, or attitude while on the job
- desire to provide job enlargement

Can security directors creatively apply protection programs to individuals and/or corporations?

Goal ambiguity is probably one of the more critical problems in developing effective security programs. If you cannot precisely state the protective goals, it will be very difficult to determine if they have been attained. This is particularly true in the development of security programs since they deal with basic values, beliefs, and attitudes that the employees bring with them to an organization.

In other words, to have a secure company environment, it must, in the estimation of those participating in it, first be worth securing. Security is often spoken of as a program to conserve economic resources for an organization. These resources can be tangible physical objects or the services of individuals who are employed by it. Security programs are established out of economic necessities.

Public and organizational policies on security are controversial. Citizens or employees hold a wide variety of views about what should be done in security programming, how it should be accomplished, and what amount of either public or private resources should be allocated to the development of the programs. It has been suggested in current literature relating to criminology and crime prevention that a significant portion of the work of the security profession is in the development of the perception (feeling) of security by the users of the service.

One of the primary characteristics of creating that perception is that it does not require any of the security department's resources or dilute its other efforts. The promotion of the perception of security is probably one of the truly creative aspects of security administration. Creating the perception of a secure environment thus is a spin-off, a side benefit, accruing to the security director through his or her imaginative development of policies, programs, and services.

The development of security in an objective sense—the protection of lives, for example—is obviously important. But the development of security in a subjective sense—creating the perception of protection from external threats—cannot be accomplished without eliminating organizational conditions unacceptable to the majority of employees.

The development of security in a projective sense (eliminating anxiety and fear) is of little value in itself. The fear of losses and the organizational anxiety produced by it can have harmful effects on the entire organization. It tends to reduce the organizational perspective from one of goal orientation to one of defensiveness or threat orientation. Anxieties about losses or threats to the organization are reduced only when adequate information is available about real and not perceived losses or potential for losses within the organization or directed against it.

Every individual taking part in organized activities brings to that group a set of beliefs, values, and needs. As long as the group assists the individual in the attainment of his goals, it will have his active and full support and participation. However, if the group does not fulfill his goals, or attempts to impose unacceptable patterns of belief or conduct upon him, he will reject the group and its goals. Once this is done he will either leave the group or work against it internally. In organizations, the options of such an employee range from quitting to staying and causing problems.

The longer a person remains in an organization, unless he is progressing according to his expectations, the more likely he is to remain and cause problems. As he becomes less and less marketable outside the organization, the more he will seek to solidify his position internally even though he might not be happy with his job or work conditions. His personal security demands it. Consequently, efforts must be made to increase the acceptance of company goals. Each organization has specific modes of conduct that it prescribes for its members. Its goal are promoted and held up as things to be striven for by members.

A considerable number of organizational histories seem to support the deduction that the less violent the organizational changes and the less individuals are required to change basic operating patterns, the greater the probability of successfully integrating them into the organization. This in many ways runs counter to the corporate tenet that the individual must be made to comply with the organization's rules and regulations, whatever they might be.

MANAGEMENT CONCEPTS

Development of management support for security programs and systems is very important. There must be a focus by security department management on the employee's perception of importance, effectiveness, and the cost of services. Directing and redirecting the mission to deal with emerging issues affecting the company is a continual process. So whether or not the goal is the establishment of security as a profit center rather than only a cost to the company, the application of sound management principles is essential for improving and maintaining support within the company.

REQUIRED POLICY

Every security program requires a definitive statement of corporate policy to both give structure to the department and its efforts and clearly articulate the key goals and objectives of the program. Both dimensions of policy must be the result of the security professional's analysis of company requirements and vision to establish the charter for the operations of the security function within the company. Determining what is most important to protect and how it should be done provides the basis for all program activities. A sample policy guide is included in the Case Study at the end of this chapter. It identifies the major policies for information security, physical security, emergency situations, protection of assets, and a description of the company security program. A sample policy on information security is also included.

ANALYZING RISK

Determining the company's security needs involves identifying what is important for the continued well-being of the company, its employees, and shareholders. This is more than the traditional security or risk vulnerability survey. It is the integration of this basic risk information with the strategic interests of the company to focus on key business concerns, strategies, and projects that require the most attention. The development of a basic level of protection for all company requirements in information and people protection as well as physical controls is the first step in the process. Once this has been determined, the security professional can decide on specific tasks to be performed, at what level, and how they add value to the company. Figure 6–2 presents an overview of how such a system operates in the security department.

Risk has a constantly changing value for the company. It increases and decreases as new products, services, or events enter the evaluation process. In highly competitive industries, risk is strongly associated with the lead time required to get a new product to the marketplace from the research and development process. In the pharmaceuticals industry, for example, all R&D

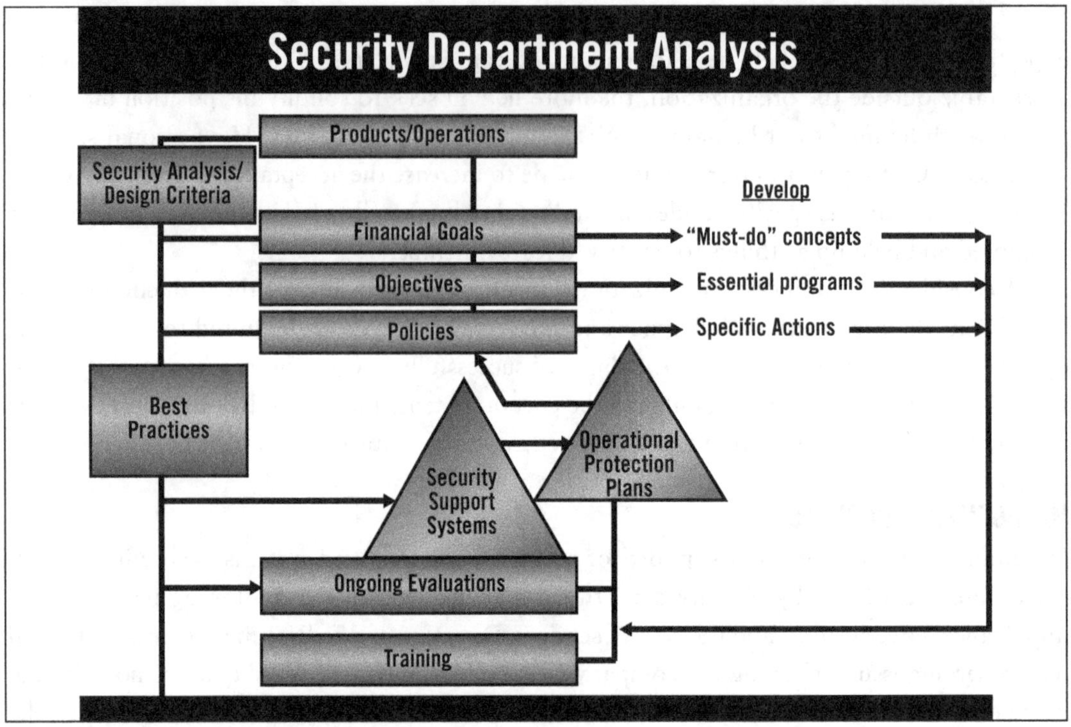

Figure 6–2. Security Department Analysis. *Source: Copyright 2009 Post & Post LLC. Used with permission.*

costs often must be recovered before the original patent expires, allowing competitors to market like products. Should counterfeit products or infringements of patents occur before an investment is recovered, there could be serious losses for the company. Consequently, risk analysis, protection for R&D efforts, and patent protection and investigations would have a high priority in determining risk levels and corresponding security efforts.

AVAILABLE RESOURCES

Determining the availability of corporate resources to accomplish the required or desired level of protection for the operations is not always simple. The budget for the security department is usually well defined. The resources of the corporation devoted to providing protection are often not so clearly apparent. Getting an inventory of the kinds of security features present in a company, such as locks, fences, access control systems, and the like, is sometimes difficult. Facility records are not always kept in ways that allow a clear picture of the state of readiness or level of sophistication of equipment for providing a desired level of protection.

Getting division or business unit management to answer survey requests can be difficult and time consuming and often will not result in anything useful being achieved. Determining the internal control systems and procedures being followed to protect internal assets is usually better, as regular audit procedures and operations analysis often touch on these areas and document existing procedures. A review of these controls can be useful in determining how good control systems might be in current operations.

The level of guarding expenditures, staffing schedules, and standing and post orders all present some description of usable resources. More problematic is the level of support for programs, policies, procedures, and controls by the people who must make them work, namely the management and employees of the company. Are programs accepted by those using them or are they perceived as making it more difficult to do business? Are programs or procedures viewed as making employees feel secure and appreciated or considered a bother or, worse, useless? Have there been any incidents where any of the programs, policies, or procedures have proven useful by stopping a loss incident or protecting an employee, or are they considered a waste of time or a joke?

DEVELOPING A STRATEGIC RESPONSE

Developing an acceptable strategic response by the company to the threats, risks, and vulnerabilities present is the most important aspect of the entire management process of the security function. This is where all of the input from the risk threat analysis, an understanding of the strategic plans and requirements of the company, and the acceptability of various controls and systems are blended, modified, and finally codified. This is where the "must dos" are separated from the "nice to dos" and the protective plan is coordinated with the strategic plan of the company.

Each company has its unique set of requirements for protection. The strategic response for one may be nothing like that for another, even within the same industry. This is because every company has a different set of internal requirements, is at a different state of corporate growth and development, and has a different corporate culture. While the basic features of company protection programs will have similarities, there can be significant differences between how one company approaches a problem and how another addresses the same issue.

Example 1. At a large multinational company several employees were killed by a disgruntled worker. The CEO decided that there should never be another incident like that again in the company. This resulted in a several-hundred-million-dollar upgrade to physical security and access control systems for the worldwide company. Another company with a similar problem reviewed its requirements and made some upgrades in its lesser protected facilities but no wholesale improvements.

Both companies did what they thought was the correct thing in response to a similar problem. Both programs were very different responses to what was viewed as a strategic problem. Was one approach better than the other? Did one company value its employees more than the other? Or did one just have more money to spend than the other?

The attitudes, money available, and a myriad of often unrelated factors can greatly impact how a strategic problem is handled.

Example 2. A large multinational company was being coerced by terrorists to pay a "tax" to keep one of its foreign operations from having problems. Local management recommended that the relatively small payments be made so the business would be protected. Corporate management determined that the payments would fall under the guidelines of the Foreign Corrupt Practices Act and could not be made. The terrorists stepped up their pressure and executed 11 security guards in an ambush and again demanded payment. The company again refused. An additional 13 were killed the following day. Negotiations were then conducted by the security director of the company and the terrorists, which resulted in another attack against the employees of the operation, killing several more people.

The company determined that, despite the local police and military providing protection, it was impossible to meet the legal requirements and still protect its employees. It quickly sold the foreign business to another foreign company, which promptly made the payments demanded. It was able to conduct normal business thereafter with no threats or attacks. This was a strategically important business but it was sold, regardless of the monetary cost, because it could not be properly protected and the human cost was too high. The foreign company had no problems in making the payments to the terrorists. It let the government worry about how to deal with the terrorists in other locations, as it had provided for its own security.

Which company had the correct strategic response to the problem?

ESTABLISHING CORPORATE ACCEPTANCE OF PLANS AND PROGRAMS

As previously discussed, selling the security plan is key to both its success after implementation and to getting all the required support during the approval process. Each company has a specific format for review of yearly plans and budget forecasts. While the required structure must be followed, selling takes place in planning meetings, project reviews, and any opportunity where senior management can be told what the security department is doing for the company. It involves understanding the business the company is in as well as lobbying senior executives and finding ways to help improve profitability through the security function.

There is also the question of input-output considerations in preparing and approving the budget. There must be a balance between what is spent for protection and the value to the company for the expenditure. The program must have a focus, and costs for protection must reflect that focus. Several additional criteria must be considered in organizing the program:

- What are the ongoing threats or risks that this program is addressing?
- What is the cost of eliminating the existing or planned program?
- What alternatives are there to the proposed program?

GETTING THE BUDGET

Obtaining the necessary budgetary approval for plans and ensuring that required capital is available is essential. Excess expenditures are hidden in every budget request, and management has the responsibility to find and eliminate them. Budget planning is an art form in many organizations, and the final budget may have little if anything to do with the merits of the programs and plans presented. Regardless, the budget that evolves through the unique process in a specific company is what the department will have to work with for the next fiscal period. It is therefore essential that the security manager learn how the system works and work the system for the benefit of the company and the department. Getting the budget is often a matter of timing. Having a sound program is sometimes not enough.

Example 3. A large multinational company headquarters was located immediately adjacent to another multinational headquarters that had a high risk and threat exposure to terrorists. The company security director received intelligence that a known terrorist group was planning a bomb attack against one of the several headquarters facilities in the area. An upgraded physical security plan was included in the budget of the next fiscal period. When the attack did not materialize, the upgrades were deleted by the budget review committee before the final budget was approved.

Just after the new budgets were put into effect, the adjacent facility was bombed. The following morning, the CEO called the security director and asked what the company was doing about upgrading security of the headquarters. He was told that there was a plan but it had not been approved as part of the budget. After the CEO reviewed it, it was approved that day and implemented within 30 days at a cost of several hundred thousand dollars. If it had not already been proposed and had to be crafted and submitted through the various review committees, it probably would never have been approved.

Just after the new program went into effect, the terrorists attempted to enter the company premises but were turned back, as the new security program was installed after they had completed the planning for the bombing of this facility and they did not know about the upgrades. They were unsuccessful and eventually arrested after another successful attack several miles from the facility.

It pays to have the contingency plan for budget items prepared, as timing is always critical to acceptance. It is always easier to get approval for a security budget after an incident than before it.

ORGANIZING THE APPROVED RESOURCES

Organize the approved resources through the application of acceptable corporate philosophy and style. Once the plan is approved, get it implemented as quickly as possible and be consistent while doing it properly. Programs that require cooperation from other parts of the company need to be properly coordinated and scheduled so that they can be accomplished in a timely manner. It is therefore essential that in the planning process all required internal approvals be collected and funds allocated so that all required assistance is likewise funded.

Example 4. The security director of a large company planned for and obtained approval for conducting a corporatewide security awareness training program to be offered in conjunction with divisional human resources personnel. After the program was approved and schedules planned and materials developed, the company learned that the funds for employee training in the divisions had been eliminated for the year. Corporate security was not able to conduct the program as planned and finally canceled the effort.

Why did this happen? There was a breakdown in communication and coordination in the budget planning process, and there was not enough buy-in by human resources for the program, which resulted in security not being informed about their budget cut in time to modify the program. As a result, both departments looked bad to management and, more importantly, an opportunity to provide a necessary training and protective program was lost.

DEVELOPING AND IMPLEMENTING CONTROL SYSTEMS

Develop and implement appropriate control systems to effectively manage the approved plans. All corporations have budget and financial control systems that both report and analyze performance against budget, usually on a monthly basis. If these financial control documents are not sufficient for security management purposes, they should be supplemented with more detailed reporting from either divisional security staff or corporate staff responsible for managing portions of the overall budget. If contract guard service management is a staff responsibility, reviews of manpower assignment logs, daily reports, deficiency reports, and billings should be conducted. If numerous investigators are in use, their billing and budget should be reviewed or problems will develop unnoticed.

Example 5. A large bank hired a private investigative firm to locate a former employee who embezzled $4 million and fled to a series of foreign countries. A weekly budget had been agreed upon and a contract written. The investigation continued for several months with just enough success that the security manager did not want to give up the search. As the budget increased, the private detectives began demanding payment from the bank. At that point, the fees had reached several hundred thousand dollars. As the case entered the sixth month, the security director said the bank had to reduce the manpower on the case, as it was getting worried about the costs involved, but that a check was forthcoming for the balance due. The investigators demanded that some money be paid or they would stop work. When the security director declined payment, they went to the general counsel of the bank for payment. It was subsequently discovered that the security manager had not told bank management about the costs being incurred and, since there had not been any results, he held the bills in his desk drawer without telling anyone that they were well over half a million dollars.

Fortunately, the investigation had continued during this controversy and the suspect was apprehended and the money returned to the bank. A settlement was made with the bank and the security manager was reprimanded. He had decided that although the cost was far in excess of his proposed budget to the bank, if the investigation were successful, the bank would not mind paying a high fee. He therefore held off submitting the invoices until there was a success. It unfortunately took longer than he had planned and caused serious problems for all parties concerned.

Always ensure that all company polices are followed in both planning for an investigation and determining any budget or plan associated with it.

STAFF PLANNING

Ensure that adequate levels of properly trained and experienced staff are available to provide required services promised in a plan. Recruitment and selection of staff, even given the current oversupply of highly qualified professionals in the job market, can be a time consuming and tedious process. Getting the correct match in both skills and corporate style is essential if programs are to be implemented successfully. A lead time of several months is not uncommon, especially if the candidate has a current position and must give notice and relocate. If these factors are not taken into consideration in developing the plan, successful implementation could be jeopardized.

REPORTING AND DATA MANAGEMENT

Implement the required reporting and data management system so that timely information for adequate communications can be provided to all levels of management. Depending on the type of organizational structure, the requirement for regular reporting on incidents or operations can vary considerably. These factors, while they are all part of the planning process, must be carefully managed to ensure that all required and "need-to-know" information reaches the correct levels of management.

The establishment or continued operation of a security department data management system must be secured against access by nonsecurity personnel, and employee-related data must be managed in ways consistent with federal and state employee information privacy guidelines in each state where the security department maintains an office.

Communication of security-related information, political risk reports of a sensitive nature, and reports on sensitive investigations must be transmitted, stored, and protected at the highest level of sensitivity allowed by company data systems. Separate and stand-alone data management systems are preferable for all security-related data.

Timely communications via e-mail or any other form of easily accessed data networks should be discouraged except for the most normal information and routine management communications.

KEY AREAS OF DIRECT INVOLVEMENT

There will always be several select key areas of responsibility for direct involvement by corporate security staff. These activities often include special investigations into sensitive matters within the company, often involving senior executives or highly sensitive commercial activities. Executive protection for top executives, political risk reporting and international security, policy development, and ethics violations are most often on the list.

Whether there is a business unit security department or only a single corporate security manager, certain selected issues, approved in the plan by senior management, are reserved for handling by corporate staff only. While assistance might be provided by consultants, reporting and control over these activities is held at corporate headquarters.

Example 6. A large multinational company suspected that a senior executive of a foreign operation was involved in kickbacks. Because of the sensitive nature of the executive's position, corporate security was authorized to hire private investigators to conduct an outside investigation into his activities. Outside counsel was also hired so that internal corporate counsel was kept out of the investigation until the case was completed.

Based on the information collected and legal advice, the company charged the executive with violations of the federal RICO statutes. He was arrested, charged, and eventually convicted and forced to make restitution to the company. Division management was not informed of the case until the morning he was arrested.

In this situation the corporation wanted corporate security to manage the investigation but do it outside of the company so that there would be no chance of a leak occurring while it was being conducted.

ENSURING ADEQUATE BACKUP

Ensure adequate backup support to protect continuity of service in the event of crisis situations or unusual workloads. There are always peaks and valleys in the work load of a security department. Usually when it is busy, more work appears as if by magic! These contingencies must be planned for so that there will be no loss of effectiveness when they occur.

Most security departments have a file of consultants available to provide a variety of services in required areas. These files include the specialty, location, rates charged, references, and confirmation of proper licenses or credentials. Unless the quality of the work is personally known to the security department, interviews or trial assignments are given to verify quality before consultants are required for a serious matter.

Succession planning within security departments, except in the largest corporations, is almost unheard of today. It is generally not possible to have enough internal security staff to handle a significantly increased work load without contracting for the service with consultants or service providers. These contracts need to be in place, retainers paid (if required), and contracted staff knowledgeable about company operations so they can go to work immediately without lengthy training periods.

GETTING EMPLOYEES INVOLVED

Develop plans and strategies for the active involvement by all levels of management and employees in asset protection activities. If strategic issues have been properly identified and sold during the planning and budget process, management and employees should be convinced of the value of their participation in security programs. It is possible, however, that an active program of security awareness training might be required to raise their level of sensitivity to the value of taking personal responsibility for asset protection. This can take place though focused training sessions related to job performance, profit improvement, or personal protection.

As noted elsewhere in this chapter, obtaining voluntary participation from employees is a valid corporate function and essential for an effective security program. It is better to have 1,000 highly motivated and actively involved employees protecting company assets than one highly trained security manager who cannot get anyone to listen.

EVALUATING PROGRAMS

Ensure that program results are evaluated on a regular and consistent basis following approved management guidelines. The impact on clients of the departmental program is often easy to measure. Do people call and ask you to do things for them? Do they refer other people to you for answers to questions or solutions of problems? Have known incidents been prevented by programs or has liability exposure been reduced by department efforts or initiatives? In other words, are the customers happy?

If operational tests are performed to ensure adequate levels of preparation are present, how does staff respond? If improvements are required, are they completed in a timely manner? There are also quantitative measurements that can be applied to other management functions. Measure whether program goals have been attained, if expenditures are within budget, and if losses are kept under control.

THE PLANNING PROCESS

Ensure that the improvement process is coordinated with planning for future programs and future budget planning. If risk or threat levels have changed from the original planning assumptions, programs and current operating plans should be changed immediately.

Improvement is a continuous process involving all company personnel responsible for asset protection. Where shortcomings or failures are noted, improvement plans, which may include immediate revisions to procedures, may be necessary. Well-trained and motivated staff will understand the need for continual improvement. The more common use of TQM, ISO, and other quality improvement and assurance techniques makes the institutionalization of the improvement process something more people are comfortable with in the company. Improvement planning is no longer seen as an expression of displeasure with personal performance but rather an ongoing requirement in business to stay current with changes in the business environment.

MANAGEMENT CONTROL SYSTEMS

Several specific management techniques have been used in the security function. They include cost-benefit analysis and benchmarking, and these require some additional comments.

COST-BENEFIT ANALYSIS

A complex series of issues relating to organizational, operational, and cost considerations must be managed within each organization. Each security manager is responsible for directing the activities of her department within the constraints imposed by her company. To the extent that she has identified risks, developed programs, and anticipated problems, she can be considered successful.

There are numerous techniques for evaluating the costs and benefits of a security program. Their effectiveness in revealing absolute values from either spending or not spending funds for specific protective measures has yet to be demonstrated. It is still not possible to quantify the complete cost of having a criminal attack except to calculate direct and indirect costs associated with an occurrence. While helpful, such calculations are no more specific than financial models, which utilize return on investment models for making judgments about the efficacy of a specific security purchase or budget request. Criminological theory is not yet developed enough to make more than educated guesses about what will or will not have a monetary value or financial return for a company making an investment in security.

However, it is possible to apply uniform financial criteria to a security program, as can be done for any other company activity, and make a judgment about cost versus value. The more accepting of risk the company might be, the more likely the financial evaluation model will emphasize lower expenditures and rely more on insurance coverage to settle loss claims rather than spend money to prevent an event that may or may not occur.

Developing metrics for security operations measurement is an ongoing process within the security industry. The Security Executive Council published the most recent attempt to establish these measurements in George K. Campbell's book, *Measures and Metrics in Corporate Security*. The American Society for Industrial Security has several publications that include company-specific models for establishing baseline measurements. The important thing is to identify what is important to the company and track performance against the goals you have set.

BENCHMARKING

The development of best practices for providing protection is a growing trend in security management. Review of operations, to establish the most effective, measured methods of providing service consistency, is a long-term strategy that adds value to security.

The uses of benchmarking include both operational and administrative, including:

- improving security officer performance
- upgrading physical security practices
- validating budgets and plans
- evaluating salary data
- determining risks and vulnerabilities
- verifying supplier information

Benchmarking groups have developed in many major industrial groups, and clearinghouses have been established to assist in developing a broad database from which to compare and analyze existing company practices against the established standards.

The budget for the security operation is another form of planning. The final budget is a decision package of the approved programs and methods accepted by company management based on the security professional's risk-cost evaluation. This budget, unless it is subject to a zero-based system, increases or decreases from prior years' program funding. Zero-based systems require managers to review risks and threats and establish programs to deal with current issues without regard to long-term programs, staff, or prior commitments.

SUMMARY

Security is a management function with all of the same accountability issues and responsibilities as any other company activity. The key requirement is that the company has a thoughtful process of considering risks and responses. Programs and systems have been developed and implemented to handle the day-to-day emergencies as well as the unusual emergencies and crisis situations. Protection is a result of being prepared. The old adage "chance favors the prepared mind" is absolutely the case in protective planning and management of security issues for the company. Thinking about unthinkable events, preparing for such disruptive events, and managing the crisis effectively are the true tests of the security program. How the assets and resources of the company are organized and managed depends in no small part on the vision, creativity, and salesmanship abilities of the security professionals in charge.

Analyzing the risks and threats, obtaining management and employee support, and responding to threats are the hallmarks of an effective manager. Ensuring that losses are kept to a minimum, within the agreed-upon budget constraints, and that controls and procedures are accepted by the majority of employees, are the goals of any security program. An effective security program is what works within a particular organization.

One program style does not fit all companies. Like any other management goal, the correct blend of cost-benefit-performance is an ongoing and constantly changing target. This is both the opportunity and challenge for the security manager.

CASE STUDY: CORPORATE POLICY ON SECURITY

Case Objective
Letter from CEO for Policy Statements
Corporate Policy on Security
Security Program Summary
Analysis of the Security Department

Case Objective

The objective of this case is to outline the basic policy information concerning the security of Company assets.

Letter from CEO for Policy Statements

Month, 20XX

To: Key Management List

Protecting our assets is the responsibility of each and every employee in the Company. To assist in this effort, I am providing you with the attached Corporate Policy on Security. This policy and the five Corporate Security Department policy guides that flow from it are intended to establish the working relationships between management, employees, and the departments in the Company responsible for assisting us to protect our assets from theft, loss, and risk, and to provide support to all of you in dealing with emergency and crisis situations. The design of the new Corporate Security Program is summarized in the attached Program Summary.

In managing our business effectively, it is critical that our assets be safeguarded through sound management practices. We are showing, by the scope of the policy, our continued resolve to take effective management action and to highlight our responsibilities to protect our assets and increase profitability.

It is essential that each of you review your individual security requirements and develop an appropriate strategy. The Corporate Security Advisor is available to assist in the identification of risks and in the development of strategies. The advisor is also responsible for policy guidance and such protective functions as investigation, risk assessment, and emergency operations most efficiently performed as a corporate function.

This policy will be in effect throughout the Company and is to be used for your guidance in developing your business unit and facilities to comply with it. Please ensure that it is communicated throughout your organization.

—YOUR COMPANY CEO

Corporate Policy on Security

Need for a Policy

The protection of Company assets, personnel, and information is a primary concern of corporate management, as well as all employees. Without a clear understanding of the corporate philosophy for protection, management and employees cannot fully participate in the program. This policy sets forth the goals, objectives, and methods by which the Company can protect assets and profitability.

The development and maintenance of a security program is the responsibility of each business unit. The application of corporate security policies is a line responsibility of business unit and facility management. It is also a job requirement for all employees within all YOUR COMPANY companies.

The Corporate Security Advisor will provide implementation guidelines and technical assistance to assist business units, facilities, and departments to develop individual security operating plans and assist in the implementation of programs.

Specific protective policies, as well as the activities of the Corporate Security Advisor, are delineated in the Corporate Security policy statements.

Statement of Policy

It is the Policy of YOUR COMPANY that:

1. Individual managers and employees are responsible for the protection of assets under their supervision, control, or use.
2. Protection of life, property, and assets of the corporation is a key element of sound management practice.
3. Each business unit and facility is responsible for the development of a comprehensive security program.
4. The Corporate Security Advisor provides operational policy guidance and support to all business sectors, facilities, and operations of the Company so that they may develop sound asset protection programs at the lowest operational level within the organization.

5. Each business unit executive, manager, or employee who suspects or has knowledge of security-related incidents involving YOUR COMPANY operations must report such information through appropriate channels to the Corporate Security Advisor or General Counsel.
6. All losses with security implications involving YOUR COMPANY property or personnel will be investigated by the Corporate Security Advisor.
7. The Company will, in appropriate circumstances, report to public law enforcement authorities apparent violations of the criminal law with respect to Company personnel, property, or proprietary information.
8. Any decision to initiate prosecution of a criminal incident shall be made by a business unit executive officer with the advice and consent of the Chairman and General Counsel of the Company.
9. Regular facility procedural and operational reviews will be conducted by the Corporate Security Advisor (and the General Auditor) to assist managers in maintaining compliance with all aspects of the program which they have developed or for which they have responsibility.

Application of Policy

This policy will apply to all personnel in all YOUR COMPANY business units, facilities, and operations. Each manager will be responsible for ensuring that those under his or her direction comply with all provisions of the policy. Support for sector and facility management in the development of implementation procedures for the physical protection of life, property, and YOUR COMPANY assets will be provided by the Corporate Security Advisor.

Security Program Summary

The Corporate Policy on Security is an overall statement of objectives of the Corporate Security Program and the responsibility of all employees toward the protection of YOUR COMPANY assets. It focuses attention on the role of the Corporate Security Advisor in assisting management and employees in carrying out their protective responsibilities regarding Company assets.

Five security policies will provide detailed guidance to management and employees on the operation of an effective security program.

1. Security Policy I. The YOUR COMPANY Corporate Security Program will be outlined. It:
   (a) fixes the responsibility with individuals at various management levels in the Company
   (b) identifies the Corporate Security Advisor's responsibilities in a support role
   (c) identifies specific responsibilities of the Corporate Security Advisor in coordinating all investigations in the Company
   (d) details the type of reporting that must take place regarding security incidents against Company assets or personnel
   (e) establishes security representatives throughout the Company and its facilities
   (f) describes liaison relationships with local law enforcement
   (g) outlines procedures for prosecution of persons committing crimes against the Company

2. Policy II—Physical Security. Identifies the Company's philosophy on use of physical security measures to protect its assets
3. Policy III—Information Security. Identifies the Company's philosophy and priorities in protection of sensitive Company information
4. Policy IV—Emergency Situations. Identifies the types of emergency situations for which preplanning should be accomplished
5. Policy V—Protection of Assets. Indicates that a self-inspection program is to be initiated and a security program developed for each facility

The Corporate Security Advisor would also issue a set of Standards to supplement each of these policy statements. They are to assist each facility manager and business unit executive in implementing the policies within their unique operating environment. These detailed standards would be used to provide day-to-day guidance for local decision making and to assist in the development of the facility security programs required under Policy V.

Information Security

The long-term success of the YOUR COMPANY Companies is based on the protection of proprietary information relating to its business and research activities. It is the policy of YOUR COMPANY to take all necessary precautions to prevent unauthorized disclosure of information that could be damaging to its interests.

The Information Security Program for YOUR COMPANY consists of the following:

- a classification program to identify sensitive information
- use of patent and confidentiality agreements for employees and contract suppliers of labor and other services; provision of adequate physical security measures to protect proprietary information; records of the generation, storage, dissemination, and destruction of proprietary information
- compliance with all client company requirements for protection of sensitive information

It is the firm policy of YOUR COMPANY to protect proprietary or sensitive information regarding activities, products, and internal operations. It is to be disseminated only on a "need-to-know" basis, with strict adherence to standards developed for each category of use and/or information.

All required equipment, procedures, and safeguards are to be employed by YOUR COMPANY employees to carry out the intent of this policy. Security standards that outline procedures designed to provide protection of Company proprietary information are available from the Corporate Security Advisor and must be followed.

Analysis of the Security Department

Executive Summary

The purpose of this project is to review current YOUR COMPANY security programs, policy, and standards in order to determine the requirements for the development of security staff training programs to reduce risk to corporate activities, provide increased service levels, and improve the professionalism of existing staff. The study will include:

- an assessment of current program activities and current levels of staff training and experience
- quality of management/employee involvement in loss-prevention efforts
- baseline study of tasks/activities currently performed
- review of existing training program, plans, and course content
- review of corporate philosophy of security
- identification of key corporate protective requirements

Interviews will be conducted with key management and existing policy and program evaluations will be reviewed. A final report will be presented to management with appropriate findings and recommendations on all areas reviewed. Recommendations for a comprehensive training and staff development program will be provided. The estimated cost of this project is $XXXX. It will require 60 calendar days to complete.

Proposal

Review of Corporate Security staff training and development requirements for the development of training programs, policy, and strategy for YOUR COMPANY.

Study Objectives

The Corporate Security function is responsible for a broad range of protective activities within YOUR COMPANY. In its role of reducing corporatewide exposure to risk, it carries out multiple assignments in investigation, consultation, and program development. The decentralized and national/international scope of corporate operations requires that the staff is capable of supporting the corporate strategic objectives inherent in cost-effective security support for corporate activities.

The purpose of this project is to review existing security activities; establish current baselines for staff training and development; and review policies, standards, and corporate programs in relation to corporate objectives. All aspects of vulnerability and risk will be reviewed against existing programs, corporate management style, and future plans. This data will provide the basis for determining what training and development programs are required for existing staff.

The data collected and the recommendations and conclusions will be used for discussions with security and corporate management to develop appropriate strategic training and staff development plans for the security function. Any gaps in policy, standards, or programs will be identified and recommendations for improvement provided.

Comprehensive recommendations for a multiyear training and development program will be presented to management.

Data Collection

Collection of required information will take place through interviews, review of documents, task analysis, and on-site visits to key noncorporate headquarters facilities.

Interviews will be conducted with key management responsible for security and other functions that have protective responsibilities or management oversight for loss and protective-related issues. Appropriate line management will also be interviewed. Representative staff positions will have job audits performed.

All existing policy, standards, guidelines, operational plans, procedures, letters of agreement, and program materials will be reviewed. Corollary documents from human resources, audit, risk management, and international may also be reviewed as appropriate.

Two major operating division facilities will also be visited to meet with key management and discuss protective problems, goals, and requirements.

A complete analysis would also include a questionnaire survey of 25% of YOUR COMPANY facilities and site visits to representative facilities in each phase of corporate operations. Evaluation efforts would include the following components, in addition to the above abbreviated analysis:

Security Policy Considerations

- managerial expectations/philosophy
- national and industry standards
- program goals and objectives

Training Assessment

- key position job analysis
- current programs and plans
- perceived levels of effectiveness
- staff background and experience profiles
- review of incident reports and corrective actions

Operational Security Requirements

- physical security control

- protection of key corporate activities
- current technology/equipment use
- location(s) and environment(s)
- investigation and loss-prevention requirements
- employee training and educational plans

Present Program Components

- conformity to security program
- operational security procedures
- employee attitudes/security awareness training/controls
- criminal incidents/loss response
- law enforcement liaison/cooperation/resources
- security staffing levels

Unique Security Needs

- production/distribution facilities
- technology and process protection
- corporate strategic plans affecting security
- exploration
- retail operations

Deliverables

Upon completion of this project YOUR COMPANY Corporate Security will have a strategic plan to upgrade the training and development of all security personnel. This plan will be consistent with existing corporate business plans. Recommendations for support of operating divisions, specific corporate requirements, and security program management efforts will be included. Budget planning, resource allocation, and appropriate staff training recommendations, as appropriate, will be made.

The specific deliverables will include:

- commentary on the current program
- recommendations for organizational, policy, or operational improvements
- task database and staff training profiles
- a strategic training and development plan for corporate security
- recommendations for specific training policy, programs, and guidelines
- management discussions to finalize plans and training recommendations
- budget and resource allocation for staff training recommendations

Work Plan

A review of all existing policy, plans, programs, and related documentation will be conducted. Initial interviews with key management will be conducted at the Corporate headquarters.

Follow-up reviews of documentation and discussions will take place. Between two and six key operating division facilities will be visited for interviews, job audits, and follow-up reviews of documentation. A review session with security and appropriate key management will be conducted with draft recommendations for training policy, strategy, and programs. A final report will be issued that incorporates all comments, recommendations, and plans.

BIBLIOGRAPHY

Barefoot, J. Kirk, and David A. Maxwell. "Organizing the Corporate Security Department." In *Corporate Security Administration and Management*, 13–21. Stoneham, MA: Butterworths, 1987.

Kovacich, Gerald L., and Edward P. Halibozek. *Manager's Handbook for Corporate Security.* Stoneham, MA: Butterworths, 2003.

Kovacich, Gerald L., and Edward P. Halibozek. *Security Metrics Management.* Burlington, MA: Butterworth-Heinemann, 2005.

"Organization and Management." In *Protection of Assets Manual*, edited by Timothy J. Walsh and Richard J. Healy. Santa Monica, CA: Merritt.

Post, Richard S., and Arthur A. Kingsbury. "Organizational and Societal Constraints." In *Security Administration: An Introduction to the Protective Services.* 4th ed., 207–213. Newton, MA: Butterworth-Heinemann, 1991.

Post, Richard S., and Penelope N. Post. *Global Brand Integrity Management.* New York: McGraw-Hill, 2008.

Post, Richard S., and David A. Schachtsiek. "Duties and Functions." In *Security Manager's Desk Reference,* 201–215. Stoneham, MA: Butterworths, 1986.

Purpura, Philip P. *Security & Prevention: An Introduction.* 5th ed. Stoneham, MA: Butterworths, 2007.

Simonsen, Clifford E. "Organizational Bureaucracy: A New View of the Security Function." In *Encyclopedia of Security Management Techniques and Technology*, edited by John J. Fey, 522–525. Newton, MA: Butterworth-Heinemann, 1993.

7

Organization of a Security Program

Richard S. Post, CPP

CHAPTER 7

OBJECTIVE

The objective of this chapter is to review the major functional, structural, and organizational approaches to effective security operations in a company. While there is no set formula for determining which approach is best for a specific organization, there are numerous options to consider. These factors also permit analysis of existing organizational approaches to optimize available protective resources.

INTRODUCTION

Central to the discussion of how best to organize the security function is the question: What is the function supposed to do for the company? Once this has been decided, the overall organizational style of the company will determine how the function will fit within existing management systems and structures. This organizational fit gives form, substance, organizational status, and structural-functional identity to security within the company. It is what the function does that sets the organizational value for the function, its policies, activities, and staff.

The operations of security as a corporate line or staff function can vary as to functions performed, authority, scope of operations, responsibility, and performance measurements. The organizational location also can impact the effectiveness, degree of management support, and status of the function within the company.

Of the methods used to organize company resources to provide for protective services, the single-purpose security department is the most traditional. The corporate model of fewer layers of management, less specialization, and greater employee and managerial responsibility for security services has led to a number of ways to organize and manage the security function.

The nature of security is open to many approaches, including providing service to an entire organization as a corporate activity, servicing a single or many facilities, or overseeing only a few specific activities, such as contract-guard service contracts.

Company security cannot function without senior management support for ideas, programs, and techniques for loss prevention. This support must come from the top down; in giving their support, managers often find additional challenges to be addressed by security. Despite these additional challenges, the security or loss prevention function also can develop new programs. As with selling insurance, security must sell the intangibles of prevention and protection. Efforts are focused on making programs more effective by using benchmarking, cost-benefit, performance budgeting, and related cost-effectiveness techniques. However, required functions and goals should be reasonably attainable within the existing organizational structure.

How the company values the security function is expressed by rewards, status, and roles. In this way, both management and staff understand how the organization values the function. In organizations with strong ethics programs, or those that require strict adherence to government regulatory standards, security activities can be more valued than in organizations where security is considered just another management function to be provided at the best cost-benefit ratio possible. Simply put: no value perceived by the company (or organization), no status, no budget, and quite possibly, no security function.

The basic structural-functional form of line staff, centralized-decentralized operations, single function–multiple function department, service provider–mandatory approach, cost center–profit center, or consultant should be reviewed during the development of a company model. The size of the company also is a factor in determining how it will organize a security area: whether it hires internal staff or contracts out for required services.

Likewise, the company must address the function of security by determining which tasks are important for the company and who should do them, and deciding where the final authority for oversight and execution rests. Key tasks most often assigned to the security department include the following:

- protecting employees
- securing physical assets
- securing information
- conducting investigations
- managing guard services
- analyzing political risk
- complying with government programs
- dealing with product integrity issues
- protecting intellectual property and trade secrets
- supplying chain security
- creating ethical standards programs
- supporting litigation
- supporting mergers and acquisitions
- investigating prospective employees

The way in which the required services are performed varies, depending on how general management authority is regulated in the company. In a strong corporate staff-operating division structure, a staff specialist can have only policy development and compliance audit authority; all implementation is carried out by divisional staff. In other structures, a corporate staff function can have full policy development, standards development, and procedural guidance authority, as well as the authority to intervene directly in divisional activities to handle problems, whether division management wants the assistance or not.

In short, there is no one way to organize for protection of company assets. Variations on basic models exist to satisfy any desired organizational style, approach, or requirement. There are, however, positive and negative points for each approach. These will be reviewed so that the best approach can be selected for specific organizational requirements.

The social, economic, and cultural environment affects the ground rules for security management. Because they may be operating with fewer personnel or smaller budgets, law enforcement in the United States and around the world may be less able to meet basic protective needs. Therefore, although it is essential to establish programs of liaison, mutual assistance, and contingency planning with local law enforcement agencies to support the security department, the bottom line is that the company is still responsible for protecting its operations, employees, and products. No one else is going to do it.

APPRAISING NEEDS

The development of protective programs must be based on a realistic appraisal of needs. The better needs are analyzed, the better programs can be developed to provide the required amount and type of security for the organization. Perceptions of needs must be realistic. Priorities for security that allow for the protection of basic activities first must be established. Those issues of lesser importance should receive correspondingly less attention.

It might sound like a simple task to identify what needs to be protected, but in practice, it can be quite complicated. Are employees to be better protected than the facilities in which they work? Are the products and ideas generated by the employees more important than the facilities? Is the supply chain that produces the products more or less important than the distribution facilities that move products to customers or end-users? All have value, all need to be protected: how much, by whom, and at what cost are all questions that need to be answered so that an effective plan can be prepared.

FLEXIBILITY IN PROGRAMS

Programs require the support of employees at all levels in the company. While security programs can be stringent enough to force employee compliance with regulations, getting support and cooperation makes the program more effective. Internalized employee and management support for program goals and controls makes for a less coercive workplace, voluntary cooperation, and higher compliance with program objectives.

It is important to get all levels of management and operations to participate in protective goal setting, incorporating protective activities into job descriptions across the company, providing financial incentives through creative compensation programs with protective goals and metrics. Personal involvement and buy-in to protective goals increases support for programs and ensures multilevel participation in providing protection for the company.

CONCEPTS

To be effective, security and loss-prevention programs involve the entire company. If senior management is not involved in developing strategy and establishing an appropriate philosophy and approach to the protection of corporate assets, the program will probably fail. The responsibility of the senior security professional in an organization is to lead security program development and continuity. He can use many techniques to ensure adequate involvement, understanding, and support for required protective programs.

The discussion thus far has presented a rationale for loss prevention or security to be viewed as an ongoing process to be provided, not an end to be attained. This process is efficient only to the extent that it reflects those principles and values that are useful for the attainment of overall organizational goals. Mere immediate dollar savings are not sufficient justification for expediency in this sensitive and critical area of organizational and societal concern. Recommendations to senior management should reflect the best in loss-prevention program development and management skills for managing identified risks.

Fears about losses and the desire to protect assets are justified organizational concerns. Anxiety over unsubstantiated possibilities or the reinforcement and expansion of concern over actual losses by unrealistic assessments can prove counterproductive within the organization.

The benefits of management and employee participation in design and implementation include ease of approval for recommendations, allocation of required resources, provision of budgets, and direct personal commitment for program goals and activities. Security or loss prevention professionals must have a clear understanding of the implications for the organization when they makes these specific recommendations.

ORGANIZATIONAL GOALS

Maximizing shareholder value, providing high-quality products at the right price, growing the business with high-quality employees, and being a good corporate citizen are all commendable corporate mission statements or goals.

ORGANIZATIONAL DESIGN

What should the organization of the security area look like to help meet these goals? The model selected for a company need not look like anyone else's security department. It needs to reflect corporate style, commitment to the program, and resources committed to protection purposes. Once these factors are agreed upon, the structure should follow the corporate approach to management and include provisions to deliver all program activities at all levels and geographic areas of the company. All security guidelines are applied across the company; specific operational plans are developed for various regions of the world and for specific countries, company facilities, and activities such as data security. Each company must determine what the corporate strategic response is going to be for each type of risk or threat.

ANALYSIS

The type of organization, size and number of facilities, functions, and staffing issues, as well as security needs and goals, play a role in the organization of a security department. The following sections provide information on these issues.

Types of Security Service. Company size is one of the major factors in determining how the company will organize and operate its security activities: Will it hire internal staff person(s) or contract out for required services? The manner in which security services are structured should follow the basic management and organizational style of the company. Major approaches to security program structure include line-staff, centralized-decentralized, single function–multifunction, or service provider–mandatory structures, and internal consultant. Each of these approaches is associated with specific management characteristics that either enhance or detract from successful operation of the security function.

Line Staff. This is a very traditional approach that retains corporate control over all key policies, standards, procedures, and operational controls at the corporate level by specialist staff. They have direct authority in and at all levels of the company to handle any security-related issue. Security managers assigned from the corporate level can be located in operating divisions, reporting to, paid by, and responsible only to the corporate level. This tightly controlled, highly structured approach is falling out of favor in general management as well as in security operations because of its expense, implicit lack of trust, and changing management styles.

Centralized versus Decentralized Operations. This approach is similar to line staff except that corporate staff develops policy, standards, and procedures but allows local management to implement and manage the function at all other levels in the company. Senior management maintains a line-reporting relationship for information about security occurrences in the company, managing companywide programs, such as investigations, protection of executive staff, political risk, and so on. Local facilities are responsible for managing their own guards and physical security and following approved standards. Incidents are reported to corporate-level staff, and usually no investigations are permitted without the approval of the corporate staff.

This approach is still used in many large corporations; more autonomy is given to operating divisions to manage their issues directly without checking with corporate-level staff or necessarily reporting all incidents to them. The costs are generally lower with this system than with line staff, as fewer corporate personnel are needed to manage divisional issues, and senior management review of local problems is not required.

Single versus Multifunctional Department. While not a common approach, there are security departments responsible for only a single function, such as investigations, physical security, or information protection. They can be located at the corporate level or in an operating division and have limited or no coordination with other security personnel in the organization. They also may incorporate the only security personnel in the company, depending on what the company does or where it is located.

The common approach is to have a generalist security function that bases its programs and plans on the needs of the overall organization. This department often will manage commercial or proprietary guards, manage alarm systems, conduct theft investigations, manage facility protection, assist with background investigation, respond to emergency situations, provide for executive protection, and support all corporate activities as required.

There is also a growing trend to have the security department responsible for more than only that function; the most common additional function is safety. In a growing number of corporations, occupational safety and health, fire prevention, security, and environmental health are controlled by the same department. Multiple-service departments, while cost-effective, often produce a diluted, distilled minimum level of oversight, program management, and compliance.

Corporate Service Provider versus Mandatory Service. The current trend for security areas in organizations is to downsize and contract out. Having perhaps one internal security practitioner provide basic security support for the company is becoming the norm for corporate protection. The corporation pays for basic overhead, but will allocate charges for security services to business units or divisions that require security support to solve problems. This is the opposite approach to that of a corporation requiring incidents to be reported and then conducting an investigation or the approach that mandates an upgrade of security after a security inspection. The negative feature of this program is that, given the choice, divisional management can try to find a less expensive way to solve the problem than reporting it to corporate management. It also does not permit the corporate level to intervene in divisional problems except under unusual conditions.

Cost Center–Profit Center. The traditional model of corporate staff service allowed for the corporation to provide services that were either the result of mandatory requirements for reporting or compliance to subordinate business units. Any required services determined by corporate staff to be essential were paid for by corporate staff and often charged back to the unit for reimbursement. The budget for expenses of the staff unit was considered corporate overhead and a portion of the general and administrative expenses of the company.

Under current decentralized or semiautonomous operating styles, a minimum of corporate overhead is acceptable and business units are required to provide their own support services, some of which they can purchase from the corporate organization. Many staff units now are required to "sell" their services to business units to maintain more than a minimal corporation-supplied budget. Several more enterprising firms have spun off security departments as separate business units to sell their services both internally and externally.

Consultant. An internal or external security consultant is becoming more common in corporations. This enables organizations to maintain a minimum staff level yet have a corporate emergency response resource, provide requested services from operating divisions, and coordinate the purchase of outside services to supply security and investigative services for the company.

Outsourcing for day-to-day services for security control centers, investigations, physical security audits, and policy development is now commonplace and cost-effective.

Nonsecurity Generalist Manager. Some organizations do not have a professional security specialist on the payroll to advise, assist, or manage programs. Instead these organizations rely on human resources, legal staff, or outside consultants to provide whatever is required on an ad hoc basis. In some cases, security managers organized effective programs while they were employed, but when the organization was downsized, they were separated from the company while the programs they developed and managed remained in effect. Frequently, security programs are managed by some other corporate function, such as human resources, and staffed, as required, by consultants.

Other models exist, wherein functions were combined during a downsizing and the security function is given to the safety department to manage. Other examples include the model in which a security position is expanded to include all administrative services, including security. In these situations, security issues are managed by a skilled professional, but only on a limited basis.

Reporting Structure. The security area can report to any number of executives in the organization. This decision is often arbitrary rather than something given a great deal of thought by senior management. In large multinational organizations, the security function can report to human resources, legal, administration, audit, labor relations, or facilities management. In any of these departments, security is likely to adopt the departmental organization style and closely align its operational focus with that of the department to which it reports.

Legal department management of security can result in greater emphasis on investigations, litigation support, and violations of ethical activities than on physical security management activities. If reporting to human resources, security is more likely to have a well-balanced employee support focus, which will minimize investigations and promote prevention of violence, workplace protection, executive protection, and political risk-management efforts. Facilities management of security will have more emphasis on managing physical assets; most important will be guard-service management and physical security issues. If reporting to administrative services, security often will have a well-balanced approach to protection and preventive activities.

There are organizations in which executives have a personal interest in handling security affairs for the company. In these organizations, the function can report to the president or an executive vice president responsible for all administrative activities in the company, rather than to a functional department head. Often this is a more effective arrangement, as corporate interests are balanced at this senior level, and key corporate issues are more likely to surface with priority over interdepartmental affairs. If possible, this is the most ideal location for the security function.

COMPANY SECURITY SERVICES

Providing physical security services for the company means preparing policies, establishing standards for purchasing, evaluating various hardware and systems for company applications, and overseeing equipment installation for the company. Often this is a corporate function and is closely coordinated with purchasing and operating divisions, headquarters facility management, and local facility management.

This service can be provided within any of the functional approaches identified but is most cost-effective when provided as a corporate service based on standards developed within the company. The corporate department also can be required to approve all facility expansion plans, remodeling, or new facility construction to ensure that adequate security features are included at a cost-effective time in the building process. This is best done on a centralized basis.

SECURING INFORMATION

Many companies have information security operations or departments that are separate from any corporate security function in the same company. Other companies propose that protection of information is only one dimension of corporate asset protection, although a very specialized one. While both positions are valid, if they are maintained separately, coordination is essential between the information security and general security functions. Policies for information-protection and physical security standards applicable to protecting electronic data, work, and storage environments must be coordinated with the overall protection program for all other corporate assets and resources.

While computer systems have concentrated sensitive information and raised the risks from loss or compromise, the electronic controls, procedures, and safeguards rely on employees to follow prescribed procedures and plans. Laptops, distributed systems, e-mail, and electronic file transfer have made the problem of information security move outside the realm of a stand-alone protective requirement for a company. Considerable coordination through a centralized policy, procedure development, and audit function is essential.

PROTECTING EMPLOYEES

Protecting employees at risk because of job assignment, visibility in the company, position, or exposure while traveling on company business is best performed and coordinated by a centralized corporate security function. While day-to-day coordination of protective resources can be managed on a decentralized basis, information to support these local operations should be provided from a centralized source at corporate headquarters or through corporate-sponsored resources to ensure timely, accurate, and readily available data.

Assessing threats, analyzing vulnerabilities, and developing personal protective programs, domestically or internationally, is a specialized function. It can be cost-effectively and efficiently performed or coordinated by corporate staff and made available on an "as-needed" basis throughout the company.

Advanced preparation for trips to high-risk areas, special events, or to protect individuals is performed by the protective team assigned to the project. They can be assigned from either local or corporate resources, depending on the nature of the assignment, its location, and duration.

It is cost-effective to purchase risk-reporting services on an outsourced basis. Numerous sources (Control Risks, Kroll Inc., Overseas Security Advisory Council (OSAC), etc.) provide high-quality, timely, and country-specific information to avoid problems, identify risks to operations or individuals, and plan travel.

INVESTIGATIONS

Organizations can require different types of investigations—those dealing with internal issues affecting company operations or employees, and those for the benefit of the company directed against its legal or commercial problems. The former are generally best performed by internal staff because they tend to be more sensitive; the latter can use consultants and private investigators.

Often substantial liability issues are involved if an investigation is improperly performed. Consequently, it is often policy to involve only corporate staff or corporate legal counsel in advance. Agree to how and under what circumstances an investigation is conducted, as well as budget and investigative plans, before the investigation begins. Closely monitor results to ensure proper supervision of the investigation staff.

Investigations are best controlled as a corporate function. However, divisional staff may work alongside corporate and contract the assignment to an approved commercial private investigation agency that has experience working within the investigative guidelines of the company.

If contract investigations are to be used, a strict policy for their approval must be established. Contractors and the investigative plan should be approved in advance by the company legal department.

MANAGING GUARD SERVICES

Managing guards, either internal or contract, is a task involving considerable time and money. Guard contracts often are the largest ongoing expenditure for company security. When guard services are required only for the corporate headquarters, the headquarters staff should manage the contracts independently of other locations. In the case of a large national or regional company, the preparation and negotiation of national contracts covering all company facilities is often preferred.

When national contracts are not appropriate for geographic or other company considerations, establish a model guard-service contract that identifies specific terms and conditions all contractors must meet. The cost, terms, and conditions of the service contract must be approved by senior management before it is signed by local management.

A contract guard service can also provide flexibility in staffing and management of physical security services for a facility or for the entire enterprise. Companywide contract security services are often used to reduce the need for a security head count. A guard or physical security services manager can be provided to the company by the contract security firm at a reasonable additional cost.

ELECTRONIC MONITORING

Most companywide security functions have their single largest investment in alarm, ID, and surveillance systems. The use of wide area networking for global monitoring of security conditions at a single monitoring station is now commonplace in large security operations. The convergence of information technology, physical surveillance, and ID-access control over wide area networks, while cost-effective, must be monitored closely for failures. Redundancy in systems is still highly recommended and should be a part of the design specifications for this type of monitoring.

ANALYZING POLITICAL RISK

Multinational corporations need timely information about crime, terrorism, local governmental instability, and threats against their operations. This is best provided on a centralized basis within any company. Whether the information is provided by outside consultants, company resources, or any combination, it must be available or accessed centrally or regionally from a central source.

Coordination of internal information required for executive protection decision making, establishment of travel restrictions, evacuation of employees in high-risk areas, or crisis-management purposes must be at the corporate level. It is efficient and cost-effective to have it located there. Although e-mail access for updates can take place worldwide, the development of worldwide resources to collect and furnish data must be centralized.

A risk-monitoring center is becoming a standard part of companywide monitoring and control centers. Having real-time threat information, along with facility surveillance from worldwide facilities in the same room, is crucial to providing timely risk data to senior management. The use of incident/crime-reporting and analysis software, crime-mapping services, and regularly updated risk assessments of key operations and facilities is essential.

METRICS

Of all the things that a security manager must do, developing measurement criteria (metrics) as performance indicators is the most critical. These metrics will vary across companies as their missions, core business objectives, and risks vary. Developing a "dashboard" with a reasonable number of evaluation measures will benefit both the company and its security function.

COMPLYING WITH GOVERNMENT PROGRAMS

Various government agencies have programs to protect sensitive information given to businesses under contract terms and conditions. In the United States, these include such organizations as the Department of Defense and the Department of Energy. Contracts should be managed to ensure security compliance is an essential part of operating a successful business venture with the government.

Managing security under a government contract requires that the company establish and maintain a program ensuring that all terms and conditions required for protective standards are met in full.

Compliance can be managed at the business-unit level alone or require corporate oversight. Responsibility for protection rests with the contractor, corporate, or business unit. Extent of protection is determined by regulatory requirements.

Programs of the Department of Homeland Security (DHS), Federal Emergency Management Agency (FEMA), and Federal Bureau of Investigation (FBI) may have implications for corporate activities depending on the type of business being conducted.

PRODUCT INTEGRITY ISSUES

Counterfeiting of products, disruption of markets by parallel shipments, gray-market products, sale of reject products or seconds as first-quality products, or contamination of products all cause breakdowns in customer or client confidence. Investigations into the causes of these types of activities can be conducted either by corporate or business units depending on how the company is organized.

Counterfeiting outside of the United States is most often handled by business units with local legal counsel and investigators hired within that part of the world. If a member of the corporate staff is present, and the problem is viewed as affecting overall corporate integrity, it can be handled in conjunction with the corporate security staff. However, this approach may not be practical. International investigation agencies are more common in most parts of the world, thus local operating management and legal departments often conduct these investigations without direct involvement by the headquarters' security department.

In conducting the risk assessment for the company, the value of intellectual property, product integrity, trade secrets, and all intangible assets must be carefully reviewed. It is possible that a defined risk to these aspects of the company may require that a brand protection focus (Brand Protection Department/Brand Integrity Department) also be established. Several companies have this model, including Johnson & Johnson, Cisco, and Apple.

ETHICAL STANDARDS PROGRAMS

Internal investigations into allegations of employee ethics violations, such as kickbacks, payoffs to government officials, transshipment of goods, fraud, expense account violations, or even sexual harassment often are handled by the corporate security department. Working with company staff, such as the general counsel, general auditor, or ethics manager, investigations can be conducted discretely to determine the facts behind an allegation. This type of sensitive internal investigation most often is carried out only by corporate staff or, if extremely sensitive, by contracted investigators. Portions of investigative work can be contracted to private investigators, but in most cases, local management would not be permitted to conduct an investigation unless the issue is relatively minor.

Maintaining ethical standards developed in the company makes it important that actual or suspected violations are thoroughly investigated regardless of who is involved. Corporate staff or outside investigators generally are better prepared to deal with these issues without concern for the individuals involved.

SUPPORTING LITIGATION

Often issues in litigation involve the interests of the company. Working with outside counsel, outside investigators often are hired independent of any corporate security involvement. In these situations corporate security is viewed as having a corporate line responsibility and should not become involved in ancillary investigations not directly a part of the corporation's primary and day-to-day business activities. A recommendation on whom to use for an assignment might be made, but the hiring decision more often is made by the counsel retained by the company. Therefore, it is possible for a private investigator to be working on behalf of a major company without its security department ever knowing about it. This can happen both inside the United States and overseas.

SUPPORTING MERGERS AND ACQUISITIONS

Corporate security departments infrequently become involved in merger and acquisition-related investigations. When the security department does get involved, it is in a major transaction in which "due diligence" requires various internal specialist functions to participate on an acquisition team review of a potential partner or takeover target. Private investigators hired by the outside law firm would likely be responsible for reviewing representations made by the target firm, checking on its business and commercial history, investigating the principals of the business, and ensuring that sources of funds in a deal were not illegally obtained (drug or organized crime money).

INVESTIGATING PROSPECTIVE EMPLOYEES

Corporate security or business unit security can support human resources activities by conducting required background investigations on prospective employees. Compliance with all relevant state and federal privacy guidelines must be ensured during this process. If the volume of these inquiries is not enough to establish an effective process and a high level of quality control, they should be performed by corporate security or contracted to a private investigation agency with the appropriate experience.

The trend is toward using consultants, suppliers of services, and retainer services to provide resources previously provided by employees. Security services once only available within the largest organizations is now available to the smallest corporation. The options for providing high-quality services for any company have improved significantly in recent years. These outsourced services include expert services, handwriting analysis, chemical testing, electronic countermeasures, political risk analysis, forensic accounting, computer security, and security awareness training.

Consultants in virtually any aspect of security management, technical assistance, or investigations also are available. Many of them are fresh from positions as corporate specialists in large companies and have the skills and sensitivities to work in large corporate environments. Consequently, small companies can do what only the large ones were able to do before—have the best services available, but only when needed. The following examples show how to best use security services.

Example 1. A large Asian multinational company developed security problems at an overseas facility. Several of the key executives in the country also were personally threatened with kidnapping. The corporation does not have a security department charged with managing risks around the world. Most protection issues are considered the responsibility of local management to solve locally.

Information about the threats was transmitted back to the corporate headquarters with recommendations to hire local security consultants to study the situation and make recommendations. This was approved, a study conducted, and a report made to local management.

Among the findings of the report was that several senior managers had been compromised by local organized crime gangs and had serious risks to their well-being if they remained in the country. Several key local employees were found to be engaged in collusive behavior, selling secret competitive information on contracts, and physical security for company facilities was insufficient.

The report forwarded to headquarters included commentary on the physical security deficiencies but nothing on the other issues. They were considered too sensitive and a serious loss of face for local management. Solutions to the problems were attempted at a local level. They could not be solved and additional thefts occurred; an executive was kidnapped, recovered, and eventually returned to the home office. Local security was increased and the jobs of the thieving employees were gradually reduced in scope. They were taken off sensitive activities but not fired. The security consultants were given high praise by both corporate and local management for their efforts in protecting the company.

Example 2. A U.S. company was experiencing theft of components from an overseas production facility. An analysis of the problem was requested by the corporate security staff. The analysis identified a lack of compliance with corporate security standards and policies that provided opportunities for employees to steal at will.

An investigation was conducted that identified an employee theft ring. They were arrested, prosecuted, and sent to prison. An example was made of their convictions to strengthen the company's program to reduce losses and involve employees in protective programs.

The recommended physical security was upgraded, closed-circuit television cameras were installed, tightened access control and separation of warehousing and raw materials storage were required, and a change in trash removal and driver waiting areas was recommended. Local management reviewed the recommendations and accepted most of the key provisions in the report.

All accepted recommendations were complied with immediately. Local management was grateful to corporate security for the assistance, and costs for the services were absorbed by corporate while security upgrades were a local cost.

SUMMARY

The trend in security management is away from large corporate staff to single corporate specialists who will develop strategic plans for asset protection, develop standards and policies, and assist in auditing compliance. Where local management has staff security specialists or security departments for specific businesses requirements, they will work within the guidelines established and audited by the corporate specialist.

As more and better security management analytical tools against which to base resource allocations become available, decisions can be made with more emphasis on return on investment ratios. These can replace nonquantitative judgments used previously to measure program value. Dashboards with established metrics are becoming the norm for measuring the effectiveness of security operations.

Loss and countermeasure information is improving both inside companies and within the security industry. The development of numerous benchmarking databases will make it possible for local management to supplement corporate data with industry-specific best practices to improve local business unit security performance.

If the company is large enough, centralized data collection, policy and standards development, and coordination will enable increased cost efficiency of security within the company. If not, local security management supplemented by industry resources will become increasingly self-supporting.

BIBLIOGRAPHY

Campbell, George K. *Measures and Metrics in Corporate Security.* Security Executive Council, 2006.

Dalton, Dennis. Rethinking *Corporate Security in the Post 9-11 Era.* Elsevier, 2003.

Dalton, Dennis. *Security Management: Business Strategies for Success.* Butterworth, 1994.

Fischer, R. J., and G. Green. *Introduction to Private Security*, 8th ed. Newton, MA: Butterworth-Heinemann, 2005

McCrie, Robert. *Security Operations Management.* Butterworth, 2006.

Post, Richard, and Penelope Post. *Global Brand Integrity Management.* McGraw-Hill, 2008.

Sennewald, C. A. *Effective Security Management*, 4th ed. Stoneham, MA: Butterworth, 2006.

8

Management of Security Investigations

Francis J. D'Addario, CPP, CFE

CHAPTER 8

OBJECTIVES

The objectives of this chapter are to examine the following two investigative routines in detail and at length:

- background investigations
- incident investigations

OVERVIEW

> **in.ves.ti.gate** \in-'ves-te-g_t\ *vt* [L *investigatus,* pp. of investigate to track, investigate, fr. *in-* + *vestigium* footprint, track]: to observe or study by close examination and systematic inquiry. (*Webster's New Collegiate Dictionary*)

The truth can be difficult to determine. Once the facts are established, however, proximate or probable causes evolve, and questions of responsibility are resolved or proved. The debris of human behavior, deceptions, omissions, and rationalizations clutter the path to the truth. When the stakes are high and those involved feel the risk of termination, criminal or civil liability, or potential censure by peers or family, discovering the truth becomes a challenge. Investigation is the path to truth.

Competent investigators recognize deceptive behavior in person and through analysis of the spoken or written word. Their ability to recognize diversion and establish the real story without jeopardizing the self-esteem of the subject enables them to obtain admissions and determine what has happened. This discussion of investigations is limited to aspects of human behavior and motivation; technical crime-scene evaluation and accident reconstruction is left to the forensic scientists. All employees must be trained to protect the crime scene and to prevent contamination of evidence; they must be empowered to close an area so crime scene investigators can do their job.

The evolution of private investigative resources from public law-enforcement origins is not surprising given our crime rate and criminal justice shortfalls. Private institutions remain at risk for the cost and liability of crime. Just as the railroads hired the Pinkerton Detective Agency to pursue the James Gang in the 19th century, modern investigators help protect customers and corporate assets today.

Investigations make up a significant part of the prevention dynamic outlined in this chapter. Successful inquiries depend on accurate incident reporting and loss analysis. Fact finding without timely and accurate reporting can turn into a fishing expedition.

When combined with the other elements of prevention, investigations promote criminal risk. Proactive companies often require a code of conduct that is intolerant of crime. Required reporting, participation in investigations, and timely sanctions for misbehavior promote risk for would-be criminals. In the words of Cesare Beccaria, a noted 18th-century criminologist, "It is not cruelty or severity that renders punishment an effective deterrent, but rather its certainty" (Monachesi, 1973).

The risk and certainty of accountability are the deterrents that the competent investigator brings to the prevention discussion. Augmented by efficient loss reporting, data collection, and analysis—in the best case scenario—the criminal's perception of risk and gain is altered to influence alternative-target selection. On those occasions when prevention does not prevail, case closure and accountability help deter recurrence. Serial offenders who are brought to justice and incarcerated lose their opportunity to strike again.

BACKGROUND INVESTIGATIONS

The first objective is to screen applicants and on-site service providers through "due-diligence" background investigation. Prevention practitioners generally agree that the majority of crime losses, including fraud, are attributable to current and former employees.

EMPLOYEE INVESTIGATION

Background screens are a unique opportunity for loss control. Candidates often misrepresent their personal and professional history; fabricate credentials, experience, and education; or fail to disclose criminal convictions. The *due-diligence background investigation* can corroborate or bring into question any misrepresentation. This investigation typically occurs before an offer of employment, before individuals are assigned responsibility for other employees or have access to critical assets.

Background investigations are an extension of the access-control concept, a primary component of crime prevention through environmental design (CPTED). Access control does not just mean screening visitors to the organization's premises or adopting a badging system. It implies qualification of individuals who seek access to organizational assets. A background screen merely ensures that candidates for employment or service contracts are who they say they are. The effective screening process validates individual qualifications by verifying professional credentials, education, work experience, criminal and credit histories, and prohibited-substance use, where applicable and to the extent allowed by law. Pre-employment verifications require examination by competent legal counsel because statutory restrictions vary by state.

Certainly, history repeats itself. Counterfeit or high-risk applicants employed by the organization can later harm customers, fellow employees, or assets. Errors or omissions in background investigations can provide evidence for tort actions against the employer, including negligent hire, retention, or supervision. In 2008 the U.S. Department of Justice reported that 3.2% of the U.S. adult population, approximately 7.3 million people of the potential labor pool, was incarcerated or otherwise under correctional supervision (parole or probation) (U.S. Department of Justice 2009). Although an aggressive screening process is likely to decrease infraction frequency, most such incidents involve employees and ex-employees.

Incident investigations provide a protocol for evaluating crimes, accidents, and claims for the purposes of determining probable or proximate cause.

Probable cause is a legal term that establishes reasonable grounds for a presumption of guilt that is less than proof—more than probable but less than certain. It is a degree of probability used to determine whether an individual or corporation is chargeable with a criminal offense. A determination of probable cause is required in most cases before warrants can be issued or an indictment charged. That is the first step in establishing the accountability of a person for a crime.

Proximate cause refers to a prerequisite condition without which an accident or injury would not have occurred. Incident investigations can establish whether individuals, loss, accident, or contributing causes are responsible for a crime. Incident investigations require that all crimes, accidents, or significant losses be investigated for cause in a timely manner.

REQUIREMENTS

Pre-employment screening restrictions vary by state. Because of the variety of state laws governing employment issues, getting the advice of competent legal counsel is recommended before adopting any investigative protocol or employment condition.

Pre-employment and post-employment polygraph examinations, commonly called lie-detector tests, have been restricted by the Federal Polygraph Protection Act. National security, law enforcement, and a few private-sector industry employers are exceptions. Most employers can use a polygraph examination as part of an incident investigation if:

1. A specific crime or loss has occurred.
2. The employee had access.
3. A reason other than access causes the employer to suspect that a specific employee can be involved.

Polygraph tests must be voluntary. They are nearly always inadmissible as evidence in a court of law.

Credit inquiries must conform to the Federal Fair Credit Reporting Act as well as to restrictions by individual states. Employees have a right to know whether their employer has access to their credit history. Employees or applicants can request a copy of any credit report that was used to qualify them for a job. Credit reports include payment histories, bankruptcies, mortgage information, previous addresses, employers, and even spousal information. Credit histories can be routinely released in compliance with the Fair Credit Reporting Act on employment applications.

Although criminal-conviction histories are generally considered public information, arrest histories may not be. Prohibited-substance or drug-testing prerequisites also vary by state. Employers can be guided by competent legal counsel about what releases are considered routine for credit, criminal information, voluntary prohibited-substances testing, or other employment conditions.

When pre-employment restrictions are pervasive, post-employment screens including conditional offers and bonding, can be used. (These are examined later in this chapter.) Generally, it is advisable to warn the candidate of any conditions (such as code-of-conduct accountability, or prohibited substances testing) on the application.

The organization's employee help line or incident-reporting number also can be presented in the application with a statement of zero-tolerance for crimes, threats, harassment, discrimination, unfair treatment, and prohibited substances. Reporting and investigation-participation requirements should be detailed.

CONCEPTS

Key hires and managers responsible for subordinates, assets, or proprietary information can be subjected to a post-employment bonding process. *Bonding* is a traditional means of qualifying individuals for additional insurance when they handle money or significant assets. Today most organizations are, to a large extent, self-insured.

The bonding protocol is invaluable for early detection of high-risk or counterfeit candidates in the employment or promotion process.

The bonding form (such as the sample in Figure 8–1) itself can be redundant to the preemployment application in many respects. It asks for the prospective employee's name, address, employment history, education, and other data. As a post-employment device, the bonding form also collects information that is *not* prohibited after an offer of employment or promotion is extended. Date of birth, previous addresses, credit information (in conformance with the 1987 Fair Credit Reporting Act), and motor vehicle information, as well as whom to notify in the event of an emergency, can be required.

Individuals who have been less than forthcoming on their pre-employment application just days or weeks before completing a bonding form will find it complicated to provide the same erroneous professional credentials or education histories. The application and bonding forms, once compared, often yield a number of inconsistencies.

Often, when faced with the prospect of a detailed due diligence background investigation, candidates will disclose criminal convictions and employment terminations purposefully misrepresented or left off the application. This disclosure after the fact constitutes application fraud in most organizations. It typically entitles the employer to terminate employment immediately for cause or to withdraw a conditional offer. Bonding affords a low-cost method for disqualifying counterfeit candidates before a costly background investigation and before exposing personnel or assets to a high-risk employee.

The bonding form can be adopted as a prerequisite for any incident investigation. Once a subject's biographical information is in hand, investigators can be stymied when attempting to find former employers or other sources who will corroborate information. Deceptive applicants often claim self-employment or military service to account for omitted work history or time spent in prison. The employer can be impressed by a candidate whose resume includes a Heisman Trophy and the Medal of Honor. In nearly all instances public record is a ready resource. Military service can be routinely corroborated by writing to the National Personnel Records Center, 9700 Page Ave, St. Louis, MO 63132-5100, or by e-mail at mpr.center@nara.gov.

Criminal histories, including conviction information, are readily accessible from most county records-retention centers (National Employment Screening Services, 1994). Credit history provided by the candidate in conformance with state and federal guidelines often reveals former employers and residences. Counterfeit applicants omit residences or employment that could yield derogatory information. Such omissions highlight grounds for initiating the verification process.

Similarly, social security numbers are assigned by state. Individuals who proffer a social security number without complementary education or employment history from that locale require further research. Investigative services and software packages are also available to verify social security numbers.

Various agencies conduct background verifications for profit. Employers are advised to take bids and check references, including the Better Business Bureau. Require acts-and-omissions insurance naming the employer as an "additional insured" on the service provider's liability insurance. Insurance brokers and reputable investigations or credit companies readily provide this proof of insurance.

Many sources of public information are free. Secretary of State offices routinely carry historic information on incorporations. They can be located through the telephone information office in any state capital. County departments of taxation also list business information. Job candidates who claim to have run their own businesses should be listed even if the enterprise is defunct. You can also learn that a candidate supposedly divested of a business is not.

The public library is good source of information. Often, libraries carry telephone books for

Page 1 of 4

DO RIGHT, INC. BONDING FORM

In order to control losses, DO RIGHT requires all management (new hires and promotees) and key personnel who manage or control assets (money, inventory, proprietary information) to provide background information. The company reserves the right to investigate any person or persons at any time during their employment. All employees, as a condition of employment, are required to fully cooperate with any investigation conducted by the company or any authorized agency. In addition to reporting all crimes for prosecution, DO RIGHT , within the limits of the law, will attach wages, bonuses, vacation pay or (other items of value) of any party responsible for a loss. DO RIGHT, INC., requires the reporting of all crimes against the company. Information may be reported confidentially to the Senior Director of Internal Audit at 1-800-EXAMPLE. DO RIGHT INC. offers a reward up to $1,000. for confidential information leading to the arrest and conviction of any individual(s) responsible for crimes against the company, its employees, or customers. DO RIGHT INC. retains the right to use all lawful investigative means available including prohibited substance screening and audio/video surveillance to protect personnel and assets.

PERSONAL

NAME: ____
(Last) (First) (Middle)

NICKNAME, MAIDEN NAME, OR OTHER NAME(S) BY WHICH YOUR FORMER EMPLOYERS OR SCHOOLS MAY HAVE KNOWN YOU:

SOCIAL SECURITY NUMBER: _____-____-_____

PHYSICAL DESCRIPTION: HEIGHT ____(ft) ____(in) WEIGHT ______(lb)
EYES ____________ HAIR ___________

DATE OF BIRTH: ____/____/____ month /day / year

PLACE OF BIRTH: ______________________________
(City) (State)

(Country)

CURRENT ADDRESS: ____
(No.) (Street) (City) (State) (Zipcode)

PHONE: (____) ____-______ MONTHS AT CURRENT ADDRESS: _______________

LAST PREVIOUS ADDRESS: ______________________________________
(No.) (Street) (City) (State) (Zip)

EMERGENCY CONTACT NOTIFICATION

NAME: ____
(Last) (First) (Middle)

ADDRESS: ____

PHONE: (____) ____-______ RELATIONSHIP: _______________________

Figure 8–1. Sample Bonding Form. The sample bonding form asks for much of the same information that a job-application form asks.

Page 2 of 4

DO YOU OWN REAL ESTATE IN YOUR OWN NAME? YES ___ NO ___ RENT___

DO YOU OWN A VEHICLE? YES __ NO __ DO YOU HAVE USE OF A VEHICLE?

IF YES, MAKE ______________________ MODEL __________________ YEAR
COLOR _______________ TAG NO. ___________________ STATE

DRIVER'S LICENSE NO. _____________________________ STATE

DISCLOSURES

HAVE YOU EVER BEEN CONVICTED OF A CRIME OTHER THAN A MINOR TRAFFIC VIOLATION?
YES ___ NO ____ IF YES, GIVE DETAILS. USE BACK OF FORM, IF NECESSARY.

____
(Date) (City/State) (Charge) (Disposition: Jail Time, Fines, Probation, Etc.)

____
(Date) (City/State) (Charge) (Disposition: Jail Time, Fines, Probation, Etc.)

____
(Date) (City/State) (Charge) (Disposition: Jail Time, Fines, Probation, Etc.)

____
(Date) (City/State) (Charge) (Disposition: Jail Time, Fines, Probation, Etc.)

HAVE YOU EVER BEEN DISCHARGED, REQUESTED TO RESIGN, OR CHARGED WITH A SHORTAGE OF CASH OR LOSS OF COMPANY PROPERTY? YES ___ NO ___. IF YES, GIVE DETAILS. USE BACK OF FORM IF NECESSARY.

____
(Date) (City/State) (Company Name) (Description of Violation)

____
(Date) (City/State) (Company Name) (Description of Violation)

HAVE YOU EVER BEEN REFUSED A BOND, HAD A BOND CANCELED OR BEEN REQUIRED TO FURNISH SEPARATE SECURITY TO OBTAIN A BOND? YES ___ NO ___. IF YES, GIVE DETAILS. USE BACK OF FORM IF NECESSARY.

____
(Date) (City/State) (Company Name)

DISCLOSURE DETAILS, EXPLAIN (WHO, WHAT, WHEN, WHERE, WHY). USE BACK OR ADDITIONAL PAPER IF NECESSARY.

<u>YOUR LAST EMPLOYER BEFORE DO RIGHT, INC.:</u>

NAME: ________________________________ TELEPHONE (___) ___-________

ADDRESS: ____
(No.) (Street) (City) (State) (Zip)

STARTING DATE: __________ LEAVING DATE:_________POSITION:______________

Figure 8–1. Sample Bonding Form. *Continued.*

Page 3 of 4

REASON FOR LEAVING: ____________________

SUPERVISOR'S NAME AND TITLE: ____________________

YOUR NEXT PREVIOUS EMPLOYER:

NAME: ____________________ TELEPHONE (   )   - ________

ADDRESS: ____________________
(No.) (Street) (City) (State) (Zip)

STARTING DATE: ________ LEAVING DATE: ________ POSITION: ________

REASON FOR LEAVING: ____________________

SUPERVISOR'S NAME AND TITLE: ____________________

YOUR NEXT PREVIOUS EMPLOYER:

NAME: ____________________ TELEPHONE (   )   - ________

ADDRESS: ____________________
(No.) (Street) (City) (State) (Zip)

STARTING DATE: ________ LEAVING DATE: ________ POSITION: ________

REASON FOR LEAVING: ____________________

SUPERVISOR'S NAME AND TITLE: ____________________

EDUCATION

LAST SCHOOL ATTENDED: ____________________
(Name) (City) (State)

LAST DATE ATTENDED: ___/___/___ DIPLOMA OR DEGREE: ________
month / day / year

YOUR NEXT PREVIOUS SCHOOL : ____________________
(Name) (City) (State)

LAST DATE ATTENDED: ___/___/___ DIPLOMA OR DEGREE: ________
month / day / year

FINANCIAL

ALL BANKRUPTCIES, LIENS, GARNISHMENTS, (Within the Last 7 Years)

BANKRUPTCY____________________ Chapter 7 / 11 / 13
(City) (State) (Date) (Circle One)

GARNISHMENT OR LIEN ____________________
(Circle One) (Total $) (Date) (City) (State)

TYPE OF DEMAND (child support, debt, probationary agreement)________

IF YOU NEED ADDITIONAL SPACE, USE BACK OR CONTINUE ON A SEPARATE SHEET OF PAPER.

Figure 8–1. Sample Bonding Form. *Continued.*

Page 4 of 4

PLEASE REVIEW YOUR ANSWERS TO ALL QUESTIONS. SIGN AFTER YOU READ AND UNDERSTAND THE FOLLOWING STATEMENT:

I claim that the information on this Bonding Form is complete and correct. I understand that any misrepresentations, lies, or omissions can be grounds for terminating my employment. I authorize representatives of DO RIGHT, INC. to investigate my background and to verify the information that I have provided. I hereby release all persons who furnish information to DO RIGHT, INC. from any liability in connection with providing accurate information about me.

AUTHORIZATION AND RELEASE

I, ______________________________, hereby authorize any individual, organization, or
(Print Name)
governmental agency to furnish to the Senior Vice President of Human Resources of DO RIGHT, INC., or his/her designee, any and all information (including documents) regarding me. I hereby release from liability all such persons, organizations or governmental entities who provide any such information (including documents). A photocopy or facsimile of this release shall be as valid as the original.

Employee's Full Printed Name ______________________________

Employee's Signature ______________________________ Date __________

Employee's Location ______________________________
Department or Unit Number Address City/State Zipcode

Supervisor's or HR Rep: ______________________________
(Print Name and Title) Area/Dept./Location

Supervisor's or HR Rep's Signature ______________________________ Date __________

Supervisor's Telephone or Pager __________________

Figure 8–1. Sample Bonding Form. *Concluded.*

each year. Closed or sold businesses should appear listed in the years they operated. Reference librarians can readily verify most applicant claims. For example, Medal of Honor and Heisman Trophy winners are easily verifiable. There are few claims that can be made by an individual that cannot be validated in the public record.

Once conflicting information is developed, the employer can choose to adopt an affidavit process that informs the applicant and requires an explanation. For instance, "A routine investigation revealed an arrest and conviction for aggravated assault on 6/2/96 in Dade County, Florida. Please affirm or deny this conviction. Circle one. Provide a detailed account of what happened." A written statement from the applicant should confirm or deny the investigative information and detail the subject's side of the story. Serious security risks can include open warrants and undisclosed felony convictions. If the organization has adopted such a policy, such negative information can warrant imposition of investigative leave before resolving the issue.

The affidavit process allows the employee to deny, confirm, or add mitigating circumstances to the report. Errors in public records make the affidavit process particularly reasonable when the record is wrong. Misinformation can be corrected to an applicant's advantage when a school has transposed a social security number at the registrar's office and denies having conferred a degree, or when a "junior" is erroneously cited for felony convictions that belong to his father.

Information discovered by routine background checks ranges from outstanding warrants to fraudulent use of a social security number. Professional certifications—including diplomas, degrees, and grade-point average—can be embellished or fabricated. Work history can be false or omitted. These are all signs of potential problems that require appropriate disciplinary accountability if corroborated.

The affidavit is one type of written statement that can be solicited from an information provider, victim, witness, or suspect, to provide additional facts. Routine disciplinary action, up to and including termination, can protect the organization from employment-negligence actions. Interestingly, the affidavit process affirms application fraud in 90% of the cases. Because no action is taken before investigating the issue, no litigation is reported as the result of faulty terminations or withdrawal of contingent offers.

WRITTEN STATEMENTS

Regardless of the facts discovered in an investigation, witnesses and potential suspects ideally are required to complete a detailed narrative statement regarding their activities immediately before and during the time of the crime or incident. (See the sample written statement in Figure 8–2.) The investigative protocol ensures equal treatment for all information providers. It also "locks in" witnesses and suspects on their version of events early in the investigation.

Written statements offer the investigator the writer's version of the event. Investigators who are trained in scientific content analysis (SCAN) are qualified to analyze the linguistic code of the statement for significant omissions of fact and for deception (Sapir, 1987). Founded on the simple principle that if it is not written, it did not occur, the process allows the investigator to identify a truthful statement and clear an investigative subject in a timely fashion. Conversely, scientific content analysis also allows the investigator to identify suspects by analyzing pronouns, tenses, and linguistic codes. For example, when the alleged victim of a robbery wrote the following statement, his admission followed in an interview that was expedited by scientific content

Written Statement

Please print in ball-point pen. Use as many pages as necessary.

Date__________ **Unit #**__________

Name____________________ **Work Phone (___)**__________

Address____________________ **Home Phone (___)**__________

Street

____

City **State** **Zipcode**

Job Title__________ **Social Security #**__________ **Date of Birth**__________

Incident ____________________ **Date and Time** __________

Details (Who, What, When, Where, How, and Why)

Page____ **of**____ **SIGNATURE**____________________

Figure 8–2. A written statement confirms the testimony of witnesses and potential suspects regarding their activities before or during the time of an incident.

analysis: "Was opening the restaurant early when suddenly a voice from behind said this was a robbery. Went into the store, opened the safe, was told to go in the freezer and count to 500 and not to tell the police."

The writer conveniently omitted the "I" from his statement. No third-person pronoun—he, she, or they—was ever introduced either. Technically, neither the victim nor the robber was on the premises as described in the written statement. The second written statement taken after the interview revealed that the employee had fabricated the robbery and embezzled the missing funds. The suspect hid the money on the premises until he thought it would be safe to take it.

Requests for written statements should take place before the interview. Avinoam Sapir, "father" of scientific content analysis, reminds us that the premature interview may only teach the subject to lie once he or she determines what is important to the investigator. If the information provider is unable or unwilling to put a statement into writing, a detailed account of the event can be recorded or transcribed verbatim from the investigator's notes. This should occur uninterrupted for the purpose of obtaining the "pure version" of the storyteller. Irregularities and clarifications can be addressed later on, in additional written statements or in the interview process. All varieties of statements can be "SCANned," including incident reports, affidavits, depositions, and even suicide notes.

INCIDENT INVESTIGATIONS

The prime principle for determining a suspect is *opportunity*. *Motive* is a close second. Most crimes do not blatantly suggest motive. Property crimes including theft involve considerably less passion than conscious motive. Is there a suspect who has inordinate financial need? The suspect is often the one who, when asked to list those who had an opportunity to commit the crime, either fails to include himself or lists himself last. Those having opportunity cannot be excluded from suspicion because they are "one of our best employees" or because they were on vacation at the time of the crime. If a person had access either personally or theoretically, through collusion with a third party, he ought to be examined as a potential suspect.

The Achilles heel of amateur investigators is bias; either their own or that of someone else. A manager reports a crime and has a list of the usual suspects. The investigator who follows this lead does so at his own peril. The equal-opportunity investigator establishes, early on, who had opportunity. Good reporting systems answer the "opportunity question" in the first report process. For instance, if a safe is burgled, those having keys or combinations should be routinely listed by name, social security number, and position title. Similarly, the report taker should solicit written statements from all those having opportunity, detailing their activities during the time of the crime. Automated investigative protocols can initiate a routine criminal background check of all potential suspects.

The crime reporter or victim can often be a suspect, particularly if the crime is not witnessed. We know that up to 64% of commercial armed robberies involve current or former employees. Fabricated reports feature certain common behavioral characteristics—a lone victim, no witnesses, or a shaky co-conspirator witness. Higher-than-average losses (for example, of several deposits rather than one) are not uncommon. The investigator's attention should be drawn to a demonstrative minor injury (often a self-inflicted bruise, cut, or flesh wound). Reporting lapses (for example, notifying a manager and a supervisor before the police) or a coincidental chain-of-command lapse (the suspect's supervisor being on vacation or unavailable at the time of the incident) should cast suspicion on the reporting victim. When the "victim" manages to "save" some money (heroically reducing the robber's take), or description particulars are either very vague or unusually detailed, there may be cause for further inquiry. Typically, several of these red flags present themselves simultaneously in fabricated cases.

In addition to collecting detailed written statements from all victims and witnesses, a re-enactment of the suspicious incident may be warranted. Situations involving self-inflicted wounds are re-created with difficulty. Often impact angles would be the opposite of those the victim describes. Videotaping is a good idea. This allows suspicious events and injuries to be objectively documented.

Human behavior is repetitive. Individuals who are generous seldom give only once. People who score in the top 5% on a scholastic aptitude or intelligence test are likely to retest in the same range. Suspects who injure, damage, or convert assets to their own personal use are likely to have stolen, assaulted, vandalized, or victimized others in the past. Automated reporting systems will search for a suspect's name or social security number in a database and will detail every previous incident in which she was a reporter, a victim, a witness, or a suspect.

Behavior traits demonstrate common denominators. Suspects who may be potentially violent share a checklist of potential antisocial behaviors. Embezzlers and property criminals often are arrogant; they rationalize that they are undervalued or underpaid, even when they achieve director or officer status and are among the most highly compensated employees. Analysis of expense records of highly compensated embezzlers often provides multiple proofs of deceit. Investigative protocol requires analysis of at least 90 days of expense-account requests, company-vehicle mileage reports, receipts, business telephone records, a business calendar, and any company credit card activity. Although the suspect may be an investigative target for hundreds of thousands or even millions of dollars of theft, expense records can readily prove consistent patterns of petty embezzlement that will prove instrumental in defining the suspect's character.

COST ANALYSIS

Embezzlers can use an inordinately high amount of gas in company vehicles, well beyond average usage for their peer group. The explanation is simple. They feel entitled to fill up every vehicle in their household. Receipts from local gas stations readily conflict with confirmed out-of-town travel. Similarly, telephone credit calls and business credit card purchases are made in two or more towns simultaneously. The embezzler can feel entitled to share company credit cards with everyone in his family. These irregularities are not typically discovered because we trust people in positions of trust. Expense account approvals tend to be automatic.

Few occasions short of suspected theft endanger examination of a series of expenses over a large time period. Inordinate cash receipts, when reviewed over months of activity often show that the embezzler has his or her own receipt book because receipts are consecutively numbered. Suspicious cash receipts for entertainment check back to carry-out establishments or lists guests that were never present (easily proved when fellow employees have conflicting expense and travel records). If dishonesty is established via expense accounts, always investigate the management suspect's cost center. Thieves seldom restrict their activity to one method.

Cost analysis is paramount for all property crimes involving theft. Suspects typically report significantly higher costs or lower operating profit than their peers. Less-than-honest construction directors often feature higher building costs. Unethical marketing directors can show higher promotional costs. Risky real estate professionals rationalize higher site costs based on a limited window of opportunity. Facilities managers show higher repair and maintenance costs. Dishonest retail managers report lower sales or higher inventory or labor costs. When compared to others, the suspect's transactions nearly always yield lower returns on investments.

Inordinate-cost investigations regularly lead to suspects who misrepresent expenses or pad their inventory in an attempt to avoid detection. Occasionally, a cost inquiry may only prove that expenses were not controlled or that a manager was incompetent. The objective inquiry identifies procedural lapses and training shortfalls that, once remedied, reduce costs.

Excessive cost is often the result of the trusted managers paying too much for service in order to receive kickbacks or from theft of revenue. Inventory and labor costs are expressed as a percentage of sales on profit-and-loss statements. Theft of sales always escalates controllable costs. Receipts or invoices for service can be fabricated, routinely approved for double payment, or paid without verification of delivery of goods or services. Your organization may often supply the lion's share of the suspicious vendor's revenue. This can be routinely corroborated through audits and credit checks. Investigation often reveals a previous relationship between a supplier and a suspect.

The public record, including that of the Secretary of State's office, can prove invaluable. Discovering that your vice president is on the board of directors of a suspicious supplier is a bad sign. Similarly, a building permit registered in the county by your construction service's supplier for your director of construction's personal residence may not be coincidental. A title search that reveals a newly purchased lot was recently owned by an in-law, a business partner, or a friend of your real-estate manager can be cause for concern.

Greed and arrogance are a powerful combination. Gone are the days when the cash kickback was nearly impossible to trace to the luxury car sitting in an investigative target's driveway. Today, the third party in a conflict-of-interest investigation is probably leasing the car to take full advantage of tax treatments. An employee's exotic vacation travel is directly billed to a vendor's travel department as a promotional expense to "legitimatize" the expense. Illegal payments and conflicts are auditable.

Crimes can vary by motivation. Opportunity always defines the probability. Disgruntled individuals can steal retail cash receipts, divert inventory from a warehouse operation, contaminate a process in a manufacturing environment, or introduce a virus into a critical software application. Regardless of the motivation, suspects have opportunities that can be proved from schedules, access-control systems, telephone records, journals, eyewitness accounts, or even forensic evidence.

INVESTIGATIVE PROTOCOL

Many employers recognize that up to 80% of the criminal activity with which they must deal involves current or former employees. The investigative process tends to look outside, eliminating internal involvement as a possibility. Efficiently run organizations adopt a protocol that is timely, professional, objective, and fair.

If investigative resources are not available within the company, they can be purchased from a reputable provider. Ensure that investigative services are licensed and insured. Providers must be willing to provide a certificate of insurance naming the organization as an additional insured in the amount of at least $1 million per incident for liability acts or omissions. Investigation firms should feature certified private investigators who have specific experience and references for the specific type of investigation required. Ideally, investigators should be certified. *Certified Protection Professional (CPP)* and *Certified Fraud Examiner (CFE)* are two designations that indicate a practitioner has accomplished a basic standard of experience and education. Always verify credentials with the appropriate professional society. Contract services with the oversight of competent legal counsel.

Effective investigative protocols are anchored in a comprehensive prevention program that features a *code of conduct*. Codes typically prohibit crimes and misbehaviors such as violence, threats, theft, fraud, prohibited substances, unethical conduct, harassment, discrimination, and so forth, which will disrupt the mission of the enterprise. Prohibitions can be broadly communicated via payroll checks, internal memoranda, policy statements, magazines, or newsletters. Figure 8–3 shows a typical investigative protocol.

A *duty to report* emphasizes notifying the chain of command of any prohibited conduct. Confidential options can include a toll-free help line, a facsimile-transmission number, or an address. The *commitment to protect* encourages whistle blowers or those threatened with lateral or alternative employment where the company is at less risk. *Fact finding* should typically extend the benefit of the doubt to all parties (victims, witnesses and suspects) via written statements and follow-up interviews. *Consistent treatment* (including suspension, reinstatement, disciplinary action, prosecution, or assignment of responsibility) must be fair. (See the sample investigative protocol in Figure 8–3.)

Zero-Tolerance Code of Conduct
Violence, Threats, Harassment, Discrimination, Inappropriate Personal Relationships, Ethics, Prohibited Substances, Fraud

Duty to Report
Chain of Command, Human Resources, Employee Help Line, Law Enforcement, or Government Agencies as Required

Intervention: Move "At Risks" Out of Harm's Way

Fact Finding, Collection of Evidence, and Documents
Benefit of the Doubt for All Parties
Written Statement Solicitation
Complainants, Victims, Accused, Respondents, and Witnesses
Open Statement or SCAN View Questionnaire™
In Person, Faxed, or by Registered Mail

Interviews
John Reid Technique
More Written Statements

Personnel Action and Appeal
Mediation
Arbitration
Prosecution
Litigation

Justice
Fair Treatment

Figure 8-3. Investigative Protocol. Investigative protocol ensures equal treatment for all information providers. *Reprinted with permission from D'Addario.* The Manager's Survival Guide, *p. 55.*

INTERVIEWS

Interviews should not be undertaken before the first report has been made or detailed written statements have been collected, if possible. Amateur sleuths are cautioned not to undertake an interrogation. An *interview* is an exchange of information that should supply details of the event. An *interrogation* involves accusing an individual of wrongdoing. Mistaken accusations can carry harsh penalties if slander or libel is involved. Get the facts before the interview.

Interviews often lead to revelation of previously unknown information and sometimes an admission. Always conduct an interview in a private space, where the conversation is unlikely to be overheard or interrupted. Allow interviewees to sit next to the door or in a manner that does not block their exit. Take notes (including the date, time, and location of the interview) along with the contact data of interviewees (including their names, addresses, and telephone or pager numbers).

Never provide notice of the interview; particularly if the subject appears to be involved. Background information and a review of the interviewee's personnel file can prove helpful. Very skilled interviewers can observe the posture and attitude of the interviewee as signals to consider. A defiant facial expression or even such body language as leaning back in the chair, general inattention or sleepiness or crossed legs or arms sometimes can connote resistance to the interview when considered in light of other suspicious observations.

Always ask interviewees whether they know why you have asked to see them. If you have information in hand that implicates interviewees, you can inform them that your investigation shows they are involved. Probable-cause information can range from an anonymous tip to an access-control report that indicates presence at the location at the time of the crime. Video, eyewitness information, a lengthy criminal history, or a deceptive written statement detailing the crime could be involved.

A potential lead for the investigator can be "Our investigation suggests that you have not told the whole truth." Many people become involved in something unintentionally because of a mistake in judgment. If this is one of those situations, it should be cleared up in a businesslike fashion before the police are required to handle it. Some people who make a mistake in judgment compound it by denying what happened in face of the evidence. They lose the opportunity to control their own destiny. The investigator can say, "If this was a case of bad judgment on your part, the time to clear it up is now. Are you willing to help me clear this up today?"

Even before the interviewee agrees to be forthcoming, the following subtle signs of submission can appear:

- uncrossing arms and legs, suggesting openness
- a neutral facial expression
- leaning forward in a slumped position
- attention to the interviewer's words with occasional nodding, weeping, or a generally defeated look

Caution: tension or personal problems could produce similar responses. These may be significant only in concert with more substantial evidence.

Vince Cowdry, an interview and admission specialist, emphasizes the architecture of the "golden bridge," in a 1996 piece, "A Short Course on Interviews." Bridge-building involves establishing good and bad options or alternatives that may allow the suspect to admit wrongdoing. The interviewer can offer softening alternatives. For example, the John E. Reid behavioral interview technique refers to stealing as "borrowing." An alternative offering will posture a good reason versus a bad reason to confess: "Did you borrow the money to pay bills for your family, or was it for a drug habit? Did you plan to take this money for a long time, or did it just happen on the spur of the moment? Did you take the money to get even with management, or did you have a personal problem that forced you to do it?" (Inbau et al., 1986.) Be careful not to be heavy handed in employing this approach, however. You must never appear accusatory.

Many admissions are preceded by the question "What will happen if . . . ?" Avoid promising not to prosecute a suspect in return for cooperation. Emphasize that cooperation enhances the ability to maintain control of one's own destiny. Collect a written statement that tells the whole story from the interviewee's point of view. Sometimes deception and misrepresentation fall short of involvement in a crime. For instance, a manager who failed to secure funds or assets, might lie so she does not appear inefficient or lacking in foresight, or to avoid personnel sanctions for breach of policy.

In all cases, set the record straight with an additional written statement. Document that the story has changed to support prosecution, disciplinary action, or your case against another subject. Some suspects admit to something orally but choose not to write the admission statement. A letter of apology, a tape-recorded admission, or even an apology by voice mail will suffice to corroborate complicity short of the full written statement.

The interviewer must always convey a posture that is both professional and empathetic. If an interviewee strongly and consistently denies involvement the investigator must recognize that such behavior can suggest that the interviewee is telling the truth. The investigator's responsibility is to establish the facts.

Restrain unapproved employees from signing warrants without investigative or legal review. The civil liabilities associated with malicious prosecution and false arrest are potentially enormous. Most law enforcement jurisdictions review a case for a probable-cause warrant. Others submit investigative findings to a grand jury for indictment consideration. Always report crimes. Weigh a policy to prosecute all crimes where the prosecuting authority determines probable cause. Avoid letting managers pick and choose who is to be prosecuted. Consider paying employees for time required of them for the investigation and court appearances.

SUMMARY

Several factors contribute to conducting effective background and incident investigations. These include:

- preparation—be alert to potential problems in prospective employees and in actual or suspected problems in current personnel
- state requirements and restrictions for screening and incident investigations
- investigative procedures—ideally acquired formally through a training program—and protocol
- concepts, administrative routines, and standard forms
- resources—relevant training programs and reading material

In detailing and providing examples of all of the foregoing, it is hoped that this chapter can assist employers in conducting investigations and, by showing foresight *before* hiring new employees, employers can avoid conducting investigations later.

BIBLIOGRAPHY

The following programs and relevant reading material are recommended.

Background Investigations, 6th ed. Tulsa OK: National Employment Screening Services, 1994.

Basic Guidelines for Security Investigations. Arlington, VA: American Society for Industrial Security, 1991.

Cunningham, W. J. S, and C. Van Meter. *The Hallcrest Report II: Private Security Trends 1970-2000*. Newton, MA: Butterworth-Heinemann, 1990.

D'Addario, F. J. *The Manager's Violence Survival Guide*. Chapel Hill, NC: Crime Prevention Associates, 1996.

Inbau, F. et al. *Criminal Interrogation and Confessions*, 3rd ed. Baltimore MD: Williams & Wilkins, 1986.

The Laboratory for Scientific Interrogation, Scientific Content Analysis: SCAN Reports, P.O. Box 17286, Phoenix, AZ 85011. Online at www.lsiscan.com; e-mail: info@lsiscan.com.

Liability Consultants, Inc. Inadequate security (study). *Wall Street Journal*, Feb 1, 1993.

Monachesi, E. and Cesare Beccaria. In H. Mannheim (ed). *Pioneers in Criminology*. Monclair, NJ: Patterson Smith, 1973.

O'Hara, C.E., and G.L. O'Hara. *Fundamentals of Criminal Investigation*, 6th ed. Springfield, IL: Charles C. Thomas, 1994.

Reid, John E. and Associates. The Reid Technique of Interviewing and Interrogation, 209 W Jackson Blvd., Suite 400, Chicago, IL 60606. Online at www.reid.com; e-mail: info@reid.com.

Sapir, A. *Scientific Content Analysis: SCAN*. Phoenix, AZ: Laboratory for Scientific Investigation, 1987.

Walters and Associates, Inc. Practical Kinesic Interview and Interrogation® Courses. 309 Arrowhead Springs Lane, Versailles, KY 40383-9591, (800) 847-4774; Online at www.kinesic.com.

Walters, S. B. *Principles of Kinesic Interview and Interrogation*. Boca Raton, FL: CRC Press, 1996.

Part 4
Personnel Security Management

PART 4

9

Human Resources and Security Management

By Eugene F. Ferraro, CPP, CFE, PCI

CHAPTER 9

OBJECTIVES

The objectives of this chapter are to

- define modern human resource management and examine its importance to the proper functioning of an organization
- explore how and why human resource management and traditional organizational security functions have blended and the implications thereof for risk management
- examine the future of human resources and security management

INTRODUCTION

Human resource (HR) management is the sum of all organizational systems that ensure human talent is used effectively and efficiently during the pursuit of organizational goals. Thus, many organizations have claimed that their ability to manage HR differentiates them from their competitors and is a key determinant of competitive advantage. Organizations as widely diverse as FedEx, Nordstrom, and Dell Computer have focused on human resources as having special strategic value for the organization (Mathis and Jackson 2006, 9). In this chapter we will examine this concept in more detail and the importance of HR management and security management's relationship to one another in the modern organization

THE NATURE OF MODERN HUMAN RESOURCE MANAGEMENT

Most observers will agree that for several decades the field of HR management has undergone a significant transformation. In addition to the traditional functions of employee recruitment, selection, training, and management, today's HR professionals are increasingly involved in organizational regulatory compliance, compensation administration, employee safety, and risk management. The enormous breadth of oversight and responsibility of the modern HR professional has made HR management one of the most important functions in all but a few organizations.

The role of HR management is expanding, as evidenced by the frequent placement of traditional organizational security functions under the umbrella of HR. In these organizations we see a blending of HR and security activities and the melding of once distinct and very different missions. The resultant product is an HR function that is more security minded and a security function that is more humanistic. This holistic approach has birthed new organizational alliances and has brought new efficiencies unattainable a mere decade ago. Another driver behind this sea change is the mutual recognition that the modern workplace and the people in it have also changed. For our purposes, there are four noteworthy changes.

MORE SOPHISTICATED CRIMES AND PERPETRATORS

As our workplaces become more sophisticated, so do the crimes and their perpetrators. With the addition of advanced processes, systems, and technology, the opportunity for exploitation has advanced as well. It is for this reason that today's approach to these challenges must also be more advanced. Those who reject the modern workplace and all of its modernity are doomed. The security professional must embrace the new and ever-changing workplace, for it is impossible to solve modern workplace problems without a fine understanding of the environment in which they occur.

The capacity of the workplace criminal has also changed. With the availability of new and more sophisticated technology, today's workplace criminal is capable of causing more damage while using less effort and skill. The Internet provides the workplace criminal with education, new ideas, and weapons. Take, for example, identity theft. Every person has always had an identity. Yet, it is only recently that the crime of identity theft has caught our attention and entered our lexicon. Why? The answer is simple: Changes in organizational processes (storing large quantities of personal information in electronic databases) made accessible with more advanced technology (the Internet) to a more sophisticated (educated and technologically savvy) criminal have made devastating new crimes possible.

GREATER USE OF AND DEPENDENCE ON TECHNOLOGY

Technology by definition promises convenience and economy. When technology is properly crafted, the convenience and economy it provides create solutions. Technological solutions improve efficiency and profitability. Intentionally, they also reduce head count. Engineers are constantly looking for ways for machines to replace human beings. Unfortunately, however, from an asset protection and loss prevention perspective, two things happen: (1) The technology provides new opportunities for exploitation; (2) the quantity of supervision decreases. There is a subtle dichotomy here; as exposure increases, protections decrease. Look around your own organization. Do your employees not have more access to just about everything, and is your organization not operating with less supervision?

Furthermore, organizations have been tricked into thinking more can be done with less supervision. Consider for a moment a few of the revolutionary, often enterprisewide leadership initiatives the big-name thought leaders have sold to America's cutting-edge organizations designed to empower employees and create world-class organizations that are built to last (Collins 2002). For more than two decades, corporate America has been looking for the next "new thing"; it has yet to find it (Lewis 2001). Although many of these new-era corporate messages are about embracing employee relations and empowering individual contributors, the sad fact is that most employees do not care.

The truth is most employees do not want to be empowered, most do not want more responsibility, and most are neither prepared nor willing to make tough organizational decisions. Largely, employees want a safe, fulfilling job and a work environment in which they are appreciated and treated with respect. Consider that if the average line employee or staffer wanted more responsibility or risk, would he not have started his own company? What has been created in many cases are work environments in which management has involved its employees in running the business, something the typical employee does not want to do. Then, in order to capitalize on the dividend of reduced head count, management has slashed the amount of supervision in their organizations. Rarely have such initiatives worked as anticipated. Instead they serve as expensive experiments. By reducing the quantity of supervision without necessarily improving quality, many of these well-meaning organizations created nothing more than fertile grounds for employee theft, dishonesty, and misconduct.

As management reengineered itself, it often simultaneously embraced advances in technology. In most industries and workplaces, technological advancements delivered on the promise of improved efficiency and profitability. However, as previously mentioned, new technology also produces new vulnerabilities. As fast as the technology is deployed, villains may be able to figure out how to exploit it to some criminal advantage. In some cases they will steal the technology and also the product it produces, as in the practice of illegally downloading music from the Internet (Ferraro 2006).

A MORE LITIGIOUS WORKFORCE

The United States is the most litigious industrialized nation in the world. For all its greatness and potential, our society is also one wrapped in rules and regulations. Be it workplace safety regulations or the amount of vacation pay a terminated employee must be paid, many workplace activities are regulated. In the post–World War II era of regulation promulgation, we have passed more laws and regulations than were implemented during the 175 years prior. This tangled morass of increasingly contradictory rules, laws, and regulations is rapidly reaching critical mass. Even lawmakers are beginning to recognize the trouble they have created (U.S. House of Representatives 1995). However, little has been done to solve the problem. Following is just one example of such perplexing conflicts.

The Americans with Disabilities Act (ADA, 42 U.S.C. 12101, 1990) stipulates employers must reasonably accommodate those with recognized or perceived disabilities. It also imposes an obligation to reassign work or make other accommodations to assist persons with disabilities or those recovering from injuries. This makes sense because early return-to-work policies reduce costs. However, if an employee is injured and, in an effort to reduce workers' compensation costs, an employer attempts to accommodate the employee and return her to an alternative or a light-duty position, the employer may violate the Family and Medical Leave Act (FMLA). The Department of Labor has taken the position that if an employee is entitled to leave because of a "serious health condition," which may include a workers' compensation injury, the employee is entitled to the same or equivalent position upon return. So therefore, while the ADA and the employers' workers' compensation law may encourage rapid return to light-duty assignment, the FMLA allows the employee to refuse to accept it and sue if forced to accept it.

Other employer liabilities exist as well. Since its creation in 1970, the Occupational Health and Safety Administration (OSHA) has been hard at work creating a safer workplace for everyone. In doing so, OSHA has published more than 4,000 regulations, dictating everything from the height of a railing to the thickness of carpet. Navigating the maze of rules and regulations (such as how far a plank can stick out from the edge of a temporary scaffold) requires more than 2,000 inspectors and tens of thousands attorneys who prosecute and defend against the litigation those regulations have precipitated. Attempting to toe the line, industry has spent several hundred billion dollars on compliance alone (Howard 1994, 12).

Arguably many laws and regulations, such as Title VII of the Civil Rights Act, for example, have made a safer and more fair workplace for Americans, but the net effect on the employer is more liability and litigation. The litigious trend can be seen outside of the workplace as well. In Los Angeles county alone, more than 3,000 civil lawsuits are filed each day. Only a mere fraction of these are employment related, but the fact remains that Americans like to sue. Most disturbing is the trend shows no sign of reversing.

EXPANDED PROTECTION OF EMPLOYEES AND THEIR JOBS

Some, like Philip Howard in his compelling 1994 best seller, *The Death of Common Sense,* have argued the expansion of rights is a zero-sum game. That is, as the rights of one group expand, the rights of another contract. Zero-sum theory applies well to aspects of economics, but as we look at the expansion of worker rights, it seems to apply perfectly. Legislation such as Title VII, the ADA and FMLA, and more recently, the Health Insurance Portability and Accountability Act (HIPAA), has certainly provided much good. Few would argue against improving workplace safety or improving the treatment of employees. However, the cost has been the erosion of employer prerogatives and an exponential increase in compliance costs. It is estimated that HIPAA alone has cost employers more than $45 billion. The trend has eroded employer doctrines such as "at-will employment" to the point of meaninglessness. Few employers can terminate at will, much less hire whom they wish, without contemplating the litigation potential of perceived inequality or unfairness.

Sadly, the employees do not win. Increased employer costs translate into fewer resources for research and development, fewer process and technology improvements, less profit, fewer investments, and ultimately, the creation of fewer jobs. Even the government loses. With less money in circulation and fewer people working, fewer taxes are collected. The net effect is a less competitive and weaker economy.

HUMAN RESOURCES AND SECURITY UNITE AGAINST A COMMON ENEMY

If human resource management is the sum of all organizational systems that ensure human talent is used effectively and efficiently during the pursuit of organizational goals, what then is its relationship with security management? As discussed, there has been a blending of HR and security activities and the melding of their missions in the last decade. In many ways the result has changed how we look at organizational resilience and risk management. Prior to this change, most organizations believed the greatest threats they faced were from external sources.

For most of the 20th century, it was widely held that the people and forces outside of an organization posed the greatest threat to it. Assuming this logic, the security departments of these organizations embraced the notion that the protection of assets and loss prevention involved the use of locks, gates, and guards. Organizations shrouded themselves in a protective envelopment of impenetrable barriers, alarm systems, and monitored cameras. All the while HR departments went about business looking for new ways to improve employee retention, super-size benefit plans, and create a kinder, more gentle workplace. Then something changed—substance abuse entered the workplace.

In the early 1980s employers learned that employees who abuse drugs tend to be absent from the job up to 16 times more often, claim three times as many sickness benefits, and file five times as many workers' compensation claims as nonabusers. Abusers are also more likely to be laid off or fired. Additionally, drug abuse breeds dysfunctional relationships, as is often evident in drug abusers' relationships with managers and co-workers. They tend to withdraw from friends and be more secretive, spending less time at home and at work. For the employer, substance abusers are less productive and frequently represent the 20% of employees who consume 80% of their disciplinary time. Substance abusers have attendance problems, usually with recognizable patterns, such as being absent on a Friday or Monday. They appear less focused and have more personal problems. Ultimately, many abusers begin to use drugs while on the job. At some point, employers realized that one of the greatest threats to successful business operations was not external but internal.

THE NEW ROLE OF HUMAN RESOURCES AND SECURITY MANAGEMENT

A candid examination of the employee substance abuse problem facing employers rapidly led to the conclusion that the magnitude of a workplace substance abuse problem was inversely proportional to the degree to which an employer screened its job applicants, the extent to which it trained its supervisors, and the strength of its substance abuse prevention policy. In other words, the risk of substance abuse appearing in the workplace was a function of:

- the quality of the employees an organization hired and employed
- the skill and ability of supervision
- the environment and culture that the organization chose to allow or created

This discovery struck a chord not only with HR professionals, but also with security professionals. Suddenly, the thought leaders of the security industry realized that asset protection and loss prevention could be attained by better managing the enterprise and the people within it. While locks, gates, and guards could still play a role in the organization's overall security strategy, internal human resources would be used as a force multiplier. In other words, combining the disciplines of security management and HR management, the organization could protect itself from both external and internal threats without the addition of incremental cost. The concept was not new to those responsible for workplace safety. For decades safety professionals knew that industrial safety was largely a function of human attitude and awareness. Thereby, blending the three disciplines of HR, safety, and security where overlap was most obvious, a return on investment was possible. The resultant model is shown in Figure 9–1.

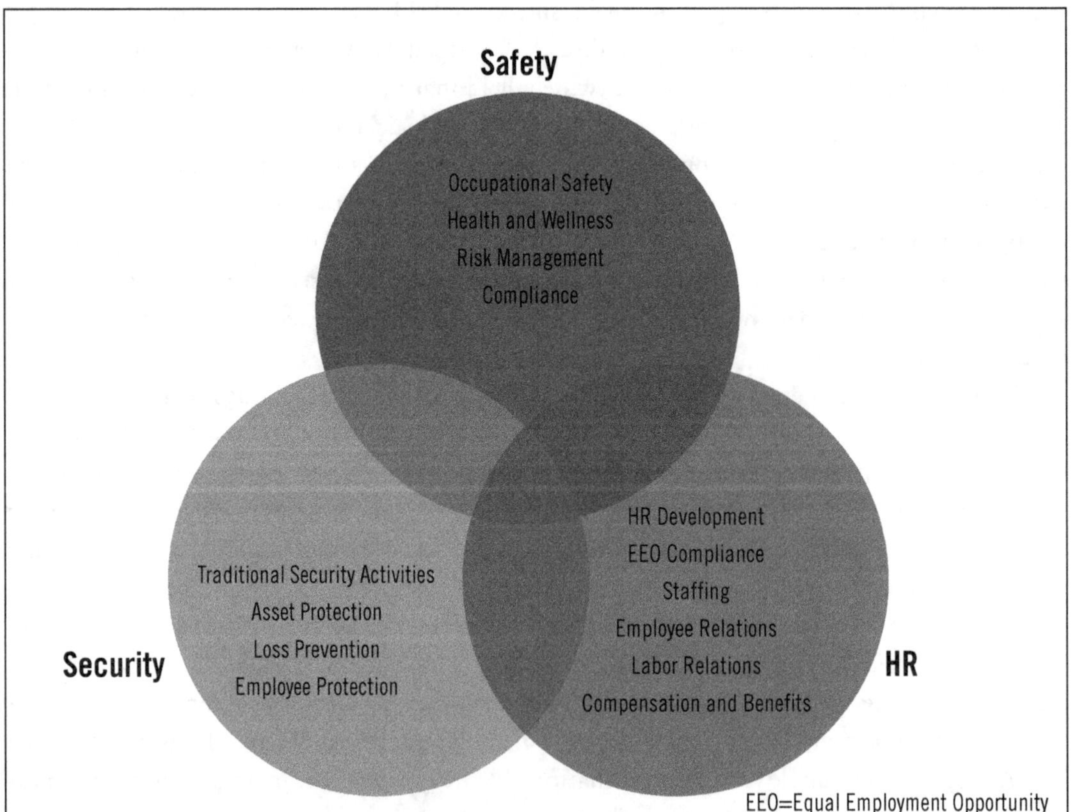

Figure 9–1. Model for Protection against Internal and External Threats to Security

Today, numerous organizations have successfully adopted this model (Mathis and Jackson 2006, 4) for protection against internal and external threats to security. Although permeations of the model are as varied as the organizations that have adopted it, there is a recurring theme. At its core is the recognition that by properly managing employees, supervisors, and the environment into which these people are placed, the organization is better able to manage risk. Thus, enterprise risk management is a thread that runs through the entire organization. It inextricably links every function and activity. Because the first component of this new enterprise risk management solution involves people, Chapter 10 Personnel Security and Employer Liability focuses more on the topic of personnel security issues.

SUMMARY

It has been said that human resource (HR) management is the sum of all organizational systems that ensure human talent is used effectively and efficiently during the pursuit of organizational goals. Security management is therefore the sum of all efforts relative to organizational asset protection and loss prevention. In this chapter we examined how these disciplines can be combined to improve efficiency and organizational resilience.

REFERENCES

Collins, C. 2002. *Built to last.* New York: Harper Business.

Ferraro, E. F. 2006. *Investigations in the workplace.* New York: Auerbach.

Howard, P. K. 1994. *The death of common sense.* New York: Random House.

Lewis, M. 2001. *The new new thing.* New York: Penguin USA.

Mathis, R., and J. Jackson. 2006. *Human resources.* 11th ed. Mason, OH: Thomson South-Western.

U.S. House of Representatives Committee on Economic and Educational Opportunities. 1995. *Conflicts and inconsistencies in workplace regulations.* Washington, DC: Government Printing Office.

10

Personnel Security Issues and Employer Liability

Eugene F. Ferraro, CPP, CFE, PCI

CHAPTER 10

OBJECTIVE

The objectives of this chapter are as follows:

- identify issues and legal risks associated with hiring, supervising, and managing employees
- discuss the legal aspects of workplace harassment and discrimination claims
- review what constitutes an unfair labor practice claim and how to avoid one
- identify and discuss some of the common legal issues associated with employee substance abuse screening and honesty testing, as well as workplace searches conducted by employers

INTRODUCTION

Although the courts have long held that, generally, employers have the right to run their business as they see fit, the Occupational Safety and Health Act of 1970 (OSHAct) and its state counterparts, as well as a large body of case law, require employers to provide employees with a workplace "free from recognized hazards that are causing or are likely to cause death or serious physical harm to . . . employees" (OSHAct1970, Sec. 5 Duties, 29 USC 654). In 1995, California went further and addressed the security aspects of violence in the workplace by producing a comprehensive report and a Model Injury and Illness Prevention Program for Workplace Security. Similarly, a number of state and federal laws require that the workplace be kept free from the sale and use of illegal substances (Title V 1988). So while a host of personnel security issues confront employers, the most significant issues are the potential legal liabilities associated with managing the workplace and keeping those within it safe.

In this chapter we will examine some of these liabilities as they relate to the work force and the overall management of security within it. This chapter will also address federal and state mandates, as well as some of the case law on the federal and state levels relative to the management of personnel security issues.

THE RISK OF NEGLIGENT HIRING, SUPERVISION, AND RETENTION

Research suggests that assaults while on duty at work represents approximately 10% of all violent crimes in the United States. A sampling of 480 members of the Society for Human Resource Management revealed that one-third of their workplaces had experienced at least one known violent crime during the past five years (Cal-OSHA Guidelines 1995). Homicide is the leading cause of death in the workplace in a number of states and the District of Columbia. While the numbers represent a very small percentage of all workers, the liability associated with workplace homicide can be devastating both to an organization's finances and, of course, to its reputation.

Security professionals know that any threat of violence must be taken seriously. A threat or a seemingly minor assault, unchecked, may be a precursor to something far more serious. Experience tells us that most employees who kill a co-worker or supervisor did not "just snap." In nearly every known case, the aggressor provided significant clues regarding his instability prior to the incident. His ultimate course of action and the resultant tragedy could have probably been averted had those around him considered the possibility that the threats were serious and taken appropriate and timely action.

The cost of these errors can be staggering. The average cost of defending an inadequate security lawsuit exceeds $38,000 *when the defendant business prevails.* The cost is significantly greater if the case goes to trial. The average (nationwide) reported verdict for plaintiffs in wrongful death cases is $2.2 million. The average verdict in aggravated assault cases is approximately $1.2 million. Absent a verdict, settlements average $750,000 (Chaney et al. 1997). Victims (and survivors) of workplace violence often allege negligent hiring. For the claim to prevail, the plaintiff must typically prove:

1. the existence of an employment relationship
2. the employee's incompetence
3. the employer had actual or constructive knowledge of this incompetence
4. the employee's act or omission caused the injuries
5. the employer's negligence in hiring or retaining the employee was the proximate cause of the injuries

An action for negligent hiring may also be based on the extent of the background investigation conducted by the employer (for more information about preemployment screening and background investigations, see Chapter 11, Employment Policies and Procedures); the employee's position and responsibilities; and the risk of harm to third parties posed by the employee (*Henley v. Prince George's County* 1986). However, in California, as elsewhere, significant obstacles stand in the way of conducting thorough employment background investigations. In such locales, employers other than those hiring for a specific position which *statutorily requires* criminal background checks are generally allowed access only to public records (*Pittman v. City of Oakland* 1988).

However, merely failing to conduct a thorough background investigation is not sufficient grounds to hold an employer liable. The plaintiff must be able to produce evidence that there was negative information readily available that would have shown that the applicant posed a risk to the safety of others. If no such evidence exists, there can be no liability imposed on the employer (*Wasson v. Stracener* 1990). Similarly, victims can sue for negligent supervision or retention once it was known or should have been known that the perpetrator was dangerous or otherwise unfit. For an employer to be found liable for negligent hiring, supervision, or retention, the offense need not be within the scope of employment, but merely somehow related to the employment. In *Yunker v. Honeywell*, 496 N.W. 2d 419 (Minn. Ct. App. 1993), the defendant rehired a former employee after he returned from a five-year prison term for a conviction involving strangling a female co-worker to death. Not long after he was rehired, the employee was transferred twice due to his inability to get along with co-workers. After the second transfer he threatened a female co-worker with death for rejecting his romantic advances. He voluntarily resigned after the woman reported the incident to her supervisor and requested a transfer. The day after his resignation, he went to the woman's home and shot her to death. Although the man was no longer an employee at the time of the murder, and the incident did not occur on the defendant's property, Honeywell was still found liable for negligent retention. This finding was based on the assumption that Honeywell supervisors were well aware of the man's violent propensities yet exposed the victim to danger by failing to terminate the man after two prior confrontations with co-workers.

THE DOCTRINE OF *RESPONDEAT SUPERIOR*

Another legal theory that has been widely used by victims as the basis of lawsuits against employers, and which is recognized in virtually all states, is the doctrine of *respondeat superior*. Under this theory, the employer can be held responsible for the acts of employees when the offense was committed within the scope of employment. Broadly, this doctrine asserts that the offender used his position in the commission of a misdeed, or that the employer in some way benefited from not addressing the misdeed. A good example would be when a hypothetical nightclub manager directs a bouncer to eject a patron and the patron is injured. While the bouncer clearly may have overreacted, his actions were the result of the instructions from the manager. Another example might be the employee substance abuser who sells illegal drugs to his employer's customers. If done in the course of his employment, and the use of illegal drugs ultimately caused the injury of another person, liability may exist.

AVOIDING DISCRIMINATION LAWSUITS

In July 1994, every American business with 15 or more employees became subject to the Americans with Disabilities Act (ADA) USC 42, §12101. The ADA allows disabled workers to bring legal action against employers and seek damages up to $300,000. The Equal Employment Opportunity Commission (EEOC), charged with enforcing the ADA, reports that approximately 87% of the ADA-based complaints it receives are from people who are already employed, as opposed to applicants claiming hiring discrimination (Shapiro 1998).

The ADA provides protection to individuals meeting the following criteria:

1. physical or mental impairment that substantially limits one or more of the major life activities of the individual
2. a record of such impairment
3. the perception of having such an impairment

The ADA prohibits employers from discriminating against a "qualified individual with a disability" relative to all aspects of the employment application process; advancement and discharge; compensation; job training; and other terms, conditions, and privileges of employment (USC 42, §12112). A "qualified individual with a disability" is one who, with or without *reasonable accommodation*, can perform the essential functions of an assigned position. Employers are allowed under the ADA to test for drug use, but they may not discriminate against people who have successfully completed a drug rehabilitation program and who no longer use illegal drugs. Current drug users are not protected.

The ADA is less specific about alcohol use and alcoholism. Thus, during the preemployment screening process, the employer should steer clear of certain questions or inquiries. Here is a list of questions that may be unlawful (Equal Employment Opportunity Commission 1994):

- How often have you used illegal drugs in the past?
- Have you ever been treated for drug abuse/drug addiction?
- How much alcohol do you drink each week?
- Have you ever been treated for alcohol problems?
- Are you an alcoholic?

- Have you ever been treated for mental health problems?
- Have you ever sought treatment for your inability to handle stress?

There have also been cases in which an employee was alleged to have made serious threats and/or actual assaults against a co-worker, then later sought protection claiming a mental disability. In *Adams v. Alderson*, 723 F. Supp. 1531 (D.D.C. 1989), *aff'd* by 1190 WL 45737 (D.C. Cir. 1990), the court determined that under the federal Rehabilitation Act of 1973, an employee who physically assaulted his supervisor because of an adjustment disorder was not "otherwise qualified" and entitled to an accommodation from his government employer. It seems the danger posed by one's inability to control violent urges would outweigh the need for job protection because of a perceived mental disability.

TITLE VII OF THE CIVIL RIGHTS ACT OF 1964

The Civil Rights Act of 1964 attempted to eliminate gender and race discrimination in the workplace. Title VII of the act criminalizes discrimination in hiring, firing, promoting, or giving raises or other benefits because of an employee's race, gender, color, religion, or national origin. While this federal law is limited to employers with 15 or more employees, most states have similar laws affecting all employers.

To qualify as a discrimination case, one of two factors must be present. The first is known as disparate or different *treatment* of a member of a protected class based solely on that difference. The second factor is disparate *impact*, which most often occurs when a company has in place a policy or test that, usually unintentionally, results in a negative impact on an entire protected class of applicants or employees. A simple example that has been recently battled in courts is height requirements for fire fighters, which have been held to be discriminatory against women in general, and Asians of both gender (who are, on average, shorter than white males). Similarly, a freight delivery company that requires all drivers to be able to lift and carry 200-pound (90.7-kg) packages could conceivably be sued under Title VII for discrimination against women.

THE CIVIL RIGHTS ACT OF 1991

When challenged under the Civil Rights Act of 1991, employers must show that their workplace practices and rules are sufficiently *job related and consistent with business necessity*. For example, the presumption that most young Hispanic or African-American male warehouse workers tend to stick to themselves, speak differently than other workers, and wear a popular style of clothing sometimes associated with gangs would not constitute reason enough to establish a practice or work rule banning such individuals. In contrast, the presence of gang graffiti on company property, evidence of a theft or illegal drug use, or credible reports of gang activity may be legitimate reasons to investigate but not exclude individuals who *appear to be involved in such activity* from consideration of employment. The following checklist provides guidelines for assessing whether your organization has taken solid steps toward safeguarding against discrimination claims.

1. Does the organization manage its work force without regard to race, color, sex, or national origin?
2. Has the organization informed all employees of its drug and alcohol policies and made it clear that the policies will be strictly enforced?
3. Are substance abuse problems in management ignored, while enforced elsewhere?
4. Is discipline consistently applied at all levels of the organization?
5. Are opportunities for rehabilitation consistently available to all employees?

SEXUAL HARASSMENT

According to a survey conducted several years ago, more than 90% of Fortune 500 companies have received sexual harassment complaints; more than 33% of them have been sued; and almost 25% have been sued more than once for sexual harassment (Sanaroff 1988, 69). The survey noted that an immediate supervisor or upper-level manager was the alleged harasser in two-thirds of the reported incidents, and that the human resource managers of these companies believed that about two-thirds of the claims were valid.

The EEOC has defined two categories of sexual harassment. The first is called *quid pro quo harassment*. This claim may arise when a supervisor conditions employment benefits or opportunities in return for sexual favors to him or her. The second category of sexual harassment, *hostile environment harassment,* occurs when an employer creates, or allows to be created, a pattern of conduct pertaining to sex that contributes to an unpleasant, intolerable, or hostile work environment.

Where harassment is pervasive and the employer fails to discover it, it will be held to have constructive knowledge of the harassment (see *Hall v. Gus Construction Co., Inc.*, 842 F2d 1010 [8th Cir. 1988]). If the employer is determined to have actual or constructive knowledge of the harassment and fails to take immediate and appropriate corrective action, it will be held liable (EEOC manual).

The prevention and investigation of sexual harassment claims present a unique set of problems. The EEOC requires that a thorough investigation of complaints be conducted quickly, yet the nature of a sexual harassment claim is such that it is particularly difficult to investigate adequately without infringing on the privacy rights of both the accused and the accuser. However, courts have made it clear that a prompt and thorough investigation and remedial action, if conducted properly, can limit potential damages. Employers should investigate sexual harassment complaints whether the complaint is made formally or not.

Because of the sensitive nature of sexual harassment allegations and the high potential for defamation lawsuits, each witness should be admonished to not discuss the matter with anyone other than the investigator. The investigator should emphasize that the purpose of the investigation is to get to the truth and that the best way of doing that is for everyone to feel assured that whatever they say to the investigator will remain as confidential as is practical.

In the event that the investigation leads to disciplinary action, well-written and complete reports regarding the employer's effort will support its decision should any disciplined employee seek legal action (*Whitaker v. Carney* 1986). These records will also be necessary if the investigation is inconclusive and the claimant subsequently files a Title VII sex discrimination lawsuit or retaliation claim.

Because many claims of sexual harassment involve incidents that allegedly occur between two people without the presence of witnesses, such charges sometimes boil down to one person's word against another's, and no conclusion can be reached as to the truth of the allegations. Sometimes it is just an honest difference of interpretation of events. It is up to the employer, in any event, to reiterate the company's sexual harassment policy and to decide what disciplinary action, if any, is appropriate. Even in cases when no definitive conclusion can be reached, it is sometimes appropriate for the employer to offer a lateral transfer to one of the parties, if possible, to alleviate the chance of future problems.

AVOIDING UNFAIR LABOR PRACTICE CLAIMS

Under the National Labor Relations Act (NLRA) employees have the right to form, join, and assist unions. Any interference with or restraining or coercing of employees in regard to a protected activity is a violation of the act and can subject the offending employer to heavy fines and sanctions. As an acting agent of the employer, the investigation team must be particularly careful to avoid even the appearance of attempting to interfere with union activity. Under no circumstances may any employer conduct surveillance of or otherwise monitor employees' union activities. Violations of these basic rights may subject both the employer and the investigative agency to claims of unfair labor practices.

Union employees also have the right to picket their employer's premises, but not to trespass on the premises during a strike. Nonstriking employees have the right to protect their safety by refusing to cross a picket line. Employees who strike over safety issues or unfair labor practices must be rehired when the strike is over, except for those who are convicted of crimes, including trespassing, during a strike or lockout. Rehiring of those convicted of crimes and those who strike for reasons other than safety issues or unfair labor practices is at the discretion of the employer.

WEINGARTEN

In *National Labor Relations Board v. Weingarten, Inc.*, 420 U.S. 241 (1975), the Supreme Court ruled that an employee is entitled to ask for and receive union representation at an employer's investigatory interview if the employee reasonably believes that the interview might result in disciplinary action against her termination. Thus, the right to representation must not be offered. Instead, it is triggered upon the request. An employer who disciplines or terminates an employee for refusing to participate in an investigatory interview without a union representative present is in violation of the act, section 8 (a)(1).

Once the employee has requested union representation at the interview, the employer must either grant the request or dispense with or discontinue the interview unless the employee voluntarily agrees to participate after being informed by the employer or another party that he has the right to not participate without union representation. The employee does *not* have the right to union representation at a meeting held solely for the purpose of informing the employee of, and acting upon, a disciplinary decision that was made prior to that meeting. The best and safest play in this case is for the employer to do nothing until the request for representation is made. If and when representation is made, document it and either provide the representative or allow the employee to make the decision of continuing without one. Whatever the decision, it should be documented. Also, document the time of and response to the request. If witnesses are used, have them do the same thing.

Furthermore, the *Weingarten* precedent allows that the employer can insist on hearing only the employee's account of the incident or insist that the union representative hold her comments until the end of the interview. The employer *cannot* require that the representative remain silent unless the representative is present only because of a collective bargaining agreement that allows her to be there. The Supreme Court also ruled that the employer need not reinstate the employee with back pay in response to the employer's violation of *Weingarten* rights if some independent evidence, obtained *other* than through the unlawful interview, shows that the discipline was for good cause.

The NLRB has also held that employees do not have the right to demand union representation at a meeting that is not an investigatory interview, and that the employer may discipline employees who refuse to participate under those circumstances. Additionally, the NLRB has ruled that an employer may discharge an employee who refuses to meet with the employer in a private office to discuss whether or not the interview will be held without union representation.

WORKPLACE PRIVACY

An understanding of the privacy rights of individuals is fundamental to the proper engineering of any security management program. The concept of invasion of privacy was first contemplated in the Fourth Amendment of the U. S. Constitution as a protection against unreasonable searches and seizures. Two hundred years later, implications of the Fourth Amendment are still being interpreted by our courts at both federal and state levels.

In 1989, Fourth Amendment issues entered into the debate over government-mandated drug-testing programs. *National Treasury Employees v. Von Raab*, 489 U.S. 656 (1989) involved applicants and employees seeking sensitive positions with the U.S. Customs Service who were required to produce urine specimens for drug-screening purposes. The court upheld the testing as "reasonable" under the Fourth Amendment because the government had a compelling interest in preventing drug users from holding positions involved in the interception of illegal drugs. Additionally, many of these employees were required to carry firearms. Under these circumstances, the court held that the need for drug testing by the government outweighed the need for privacy on the part of applicants and employees in these sensitive positions. The principle was laid forth that employees in sensitive positions have a diminished expectation of privacy.

The court again upheld mandatory testing, this time of blood, urine, and/or breath, when railway workers are involved in an incident or when a supervisor has a reasonable suspicion that the acts or omissions of the employee contributed to the incident. In *Skinner v. Railway Labor Executives Association.*, 489 U.S. 602 (1989), the court found that ensuring the safety of the public outweighs the employees' expectation of privacy. Both cases were decided on the basis of the 1987 case of *O'Connor v. Ortega*, where the U.S. Supreme Court presented a balancing test that mandates that the "invasion of the employee's legitimate expectations of privacy" be balanced against "the government's need for supervision, control, and the efficient operation of the workplace." While these Supreme Court decisions apply only to government workplaces, the balancing test of *O'Connor* has been applied to a number of state cases where privacy and the private employer were at issue.

The state constitutions of Alaska, Arizona, California, Florida, Hawaii, Illinois, Louisiana, Montana, South Carolina, and Washington address the right to privacy or recognize a right to freedom from intrusion into one's private affairs. The California Constitution, Article I, Section I, states:

> All people . . . have inalienable rights. Among these are enjoying and defending life and liberty, . . . and pursuing and obtaining safety, happiness, and privacy.

However, in *Nader v. General Motors,* the court noted that "the mere observation of the plaintiff in a public place does not amount to an invasion of privacy. But, under certain circumstances, surveillance may be so overzealous as to render it actionable." In 1972 the U.S. District

Court of New York reinforced the concept that legal action is appropriate when surveillance tactics cross the line and become an invasion of privacy. The case involved Jacqueline Kennedy Onassis and celebrity photographer Ron Galella, who had allegedly monitored Onassis's activities by corrupting a doorman, romancing Onassis's personal maid, intruding into her children's schools through deception, and visiting shops and restaurants to find out what she purchased. The court held Galella's activities were exclusively for the "purpose of gathering information of a private and confidential nature" and were thus actionable on the grounds that the photographer "[had] no right to invade [the] defendant's right of privacy."

Most of the laws regarding privacy that apply to the private sector are derived from case law on the state level. The majority of states recognize a common law right to privacy, with invasion of that privacy as a legitimate basis for a lawsuit.

Under *Hill v. National Collegiate Athletic Association*, 7 Cal. 4th 1 (1994), the California Supreme Court set a new standard for invasion of privacy claims that is more favorable to employers than earlier California case law, which was based on the 1987 *Ortega* "compelling interest." *Hill* established that a private employer may be held liable for invasion of privacy only if the plaintiff can establish:

1. a legally protected privacy interest
2. a reasonable expectation of privacy
3. conduct by the defendant constituting a serious invasion of a privacy interest

The justices in *Hill* noted that "[p]rivate entities pursue private ends and interests" and would virtually never be able to show a "compelling public interest" for intrusions on individual privacy. The court concluded that a rigid and absolute definition of privacy rights, to the exclusion of practical and legitimate business concerns, has no foundation in the law. Five out of the seven justices in *Hill* viewed the "compelling interest" federal standard as unduly inflexible in placing the burden of proof on the plaintiff.

Some courts have ruled that particular acts of invasion of privacy constitute a violation of public policy, or are against the public good.

FINGERPRINTS AND PHOTOGRAPHS

There are no specific federal restrictions on employers obtaining employee fingerprints and photographs, but some states have actually addressed how employers can use fingerprints and photographs. California Labor Code section 1051 prohibits employers from requiring applicants or employees to furnish photographs or fingerprints as a condition of employment if they are going to be given to a third party who could use them with adverse effects for the employee. Simply stated, employers can collect employee fingerprints and photographs for their own use, but they should avoid supplying them to anyone outside of the organization unless otherwise directed to do so.

If fingerprints or photographs are required by the employer, the employee should be requested to furnish them only after an offer of employment has been made. To request them prior to an offer of employment could expose the employer to a discrimination claim.

DRUG AND ALCOHOL TESTING

The law as it applies to drug testing is continually evolving. Because drug testing involves what may be considered an invasive procedure through the examination of bodily fluids, an employer contemplating testing should seek the advice of a labor and employment law attorney before proceeding. In regard to drug testing, some cities, such as San Francisco and Boulder, Colorado, have instituted tighter restrictions on preemployment drug testing than have their states. Regardless, drug testing by most private employers will not give rise to Fourth Amendment claims of prohibiting unreasonable searches and seizures or the requirement of reasonable suspicion. Only *public* employers (federal, state, county, or city) are subject to Fourth Amendment restrictions. However, private employers are subject to other restrictions as mentioned and should nevertheless proceed cautiously.

Drug testing may also be required. Under the Omnibus Transportation Employee Testing Act of 1991, five operating administrations under the federal Department of Transportation have instituted rules requiring alcohol and drug testing for employees under their auspices. These administrations are:

- the Federal Aviation Administration (FAA)
- the Federal Highway Administration (FHWA)
- the Federal Railroad Administration (FRA)
- the Federal Transit Administration (FTA)
- the Research and Special Programs Administration (RSPA)

New regulations imposed by the Federal Highway Administration apply to about seven million *interstate* and *intrastate* drivers. Testing for drugs and alcohol is no longer required as part of a physical exam, but testing is required before returning to duty after the employee has been disciplined for on-the-job drug or alcohol use, and following most incidents.

It is also noteworthy that the Americans with Disabilities Act (ADA) prohibits medical testing prior to an offer of employment; however, a test for *illegal* drugs is not considered a medical test under the ADA. To avoid a conflict with the requirements of the ADA, any preemployment drug test required must be designed to accurately identify only illegal drugs and cannot be given in conjunction with a physical exam. The ADA also disallows alcohol testing prior to an offer of employment.

Routine Drug and/or Alcohol Testing during Periodic Physical Exams. The ADA requires that employee physical exams must be voluntary or job-related and consistent with business necessity. Employees should be notified that drug and/or alcohol tests will be administered during physical exams if they are a requirement of continued employment. However, California's Confidentiality of Medical Information Act (California Civil Code Section 56.20–56.245) requires that an employer must obtain authorization from the employee before disclosing medical information relating to that employee. This little-known law also provides that "no employee shall be discriminated against in terms or conditions of employment due to that employee's refusal to sign an authorization" (California Civil Code Section 56.20 [b]). A California employer cannot terminate an employee solely for refusal to authorize a drug or alcohol test or any other medical test, but the employer is allowed to inquire of the physician as to whether the employee is able to perform his job duties satisfactorily.

In contrast, Texas law requires an employee to consent to a drug test ordered by her employer as a condition of continued employment, and an at-will employee may be terminated for refusing to consent to a drug test so long as the drug test policy is reasonable. Further, the state of Texas can deny unemployment compensation to any worker fired for refusing to submit to a required drug test (*Kaminski v. Texas Employment Commission* 1993). Employers in any state where refusal to submit to drug testing carries serious consequences for the employee should make certain to protect themselves from potential legal action by obtaining written consent prior to testing.

Federal courts have found that periodic drug testing is not a violation of the Fourth Amendment prohibition against unreasonable searches and seizures and thus have approved of drug testing as a part of an annual or periodic physical exam (*International Bhd. of Teamsters v. Department of Transportation* 1991; *Amalgamated Transit Union Div. 1279 v. Cambria County Transit Authority* 1988).

Reasonable Suspicion Drug and/or Alcohol Testing. Under the ADA, all alcohol testing of current employees must be limited to reasonable suspicion testing or some other legitimate business reason. With either drug or alcohol testing, limiting testing to situations where reasonable suspicion exists should minimize legal challenges. The courts have identified what constitutes "reasonable suspicion."

The Federal Highway Administration, in regulating drug testing of truck drivers, puts forth in 40 C.F.R. §382.307(a) that reasonable suspicion must be based on "specific, contemporaneous, articulatable observations concerning appearance, behavior, speech, or body odors of the driver." The code further states that observations "may include indications of the chronic and withdrawal effects of controlled substances" (see 49 C.F.R. §382.307[b]). The Ninth Circuit Federal Court, in two separate cases, held that in some safety-sensitive or security-sensitive circumstances, reasonable suspicion may be based on suspicion of *off-duty* drug use (*American Fed'n of Gov't Employees v. Roberts* 1993; *American Fed'n of Gov't Employees v. Martin* 1992). Refusal to submit to reasonable suspicion testing, when the testing policy is deemed to be appropriate, has been held to be appropriate grounds for termination of the employee (*Rawlings v. Police Department, Jersey City, New Jersey* 1993).

Post-Accident Drug and/or Alcohol Testing. The Omnibus Transportation Employee Testing Act, applying to airline pilots, air traffic controllers, railway workers, truck drivers, and other workers, requires post-accident drug and/or alcohol testing following certain types of accidents. The Federal Highway Administration defines an *accident that will trigger testing* as involving the loss of human life or an accident for which the driver receives a citation under state or local law for a moving traffic violation arising from the accident (40 C.F.R. §382.303[a]). Generally, testing required of employees suspected of being under the influence when an incident occurred has not been successfully challenged in federal or state courts and is considered by many authorities as one of the safest circumstances under which employers can test employees.

Random Drug and/or Alcohol Testing. Although unannounced random drug testing appears to be the most effective means of detecting and deterring drug and alcohol use among employees, it is the most controversial. Currently, the only states that statutorily affirm the use of random testing of employees in non-safety-sensitive positions are Arizona, Louisiana, and Oklahoma. As noted above, Texas allows drug testing, including random testing, with employee consent, and refusal to consent can be grounds for termination and denial of unemployment benefits. The California Supreme Court held in its review of *Hill v. National Collegiate Athletic Association*, 7 Cal. 4th 1, 26 Cal. Rptr. 834 (1994) that the NCAA's random drug testing program, which required student athletes to provide a urine sample in the presence of a same-sex witness, did not constitute an invasion of privacy according to the state constitution. Although the court stated that private entities are covered by the constitutional privacy guarantee passed by California voters in 1972, private businesses need not always meet the strict standards imposed on government when infringing on privacy. The government must show *compelling* reasons, but private entities in many cases need only show that the invasion of privacy is justified by *legitimate business* interests and needs.

Testing as a Condition of Returning to Work after Rehabilitation. Some states have statutes that require that the employer attempt to rehabilitate employees who abuse drugs or alcohol, prior to or in lieu of termination on the grounds of substance abuse (Cal. Lab Code §1025-1028; Federal Vocational Rehabilitation Act of 1973). Employers commonly require drug testing as a condition of returning to work during and/or after an employer-referred rehabilitation program. Particularly in safety-sensitive positions, substance abuse professionals typically recommend that the employee who has recently been in a rehabilitation program be subject to unannounced alcohol and/or drug testing as one incentive for keeping clean and sober. It is likely that the employee who has been in such a program recognizes the need for testing and will consent to future testing.

FALSE IMPRISONMENT

False imprisonment is both a crime and a tort. False imprisonment claims can arise if the employer restrains the employee by force or by fear for purposes of search or interrogation.

With the tort of false imprisonment, the employee must be conscious of the confinement or harmed by it. In 1930, in *Vandiveer v. Charters*, 110 Cal. App. 347, a California appellate court explained the nature of the basis for a claim of false imprisonment as follows:

> The restraint constituting false imprisonment may arise out of acts, words, gestures or the like, which induce a reasonable apprehension that force will be used if the plaintiff does not submit, and it is sufficient if they operate upon the will of the person threatened, and result in the reasonable fear of personal difficulties or personal injuries. . . . A voluntary display of force, intended to deprive plaintiff of his liberty, even without actual laying hands on plaintiff, has been held sufficient, but there is no legal wrong unless the detention was involuntary in the sense of being contrary to the will of the plaintiff.

In short, false imprisonment is established by proof of force, or an express or implied threat of force.

Most claims of false imprisonment arise when a police officer or a private citizen makes a false or improper arrest. Workplace interviews with accused persons can lead to claims of false imprisonment if the interviewer threatens or creates the reasonable fear that he will detain the subject through force or have the person arrested if she tries to exit the interview. To minimize this potential claim the interviewer should attempt to get the subject to sign the written statement, ideally in front of a witness who also signs, wherein the interviewee states that she was not held against her will or in any way felt threatened or coerced.

VISUAL AND ELECTRONIC SURVEILLANCE

In spite of recent court rulings that tend to favor the employer over absolute privacy rights, the safest course of behavior for an employer is to avoid behaviors that appear intrusive or invasive and to avoid surveillance of areas where employees have a *reasonable expectation of privacy*, such as private offices, restrooms, and locker rooms. However, employers have a number of legitimate business reasons for monitoring and surveillance of employees. Privacy issues rarely arise when employers monitor actual work activities for legitimate business purposes. It is the monitoring of personal and private behavior and communication that normally causes the employee to feel violated and choose to become a plaintiff.

The least objectionable use of monitoring and surveillance by employers is monitoring work activities for the purpose of giving the employer an objective record of performance and productivity that can be used for employee evaluations. As long as all employees are monitored and held to the same standards, using these procedures for evaluation purposes can actually help eliminate charges of discrimination or favoritism by employers in disciplining or promoting workers.

Providing a safe and crime-free workplace is perhaps the most compelling justification for monitoring employees and conducting surveillance when appropriate. It is widely held that monitoring and surveillance of employees while at work can detect and/or deter illegal activity on the premises, including staging of "accidents" in order to file fraudulent workers' compensation claims, assaultive behavior, use of drugs and alcohol in the workplace, and theft. Even the surveillance of government employees (a group that typically enjoys expanded rights relative to search and seizure) in work areas, conducted with their knowledge, does not invoke Fourth Amendment privacy issues because there is no expectation of privacy under such circumstances.

Interestingly, video surveillance is not regulated by most states, but states that have a state constitutional or statutory right to privacy, including California, have a basis on which employees in the private sector can make privacy claims. For example, in California a video recorder is a *recording device* according to the Invasion of Privacy Act, Cal. Penal Code §632. The act sets forth criminal penalties for anyone (except those gathering evidence of a crime) who records confidential communication without the consent of all parties. However, California has no explicit limitations on visual or video surveillance in the workplace. Other states may have limitations. The limitations on surveillance by employers, or their agent, are primarily derived from case law.

Employers and their agents should be aware that in 1997 the NLRB ruled that an employer must bargain with its employees' union in advance about the installation and use of hidden surveillance cameras in the workplace (*Ohio Employment Law Letter* 1997). The case came about when an employee reported his discovery of a hidden camera in the air vent of a restroom to a union steward who, in turn, reported the hidden camera to the union's officers. The human resource manager reported to one of the union's officers that the camera was placed in the restroom to aid in a theft investigation. An additional hidden camera was found in the employees' exercise room.

After filing a grievance over the use of the hidden camera and the failure of the company to respond, the union filed a charge with the NLRB's Region 9 office over the company's refusal to bargain about the use of hidden cameras. The case was heard before an administrative judge, who ultimately sided with the union. The company appealed, and the NLRB affirmed the decision of the judge, reasoning:

> The use of surveillance cameras is not entrepreneurial in character, is not fundamental to the basic direction of the enterprise, and impinges directly upon employment security. It is a change in the [company's] methods used to reduce workplace theft or detect other suspected employee misconduct with serious implications for its employees' job security, which in no way touches on the discretionary core of entrepreneurial control.

The NLRB's final comment on the matter was:

> The union has the statutory right to engage in collective bargaining over the installation and continued use of surveillance cameras, including the circumstances under which the cameras would be activated, the general areas in which they may be placed, and how affected employees will be disciplined if improper conduct is observed.

MONITORING PHONE CALLS

The Omnibus Crime Control and Safe Streets Act of 1968 (Title III) prohibits private individuals, organizations, and employers from intercepting wire or oral communications. This act was amended as the Electronic Communications Privacy Act of 1986, 18 U.S.C. §§ 2510-2520. The code sections do not specifically address the question of simple eavesdropping without an electronic or mechanical device; however, video cameras with sound tracks fall under the restrictions of this act.

> [A]ny person who willfully uses, endeavors to use, or procures another person to use any electronic mechanical or other device to intercept an oral communication when . . . such use or endeavor to use . . . takes place on the premises of any business or other commercial establishment the operations of which affect the interstate or foreign commerce . . . shall be punished . . . if such communication is intercepted for the purpose of committing any criminal or tortious act in violation of the Constitution or laws of the United States or of any State.

Under Title III, wiretappings are currently limited to use in investigating crimes such as drug trafficking, currency reporting, and bribery. The use must be based on probable cause and the establishment that normal investigation techniques have not been or will not be successful. The wiretaps can be authorized for 30 days or less and a warrant must be obtained from a court, with approval by the attorney general or assistant attorney general.

Fees for violations of the act include actual damages suffered by the plaintiff plus any profits made by the violator; or $10,000 or $100 per day for each day of violation (whichever is greater); reasonable attorneys' fees and costs; and sometimes recovery for emotional distress.

The three principal exceptions to the act are:

1. where there is consent of one party to the communication
2. when an employer uses a telephone extension to monitor employees in the ordinary course of business, which can include:
   (a) providing training on interaction with the public
   (b) determining whether an employee is discussing business matters with a competitor
   (c) determining whether an employee is making personal calls
   (d) monitoring of calls for mechanical or service checks by a phone company

Section 2515 of the act prohibits intercepted wire or oral communications as evidence. The section reads:

> Whenever any wire or oral communication has been intercepted, no part of the contents of such communication and no evidence derived therefrom may be received in evidence in any trial, hearing, or other proceeding in or before any court, grand jury, department, officer, agency, regulatory body, legislative committee, or other authority of the United States, a State, or a political subdivision thereof if the disclosure of that information would be in violation of this chapter.

At least 40 states have adopted electronic interception statutes similar to the Omnibus Crime Control and Safe Streets Act of 1968, making it illegal to intercept employees' conversations when participants in the conversation are not aware that they are being monitored. Employers as outside parties to a conversation are not allowed to tape conversations *between employees* unless the participants in the conversation have been forewarned that their conversations might be monitored.

Some states allow interception as long as one participant is aware of the interception; other states, including California, require that all parties give consent. Courts consistently have upheld state statutes, which impose more restrictive provisions than federal laws. In states with one-party consent, employers may tape a conversation with an employee, without that employee's consent. This, of course, is a double-edged sword since employees are also allowed to tape any conversation they have with their employer, and the tape could be allowed as evidence in a court of law. However, this does not mean that employees have a *right* to tape conversations between supervisors and themselves if the company has a policy against taping conversations or if a supervisor instructs an employee not to record the conversation.

Generally, a ban on taping meetings is the safest policy. In lieu of taping, supervisors should be taught to prepare written documentation, including a summary of the content of meetings with employees, and to keep the documentation on file (*You and the Law* 1998, 7).

The Texas Code of Criminal Procedure and the Texas Civil Practice and Remedies Code both govern the illegal interception of oral communications in that state. The Code of Criminal Procedure allows an exception when the equipment is being used in the ordinary course of the telephone customer's business, but specifies $100 per day, actual damages, or $1,000, whichever is higher, and civil liability for violations. Chapter 123 of the Texas Civil Practice and Remedies Code gives a party to a communication the right to sue a person whose communication is intercepted without that party's consent. Employers are specifically discouraged from the use of wiretaps or similar devices based on the common law right of privacy.

MONITORING E-MAIL

State and federal law is still emerging surrounding rights of privacy in regard to employees' e-mail. Generally, as an electronic communication, e-mail has been held by some courts to fall under the provisions of the Electronic Communications Privacy Act (ECPA). Section 2701 of the ECPA provides an exception for "the person or entity providing the wire or electronic communications service." The exception states that that person or entity (employer) is not liable for offenses regarding *stored* communications, which is sometimes interpreted to include voice mail and e-mail stored on equipment owned by the employer. However, the crucial issue on which there is no consensus yet is whether the employer can be said to "own" the e-mail system when it is merely a subscriber to another's system, which is used by means of the employer's equipment.

If the owner of the computer equipment (i.e., the employer) is legally considered to be *the provider of the service*, then, according to section 2702(b)(5) of the ECPA, an employee authorized to use e-mail for business purposes has no right of protection from intrusions by the employer, particularly on a strictly "in-house" e-mail system. Questions like these illustrate why it is imperative to consult with legal counsel prior to attempting to monitor communications involving relatively new technologies. Particularly with unresolved issues, it is important for the employer to diminish the expectation of privacy in e-mail by creating a monitoring policy including notice to employees of intent to monitor and, if possible, by obtaining written consent from employees.

Employers have also been held liable in sex discrimination and race discrimination cases, based in part on offensive e-mail messages to co-workers, which went *unmonitored* to the point where the court found them to be ongoing harassment creating a hostile work environment. For protection, employers should provide notice to all employees by having the employees sign an acknowledgment stating that they understand that their communications may occasionally be accessed by the employer.

HONESTY TESTING

The federal Employee Polygraph Protection Act of 1988 (EPPA), 29 U.S.C. §§2001-2009, severely restricts the use of lie detector tests. It effectively eliminates the use of polygraph testing by most private sector employers in applicant screenings and under most circumstances involving current employees. The act additionally prohibits employers from disciplining, discharging, or discriminating against any employee for refusing to submit to a polygraph or other *lie detector* test or for taking action to preserve his rights under the act. Employers are further prohibited from discriminating against an employee based on the results of any such test.

The term *lie detector* is defined by the act as including a polygraph, deceptograph, voice stress analyzer, or other mechanical or electrical device used to render diagnostic opinion about honesty or dishonesty of an individual (29 U.S.C. §2001[3]). The act does not regulate paper-and-pencil "honesty tests" or any other nonmechanical means of testing honesty.

Interestingly, the act does *not* apply to federal, state, or local government employers and employees. There are other exceptions:

1. Employers manufacturing, distributing, or dispensing controlled substances may administer polygraphs to applicants who would have direct access to manufacture, storage, distribution, or sale of a controlled substance and current employees as a part of an ongoing investigation into misconduct.
2. There is an exception for private security firms charged with protecting certain types of facilities that impact public health, safety, and security, such as nuclear power plants, public water supply facilities, shipments or storage of radioactive or other toxic waste materials, and public transportation, or providing protection of currency, negotiable instruments, and precious commodities.

In addition, an employer may administer polygraphs to an employee in an ongoing investigation of thefts or other incidents causing economic loss to the employer, such as embezzlement, misappropriation, or an act of unlawful industrial espionage or sabotage, where there is a reasonable suspicion that the employee was involved in an incident under investigation and had access to the property in question (29 U.S.C. §2006[d]).

The law requires that any employee suspected of theft and subject to polygraph testing must be given written notice that:

1. specifies the particular incident being investigated and the basis for testing that employee
2. identifies a specific economic loss or injury suffered by the employer
3. states that the employee had access to the property
4. describes the basis for the employer's reasonable suspicion

The requirement is met if the decision is reasonable and suspicion is grounded in *observable, articulatable* facts; is signed by a person legally authorized to bind the employer; and is retained by the employer for a minimum of three years.

Procedural requirements for administering polygraphs under the exceptions include:

1. written notice to employee of date, time, location, nature of test, and right to consult an attorney or employee representative before each phase of the test
2. written statement from the employee acknowledging his or her rights under the EPPA
3. employee opportunity to review all questions to be asked during the test, and
4. refraining from conducting the test where a medical condition may cause abnormal reactions during the examination

During the actual testing phase:

1. An examiner licensed by the state in which the test is to be conducted must conduct the test.
2. The examiner may not ask questions not presented in writing to the employee or applicant before the test, or that pertain to religion, racial matters, political beliefs, sexual behavior, or unions.
3. The examiner may not administer more than five polygraph tests per day. Each test must last a minimum of 90 minutes, and the employee or applicant may terminate the test at any time.

After the test, and before taking action based on the results, the employer must:

1. interview the employee on the basis of test results
2. give the employee a copy of test questions and responses
3. give the employee a written copy of any conclusions reached from the test

Limitations on the use of test results by the employer include the following:

1. Results of a polygraph test may not be the sole basis for the employer's decision regarding the employee.
2. Refusal to submit to a test may not provide the sole basis for an employer's decision regarding an employee; however, a negative inference may be drawn.

Only the employer has the right to see the results of a polygraph test. A present employer may not ask about any polygraph tests that the employee may have taken for previous employers.

Employees who feel their rights were violated by a polygraph test have up to three years to file legal claims. Employers may be liable for civil penalties of up to $10,000. Employees whose rights are found to have been violated may be entitled to reinstatement and back wages.

The EPPA is the controlling law in any business involving interstate commerce *unless* state polygraph law is more restrictive. According to California Labor Code §432.2, employers may *not* require an applicant or employee to submit to a polygraph test as a condition of employment, but employers *may* request that an employee submit to a polygraph test so long as she is advised in writing of the provisions of the federal EPPA. The federal law, in this case, is more restrictive than state law.

WORKPLACE SEARCHES

Private sector employees, including investigators, are not subject to Fourth Amendment protections against unreasonable searches and seizures unless the investigation is being conducted at the behest of the police or a government agency. In that event, a valid warrant must accompany every search. Failure to obtain a valid warrant would result in the exclusion of any evidence obtained from the warrantless search, commonly known as *the fruit of the poison tree.*

O'Connor v. Ortega, 480 U.S. 709 (1987), is currently the leading Fourth Amendment case addressing workplace searches. While its holding is not binding on private sector employees, it is useful to understanding the basis of many state statutes and case law. *Ortega* derives from a search of the desk and filing cabinets of a doctor at a state hospital amidst allegations of improper conduct. During the search the employer seized personal letters and cards, a photograph, and a book of poetry. The doctor sued on the grounds of invasion of privacy. The court ruled that a public employee has a reasonable expectation of privacy in his or her office, desk, and file cabinets that contain only personal files. However, the employer may search such areas if it has "reasonable grounds for suspecting that the employee is guilty of work-related misconduct" or if the search is necessary for a business-related purpose such as retrieving a file. No warrant is required under those circumstances.

Private employers are generally allowed to conduct searches of their premises for business-related purposes so long as they do not search areas where employees legitimately have reasonable expectations of privacy. The employer should diminish expectations of privacy by providing written notice to employees that lockers, desks, filing cabinets, company vehicles, and all other company property are subject to search without employee consent.

In order to remind employees that the company property they may use daily is not their property, employers should switch employee desks, lockers, and vehicles periodically, as well as supply all necessary locks and retain a copy of the key to each lock. Employees should also be notified that failure to cooperate with the company search policy might result in discipline up to and including termination.

Each employee should be required to sign a copy of the search policy, acknowledging consent to searches of company property for any reason. However, some courts have found that searches conducted randomly, for no *specific* reason are, by definition, not *reasonable* searches and constitute an invasion of privacy.

PROCEDURES FOR CONDUCTING SEARCHES

- In workplaces with a unionized work force, consider having a union steward present during the search both as a means of honoring agreements if applicable, and as a means of avoiding the appearance of impropriety.
- In a nonunion workplace, consider allowing the employee to choose his or her own witness.
- Ensure that the search is reasonable or in compliance with established policy for random searches.
- Get authorization from the human resources department or appropriate official before the search is conducted.
- Tell the subject the purpose of the search and limit the search to what is appropriate in the context of that purpose.
- Have at least two management officials present during the search.

- Try to obtain consent from the employee. If the employee refuses to give consent or cooperate, inform the employee of possible disciplinary consequences per company policy.
- Do not be confrontational or physically force any part of the search.
- Use professional search equipment such as metal detectors or trained dogs in appropriate settings.
- Seize the contraband in plain view of the suspect(s).
- Mark and retain any contraband in a secure location and preserve the chain of custody. In the case of illicit drugs or paraphernalia, seal contraband in an envelope for delivery to lab or law enforcement personnel (Abell 2006).

Clear notice of the company's search policy is mandatory if the employer or investigative agency plans to conduct searches of employees' clothing, purses, lunch boxes, and other personal effects. This type of search is highly invasive and thus carries with it the highest level of potential liability, other than the collecting of blood and urine samples, and frankly plays little to no role in most undercover investigations. Regardless, it is imperative that the search policy be clearly stated both in the employee handbook and in separate notices given to each employee along with a requirement that they sign the notice or face disciplinary action. These types of searches should rarely, if ever, be conducted without probable cause unless, for instance, the company has a policy of routinely searching *every* employee's lunch box upon exit from the premises.

Singling out individual employees for searches of clothing or other personal effects without the highest level of reasonable suspicion that would likely be acceptable to a judge and/or jury could open up a Pandora's box of liability. If the search of a singled-out employee fails to turn up the anticipated evidence, the employee may feel humiliated and perhaps feel that he is the victim of discrimination if the suspicions of the employer were groundless.

The employer should explain the basis for the suspicion and request that the employee consent to the search, even if the employee has previously signed a more generalized consent-to-search form. If the employee refuses, he should be reminded of the company's policy of disciplining employees who refuse to comply with searches. The overzealous employer who grabs the employee or the employee's purse or lunch box makes himself vulnerable to charges of assault and battery along with invasion of privacy. Generally, the employee who is subject to this invasive type of search should be given a reasonable opportunity to submit to the search, and the search should be conducted out of the view of co-workers other than designated witnesses or representatives. The object is not to humiliate the subject in front of co-workers, but rather to determine whether illicit activity has occurred and to take the *appropriate* steps in a professional manner to curb such activity in the future.

SUMMARY

While a host of personnel security issues confront employers, the most significant issues are the potential legal liabilities associated with managing the workplace and keeping those within it safe. This chapter examined some of the most significant liabilities and the overall means by which security management is used to address them. It also examined some of the risk-mitigation strategies used by security professionals confronted with these issues.

REFERENCES

Abell, A., et al. 2006. *Issues in workplace privacy and investigations.* Santa Monica, CA: Paul, Hastings, Janofsky & Walker.

Amalgamated Transit Union Div. 1279 v. Cambria County Transit Authority. 1988. 691 F. Supp. 898 (W.D. Pa.).

American Fed'n of Gov't Employees v. Martin. 1992. 969 F.2d 788 (9th Cir.).

American Fed'n of Gov't Employees v. Roberts. 1993. 9 F.3d 1464 (9th Cir.).

Cal-OSHA guidelines for workplace security. 1995. Revised.

Chaney, J., et al. 1997. *Premises security and liability.* Laguna Beach, CA: Workplace Violence Research Institute.

Equal Employment Opportunity Commission. 1994, May. Enforcement guidance. Manual §3279.

Henley v. Prince George's County. 1986. 305 Md. 320; 503, A.2d 1333.

International Bhd. of Teamsters v. Department of Transportation. 1991. 932 F.2d 1292 (9th Cir.)

Kaminski v. Texas Employment Comm'n. 1993. 848 S.W.2d 811 (Tex. App.–Houston [14th District], no writ).

Ohio Employment Law Letter. 1997, June. Labor board favors thieves—says union must be given advance notice of hidden cameras. 8:6.

Pittman v. City of Oakland. 1988. 197 Cal. App. 3d 1037.

Rawlings v. Police Department, Jersey City, New Jersey. 1993. 133 N.J. 182.

Sanaroff, R. 1988, Dec. 18. Sexual harassment in the Fortune 500. *Working Woman.*

Shapiro, J. P. 1998, July 6. The Americans with minor disabilities act. *U.S. News & World Report.*

Title V of the Omnibus Drug Initiative of 1988.

Wasson v. Stracener. 1990. 786 S.W.2d (Tex App.–Texarkana, writ denied).

Whitaker v. Carney. 1986. 778 F.2d 216, 222, 5th Cir. 1985, cert. denied, 479 U.S. 813.

You and the Law. 1998, March. Workplace taping ban is the best, safest policy. p. 7.

11

Employment Policies and Procedures

Eugene F. Ferraro, CPP, CFE, PCI

CHAPTER 11

OBJECTIVES

The objectives of this chapter are as follows:

- examine the importance of preemployment screening and how and why background investigations are performed
- review the legal and ethical aspects of screening employees
- review the process by which a security manager can screen and select a preemployment background vendor
- learn the essential components of modern workplace policies designed to address employee theft and fraud, workplace violence, workplace substance abuse, and harassment and discrimination

INTRODUCTION

It has been said that an organization's people are its most valuable asset. For this to be true, the organization must do all it can to recruit, hire, and retain the best people possible. Today, organizations use preemployment screening and background investigations to identify and qualify the most desirable people before hiring them. That effort, however, is filled with challenges. This chapter will explore why workplace policies are necessary and how they are developed. It will examine the assortment of legal and ethical issues that surround policy development and implementation.

PREEMPLOYMENT SCREENING AND BACKGROUND INVESTIGATIONS

Practical experience reveals that many job applicants misrepresent or exaggerate their experience and accomplishments to increase their chances of getting the job they seek. Statistics from the U.S. Bureau of Justice reveal that up to half a million criminal offenders are released from jail or prison every year, and recidivism rates can be as high as 67.5%. As most employers recognize, workplace violence, employee theft, and substance abuse remain serious problems. Never before has the need to screen job applicants been greater. Employers who currently do not have effective employment screening practices will likely hire their competitors' rejects tomorrow.

An effective preemployment background investigation should include:

- verifying the accuracy and completeness of statements made by the applicant
- developing additional information concerning the applicant to assist the prospective employer in making an informed decision regarding the applicant's overall suitability for the position

Federal, state, and local laws impact what the employer can and cannot do when conducting preemployment background investigations. Some of the provisions are complicated, and penalties for violating these laws can be stiff. As a result, many employers choose not to perform background investigations even though the price of doing nothing can be immeasurable, with consequences ranging from unwanted workplace crime, substance abuse, and violence to possible lawsuits.

RELEVANT STATUTORY PROHIBITIONS AND RESTRICTIONS

Title VII of the Civil Rights Act of 1964, the Americans with Disabilities Act, the Equal Employment Opportunity Act, the Fair Credit Reporting Act, and the Employee Polygraph Protection Act regulate how and under what circumstances an employer can conduct preemployment background investigations (ASIS International 2006, 13). For example, employers are prohibited, in most instances, from making any oral or written employment inquiry that directly or indirectly identifies personal characteristics such as age, race, color, national origin, ancestry, sex, sexual orientation, marital status, pregnancy, religion, creed, disability, or any recognized handicap. The mere consideration of any such characteristic by an employer is considered in and of itself discriminatory. Additionally, employers are barred from discriminating against an applicant who *perceives* he or she has a handicap. Both direct questions about an applicant's *protected* status and the use of indirect inquiries that disproportionately screen out job candidates of a protected group are strictly prohibited. Those violating the law may be subject to both civil and criminal penalties. Further, federal and state handicap laws (Rehabilitation Act of 1973 and Americans with Disabilities Act) regulate the manner in which an employer can conduct a preemployment drug test or physical exam. In most instances, drug tests and physicals may only be requested after an offer of employment has been made.

Although the U.S. Constitution provides no specific privacy protections, an assortment of state statutes and federal court decisions have created the *right to privacy*. This right protects applicants and employees from intrusions into their privacy by employers. Employers should not gather or collect information of a personal or confidential nature or offensively intrude into the private lives of anyone where a reasonable expectation of privacy exists. Visiting an applicant's home or interviewing an applicant at his residence could constitute an invasion of privacy. Such efforts are actionable and may give rise to legal claims against the employer and its agents.

ARREST AND CRIMINAL RECORDS

It is unlawful for most private employers to ask any job applicant to provide information concerning an arrest or detention that did not result in a conviction. Additionally, employers cannot ask about any arrest in which the individual is out on bail or personal recognizance. There are exceptions to this restriction, which are applicable to health facilities regarding arrests for sex offenses and possession of narcotics. However, the Equal Employment Opportunity Commission (EEOC) has ruled that an employer's use of arrest records, although lawful under some circumstances, has an adverse impact on some classes of minorities and that the use of such records is, in fact, discriminatory. For the purposes of preemployment screening, employers should use only actual conviction records, not arrest records.

Unfortunately (and contrary to the claims of many online database vendors), there is no nationwide public repository for criminal conviction records. Publicly available criminal conviction records are maintained either at the state or county level. Most statewide criminal databases available online (including those that offer instant criminal record searches) are often incomplete and unreliable and should be used cautiously. Many of these databases draw data from the state's Department of Corrections and may contain arrest records that have not been updated or are incomplete. Today, there is no statewide database for criminal records that can guarantee complete and accurate information. There is also no technology available to public employers to screen applicants by means of fingerprints. Moreover, many states prohibit employers from requesting or using applicants' fingerprints or photographs for hiring purposes.

WHAT CAN BE DONE

A vast amount of information is available from a search of public records and, for the most part, the information is reasonably accessible and inexpensive. Additionally, there are many private sources of information, including credit bureaus such as Equifax, Experian, and TransUnion. Accessed through designated licensees or resellers, these information providers and others can be helpful to those conducting preemployment background investigations.

The following discussion provides an overview of what can be learned or gained from these sources.

Criminal Courts. Criminal conviction records can be an invaluable source of information. For the most part, these records are public and available to anyone. In most instances these records can be obtained from the county criminal court clerk's office. Convictions for minor infractions to major felonies are found here. Files include full details of the case and disposition. As stated earlier, most statewide databases are unreliable and should be used with caution. Therefore, criminal conviction records are best searched by hand at the county level. To positively confirm identification, searches should use the full name and date of birth of the applicant.

Civil Courts. Civil court records are indexed by the name of the plaintiff and defendant. They can be obtained from the county court clerk's office. Included in these records are civil actions, judgments, notices of default, and any unlawful detainers. Many employers have learned that litigious applicants often become litigious employees.

Driving Histories. Records for driving histories and violations are maintained at the state level and are typically accessible to employers through the state department of motor vehicles. This is an excellent place to search for a history of substance abuse and other problems. For example, these records will reveal an applicant's convictions for recklessness, speeding, and driving under the influence. Suspensions due to child support issues and failure to provide proof of insurance will also be found here. Convictions for lesser violations will also be disclosed in full detail. Additionally, these records will produce vehicle registration information and accident histories.

Other Public Records. Corporations and other business entities are registered at the offices of the secretary of state. Filings include the names of all officers, street addresses, and the agent for process of service. This is a particularly good place to search if the applicant is a corporate officer or is suspected of being one. Also found at the state level are the licensing boards for the various professions. Nurses, doctors, building contractors, and even mattress rebuilders must be licensed in most states. If the job requires a license, it is often worth the effort to check and see if the applicant is properly licensed and/or registered.

Federal Bankruptcy Court. All bankruptcies are handled in federal court. These records are public and often reveal valuable information about an applicant's ability to handle personal finances. However, in the case of a preemployment investigation, the Federal Bankruptcy Code prohibits an employer from considering a past bankruptcy when making an employment decision.

WHAT SEARCHES SHOULD BE CONDUCTED

Search strategies and procedures vary and are often a matter of simple preference; however, the process should always be carefully crafted and its application consistent. Best practices suggest that an organization's strategy should include the following phases:

1. *Policy:* A policy is a general statement of a principle according to which an organization performs business functions. An organization does not need to maintain policies in order to operate. However, practices and procedures that exist without the underpinnings of a consistent policy are continually in jeopardy of being changed for the wrong reasons, with unintended legal consequences.
2. *Practice:* A practice is a general statement of the way the organization implements a policy. Good practices support policy.
3. *Procedure:* A procedure documents an established practice. Use of forms is one of the useful ways procedures are documented. For an organization that has a practice of checking past court records for criminal convictions, the procedure would be to document how the search was done, as well as to maintain the documents proving it was done (ASIS International 2006, 13).

It is important that policies, practices, and procedures be in writing. Written documentation makes it possible to independently verify from the procedure whether employees are conforming to the practice, and therefore to the policy. This kind of documentation makes it easy to perform reliable audits (ASIS International 2006, 13).

In order to avoid discrimination claims, all applicants should be screened similarly. It is suggested that an applicant screening checklist be developed and used as part of the screening process. The checklist should be attached to the other application materials and retained for future reference if needed. Using such a system helps prevent disparate treatment claims and the potential of hiring an undesirable individual.

The items and their order on the checklist should be given considerable thought. For example, not all employers need to search bankruptcy records. However, if the organization manages other people's money, the applicants' historical handling of their own financial affairs could be important. At the very least, employers should verify the following when screening applicants:

- identity
- personal history (including prior employment)
- credentials

Verification of Prior Employment. Verification of prior employment begins with the job application. It is not enough to ask for the name of the applicant's former supervisor. Screeners and interviewers should ask for the names of other people the applicant worked with or supervised. Experience reveals that subordinates will more likely cooperate and provide useful information than will personnel managers or former supervisors. Former co-workers can also be very helpful. Unfortunately, due to a growing concern about the potential of defamation claims, employers have become increasingly reluctant to provide valuable information. Today, many former employers provide little more than dates of employment and job descriptions when asked to verify employment. Unfortunately, many organizations have adopted policies that restrict what information will be provided and who will provide it.

Other Valuable Searches. For more sensitive positions, more in-depth investigations are recommended and may be tailored to the position. Exercising care up front can prevent many problems later. Other searches worth considering and information worth verifying include:

- personal credit history
- education
- current address
- business ownership
- fictitious business name filings
- real estate ownership
- vehicle and aircraft ownership
- corporation ownership or officer status
- listing on sexual offender registries

Consumer Credit Reports. Section 604 of the Fair Credit Reporting Act details the permissible uses of consumer credit reports. Many employers find consumer credit reports useful, particularly if the applicant will be handling money or has financial responsibilities. However, due to the complexity of these reports and the potential for their use to give rise to discrimination claims, they may be more trouble than they are worth for many employers.

Verification of Education. Typically, colleges and universities provide education verification over the telephone. However, many require signed releases and verify only by mail. Contact the respective admissions office or the registrar's office for information about the institution's procedures. Some larger schools have offices that do nothing but verify student and graduate information. In recent years, more colleges and universities have subscribed to third-party verification systems such as StudentClearingHouse.org (http://www.studentclearinghouse.org/secure_area/DVEV/dvev_bridge.asp). For a fee, this organization provides verification of education and other credentials.

SELECTING A VENDOR TO CONDUCT PREEMPLOYMENT BACKGROUND CHECKS

Employers have many choices in selecting a preemployment screening and background investigation vendor. Many vendors offer their services via the Internet. These vendors offer a wide range of services but may or may not offer personalized services. A vendor should not be selected by price alone. The following should be considered when selecting a vendor:

- *Licensure:* Many states require background screening service providers to have a private investigator license. Ensure the vendor selected has the licenses it needs to provide the services offered.
- *Insurance:* Remarkably, several well-known vendors do not have any general liability or "errors and omissions" insurance. Require a certificate of insurance and verify that the coverage is "occurrence," not "claims-made," coverage.
- *Resources:* The vendor should have the professional and technological resources necessary to do the job. Though many vendors claim to possess the resources to consistently provide high-quality results, some do not. Even some of the largest Internet vendors use primitive

database management systems and outdated computer technology.

- *Familiarity with the law:* The vendor selected should also be familiar with all of the current laws and regulations, especially the Fair Credit Reporting Act (FCRA). Employers should also be familiar with them to ensure that their preemployment screening practices are in compliance.

EMPLOYEE THEFT AND FRAUD

Statistics gathered by the U.S. Chamber of Commerce indicate that employee theft is a multibillion-dollar criminal industry. For years the chamber has warned that one-third of all businesses fail because of employee theft. In its 2008 "Report to the Nation," the Association of Certified Fraud Examiners (ACFE) reported that approximately 6% of the nation's business revenue was lost as a result of employee theft and fraud. When applied to the U.S. gross domestic product, the amount equates to losses of approximately $600 billion, or about $4,500 per employee, per year.

The report identifies small businesses as the most vulnerable to fraud and theft, with an average scheme causing $127,500 in losses. The average scheme in the largest companies costs them only $97,000. The report also reveals that the most common method for detecting workplace theft in 2008 was tips from employees, customers, vendors, and anonymous sources. The second most common method of discovery was by accident (ACFE 2008). Experts claim that for every $1 lost to shoplifting, employees steal another $15. In the food service industry, employee theft imposes a 4% tax on every customer dollar spent.

Employee theft is a problem that cannot be ignored. However, the problem is about more than just the dollars lost. The problem impacts morale, increases insurance costs, and decreases competitiveness and, of course, profits. Less profits means lower investor returns, less capital in the market, and lower tax receipts at all levels. Moreover, losses due to theft are costly to make up. Depending on margins, in order to recover a $100 loss, the average for-profit organization must sell nearly $3,000 in additional products or services. In low-margin industries such as consumer electronics and groceries, the recovery multiplier is significantly greater. Intuitively, the lower the margin, the more difficult it is to recover the loss.

Employees do not just steal steaks and CD players. More than anything else, employees steal time. The commodity of choice for most workplace thieves is measured in minutes. Efficiency consultants have known this for years. Attempts to improve processes and systems to increase workplace efficiency were undertaken as soon as the industrial revolution had begun. From Henry Ford's first assembly line to the implementation of modern robotics, industrialists have strived to improve worker efficiency. Since man first sold his labor to another, the theft of time has been every employer's nemesis. If each employee in a 200-person organization were to steal 10 minutes a day, the employer would loose 2,000 minutes per day. If the work year had 260 workdays, our hypothetical employer would have suffered a loss of 520,000 minutes, or the equivalent of 4.1 man-years. Effectively, the work force of 200 individuals is doing the work of 196. The cost of each employee stealing just 10 minutes a day translates to about $200,000 a year. In this hypothetical example, reducing wasted time by just 1 minute a day per employee creates a savings of $20,000 a year.

WORKPLACE VIOLENCE

Workplace violence touches everyone. It affects the way we think, feel, and behave. The threat of workplace violence affects the emotional stability and productivity of our employees and, ultimately, our profitability. Although the incidence of actual physical violence at work is relatively low, workplace violence affects the lives of thousands of innocent Americans each year. These unfortunate and often preventable crimes can destroy people, families, and businesses. Even when the act of aggression is only psychological, it can be painful and costly.

More locks and guards are not the answer, however. Employers need strong policies, effective security protocols, and a well-conceived strategy to confront the potentially violent employee and prevent workplace violence.

Employers have a moral duty and a statutory obligation under federal and state laws to provide and promote a safe and nonviolent work environment. Illustrative of these responsibilities are the requirements under the Occupational Safety and Health Act of 1970 (OSHAct) and state workers' compensation laws. Employers also have responsibilities to the public. Either vicariously or directly, employers may be liable for the harm brought to others by workplace violence. Moreover, employers have additional legal obligations to job applicants.

This intricate web of statutes, standards, rules, and regulations creates a legal minefield for supervisors and managers. Wading through that minefield is precarious at best; employers must protect employees and other parties without infringing on anyone's rights. Rarely has the challenge for employers been greater. Fortunately, there are solutions.

Intervention is the process of returning the aggressor to a structured work environment and helping the employee regain control of his life. To successfully intervene, management must have the will to redefine boundaries and overcome the aggressor's base of power. The intervention process must be well planned and executed. Nothing should be left to chance.

Upon recognizing inappropriate or disruptive behavior, management must act immediately and put the employee on notice. If a progression is identified early enough, the first warning by the organization is usually verbal. Verbal warnings often suffice to cause behavioral changes and halt further aggression. When such warnings are not enough, written warnings should follow. If the progression continues, the aggressor may be referred to the company's employee assistance program (EAP) or an outside resource for counseling. Monitoring performance and addressing behaviors is called performance-based management. It provides structure to the work environment while at the same time offering the employee choices.

In extreme cases, progressive discipline and professional counseling may not be enough. Under these circumstances, termination, temporary restraining orders, hospitalization, or prosecution may be the only solutions available.

WORKPLACE SUBSTANCE ABUSE

Employee substance abuse can affect nearly every aspect of an organization's operation. Workplace substance abusers are not only more likely to steal, but they are more likely to be involved in incidents and get injured on the job. They file more health claims and consume more than their share of the benefits provided them. More employee illnesses and injuries yield higher insurance costs to the employer. The absence of the injured or sick drug abuser may result in lower productivity for the organization, hence affecting the organization's bottom line. Their absence is disruptive and can be costly. Furthermore, an employee's absence may necessitate recruiting, hiring, and training a replacement.

While real claims cost real money, fraudulent claims cost money as well. Substance abusers are also more prone to file false claims and feign on-the-job injuries. As the abuser's burden grows and her performance begins to slip, some form of progressive discipline begins. Missed deadlines and careless mistakes are rarely terminable offenses; however, repeat offenders are not as readily forgiven and are usually punished. With each subsequent transgression, the punishment is enhanced.

The role of the employer has never been more important as organizations have come to accept that substance abuse is a workplace problem that must not be overlooked. The foundation of all doctrine and practices in the workplace is policy. For the creation of a drug-free workplace, a policy is absolutely necessary. The policy must be practical, functional, and enforceable. Additionally, the policy must be written, effectively communicated, and equitably enforced. An effective policy should:

- clearly state the company's objective to create a drug-free workplace, including why a drug-free workplace is important to all employees
- state why substance abuse (including alcohol use) at work or on the job is unacceptable
- clearly define what constitutes an infraction of the policy and describe the consequences
- acknowledge that drug problems and abuse are treatable and spell out the availability of treatment and rehabilitation options
- answer any questions that might be asked about drug abuse, the policy, or policy enforcement that might be unique to the organization or the industry

Once the policy has been established, it is the role of supervisors and managers to communicate and enforce that policy. One of the most significant deterrents against employee drug abuse that supervisors and managers have is communicating their familiarity with the policy and their willingness to enforce it. In doing so, management clearly defines behavioral and performance boundaries. Expectations of employees can clearly be defined and the consequences of not meeting those expectations can be explained.

Management must monitor performance. More than ferreting out drug abuse and employee drug abusers, management must also monitor performance. Supervisors and managers are not expected to catch employees using and selling drugs. Rather, they are expected to evaluate employee performance and be able to take remedial action when performance is not adequate or is substandard. Management that is able to do this avoids the destructive effects of enabling abusive behavior and codependent relationships. It creates a healthier and more productive work environment and forces employees to be accountable for their actions.

HARASSMENT AND DISCRIMINATION

Workplace sexual harassment and discrimination are insidious, painful, and expensive. Despite heightened awareness, better communications, and training, the problem continues to torment individuals and organizations. It has damaged lives, ruined careers, and left entire organizations publicly tarnished. To address the problem and reduce the damage when complaints are made, employers must investigate. In fact, employers have an ethical and lawful duty to investigate these complaints. However, their investigation must be thorough and impartial. The investigator must be properly trained, adequately skilled, and fair.

Human resources experts purport that harassment of all types costs the typical Fortune 500 company an estimated $6.7 million in increased absenteeism, diminished morale, lowered productivity, and employee turnover (Ferraro 2000). A total of 84,442 harassment charges were filed with the Equal Employment Opportunity Commission in 2002, an increase of 4.5% from the year prior (U.S. Equal Employment Opportunity Commission 2003).

Under federal and state law, an employer has an absolute affirmative duty to investigate a complaint or suspicion of harassment or discrimination. Employees are protected by the following federal laws:

- Title VII of the Civil Rights Act of 1964 (Title VII), which prohibits employment discrimination based on race, color, religion, sex, or national origin
- the Equal Pay Act of 1963 (EPA), which protects men and women who perform substantially equal work in the same establishment from sex-based wage discrimination
- the Age Discrimination in Employment Act of 1967 (ADEA), which protects individuals who are 40 years of age or older
- Title I and Title V of the Americans with Disabilities Act of 1990 (ADA), which prohibit employment discrimination against qualified individuals with disabilities in the private sector, and in state and local governments
- Sections 501 and 505 of the Rehabilitation Act of 1973, which prohibit discrimination against qualified individuals with disabilities who work in the federal government
- the Civil Rights Act of 1991, which, among other things, provides monetary damages in cases of intentional employment discrimination

The EEOC enforces all of these laws. It also provides oversight and coordination of all federal equal employment opportunity regulations, practices, and policies.

In spite of the vast body of law already protecting employees against discrimination and mistreatment, the U.S. Supreme Court issued landmark decisions in the sexual harassment cases *Burlington Industries, Inc. v. Ellerth*, 524 U.S. 742 (1998) and *Faragher v. City of Boca Raton*, 524 U.S. 775 (1998). Both federal and state courts have followed the principles enumerated in these cases, applying them to other types of harassment. Effectively, these sweeping decisions impose liability on employers for the harassment by their supervisors who engage in "tangible employment action(s)," and issues such as demotion, termination, and denial of benefits. In doing so, these decisions have diluted some of the employers' traditional affirmative defenses. Those traditional affirmative defenses include:

- the complainant's failure to take advantage of the employer's harassment policy and make a report
- the employer's prompt and effective response
- a prolonged delay by the employee in making a complaint

As a collateral effect, the *Burlington* and *Faragher* decisions have also imposed a new burden on employers to conduct effective and proper workplace investigations. While this burden has existed since the enactment of Title VII, the courts have become less tolerant of sloppy investigations by employers. The courts have more clearly defined what constitutes harassment and what necessitates an investigation. This list is by no means comprehensive:

- unwanted sexual advances
- the offering of employment benefits in exchange for sexual favors

- inappropriate leering, gesturing, and noisemaking
- inappropriate or obscene calendars, posters, art, graffiti, or drawings
- verbal conduct such as derogatory comments, epithets, slurs, or jokes
- written communications of any form that are derogatory or stereotypical based on gender, race, national origin, ethnicity, religion, and so on
- verbal abuse, graphic verbal comments, use of degrading words to describe an individual, group, sex, or physical attribute
- unwanted physical touching, brushing, impeding, or blocking of movements
- retaliation for making harassment reports or threatening to report harassment (*The National Employer* 2002, 148)

One can surmise that there are obvious defenses other than those listed above. Among the most useful is the employer's reasonable care to prevent and correct harassment. For this defense to be viable, the employer has limited time to act once the violation is noticed. The action, as well as its promptness, will in part determine the employer's liability. Employers should also be aware that the Civil Rights Act of 1991 expanded the remedies available under Title VII to include compensatory damages for pain and suffering, humiliation, and embarrassment in instances of intentional discrimination, and additional punitive damages in cases where the employer acted in a malicious or reckless manner (*The National Employer* 2002, 161). Never before has there been more incentive to conduct a thorough and prompt workplace investigation.

SUMMARY

To be efficient and effective, an organization must do all that it can to recruit, hire, and retain the best people possible. Today, organizations use preemployment screening and background investigations to identify and qualify the most desirable people before hiring them. This chapter explored why workplace policies are necessary and how they are developed. It also examined the assortment of legal and ethical issues surrounding policy development and implementation and how these policies are developed.

REFERENCES

ASIS International. 2006. Preemployment and background screening guideline. Alexandria, VA: ASIS.

Association of Certified Fraud Examiners. 2008. *Report to the nation on occupational fraud and abuse*. Austin, TX: ACFE.

Ferraro, E. F. 2000. *The process of investigation*. Golden, CO: Business Controls.

———. 2006. *Undercover investigations in the workplace*. Maryland Heights, MO: Butterworth Heinemann.

The National Employer. 2002. San Francisco: Littler Mendelson.

U.S. Equal Employment Opportunity Commission. 2003. http://www.eeoc.gov/stats/charges.html.

Part 5

Management of Proprietary Information

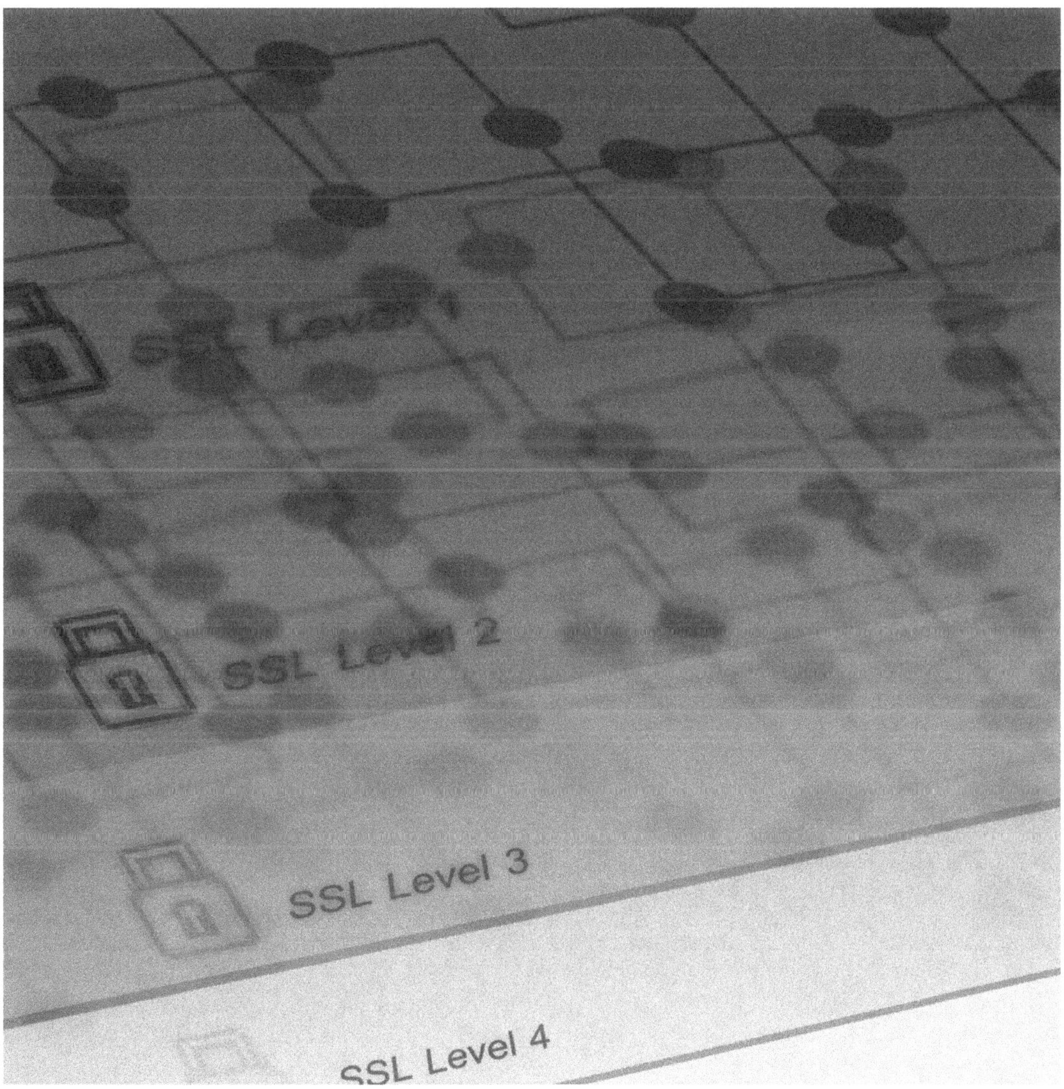

PART 5

12

Information Security Management Systems

Eugene F. Ferraro, CPP, CFE, PCI
Courtney Fischer

CHAPTER 12

OBJECTIVES

The objectives of this chapter are as follows:

- define information security
- indentify the four areas of specialized security that make up an information security program
- understand the three dimensions of an information security model and examine how they connect to build an information security management system
- understand *malware* and how it is used to corrupt and steal information
- define management's role in information security management and understand the differences between a manager and a leader
- identify the roles and functions of those responsible for information security management and the vulnerabilities that result when responsibility is delegated

INTRODUCTION

Think of a number between one and 10. Triple it. Add six. Then triple it again. Now take your answer, probably a two-digit number, and add the digits of your answer. If you still have a two-digit number, add those digits again. You should now be thinking of the number nine. The reason this works is the digits of any multiple of nine must sum to a multiple of nine. It is important to note that algebra plays an important role in telecommunication and information theory used to protect and authenticate electronic information.

Forward Error Correction (FEC) is a system of error control for data transmission, whereby the sender adds redundant data to its messages, also known as an error-correction code. This allows the receiver to detect and correct errors (albeit limited) without needing to ask the sender for additional data. The advantages of FEC are that a back-channel (or retransmission of data) is not required for authentication. Modern computer science and the systems that protect electronic information at the system level use error detection and correction techniques to ensure that data can be encrypted and transmitted without errors, even across unreliable media or networks. This is where information security begins. Today, the very survival of an organization depends on its ability to protect its information. This chapter will examine the fundamentals of information security and some of the management tools used to protect organizational information.

WHAT IS INFORMATION SECURITY?

Unfortunately, the terms *information security*, *computer security,* and *information assurance/integrity* are frequently used interchangeably and, most often, incorrectly. These interrelated fields often share the common goals of protecting the confidentiality, integrity, and availability of information; however, there are subtle differences between them. The differences lie primarily in the approach, the methodologies used, and the material that is to be protected. For the purposes of this chapter, *information security* is the means and methods of protecting information and information systems from unauthorized access, use, disclosure, disruption, modification, or destruction.

Remarkably, information security is a relatively new concept to many organizations. For some time, securing information was as simple as securing individual computers and internal networks. Information was safe as long as the organization that possessed it protected it from unauthorized access by insiders. However, as organizations began to connect their networks to the outside world, the risks to their information grew. Not only did it become necessary to protect against the prying eyes of insiders, it became necessary to keep information out of the hands of unauthorized and criminally minded outsiders. With organizations in the United States losing more than $59 billion annually in proprietary information and intellectual property, the implementation of information security systems and protocols has become vital (Nisbet 2005).

Security is "the quality or state of being secure—to be free from danger" (Whitman and Mattord 2008). However, security is not a one-size-fits-all package; the most efficient security program is multifaceted and multifunctional. In order to attain the best security program for an organization, a variety of security methods should be used in conjunction or simultaneously with each other. The principle of layering security is as fundamental as the concept of gates, guards, and locks. Whitman and Mattord (2008) expanded this definition to define *information security* as the "protection of information and its critical elements, including the systems and hardware that use, store, and transmit that information, through application of policy, training, and awareness programs and technology."

Thus, we can identify four important areas of specialized security that make up an information security program:

- physical security
- operations security
- communications security
- network security

A THREE-DIMENSIONAL SECURITY MODEL

An example of a model of information security is the Committee on National Security Systems' (CNSS) National Security Telecommunications and Information Systems Security Committee (NSTISSC) Security Model. This security model was developed by John McCumber and is often referred to as the *McCumber Cube*. In 1991 McCumber created his model for the purpose of establishing and evaluating information security programs. His security model is depicted as a three-dimensional, cube-like grid (Figure 12–1).

The concept of this model is that, in developing information security systems, an organization must consider the interconnectedness of all the different factors that impact the organization. To devise a robust information security program, the organization must consider not only its security goals, but also how these goals relate specifically to the various states in which information can reside in the organization and the full range of available security safeguards that must be considered in the design.

Figure 12–1. The McCumber Cube Model for Information Security

As mentioned, the McCumber Cube is made up of three dimensions, each of which includes key components of information security. The first dimension includes confidentiality, integrity, and availability. The second dimension includes storage, processing, and transmission. The third dimension includes policy, education, and technology. The cube consists of 27 cells, each of which represents an intersection between the dimensional components. To use this model, an organization must address the intersection of each dimensional component.

As valuable as the McCumber Cube is, some safety professionals assert that it has two significant weaknesses. First, the model does not address how to implement controls, nor does it address guidelines and policies that an organization should put in place. The second weakness occurs when an organization fails to use all three dimensions concurrently, or excludes one of the key decision-making groups discussed earlier.

THE FIRST DIMENSION—THE CIA TRIANGLE

The CIA (Confidentiality, Integrity, and Availability) Triangle is considered by many to be the benchmark for computer security in industry today. Originally, the CIA Triangle included confidentiality, integrity, and availability. However, with the changing business climate and evolution of new business standards, those three components are no longer sufficient to maintain information security. The CIA triangle now includes many more characteristics of information.

Confidentiality. Protection of confidentiality requires taking the following measures:

- classifying information
- securely storing documents
- employment of broad security procedures
- informing data keepers and end users
- cryptography

Traditionally, it is assumed that organizations will protect the confidential information of their employees, customers, or patients. When that information is not properly managed and maintained, problems can arise. Disclosure of confidential information can be deliberate or accidental. A breach can occur from failures within the organization or from sources external to the organization. The success of the organization's program will depend on the depth of the program and the sophistication of its safety measures.

Integrity. The integrity of data or information is vulnerable when it is open to corruption, damage, destruction, or any alteration of its original state. Thus, from a security standpoint, information is vulnerable when it is entered, stored, or transmitted. In addition to these internal threats to information integrity, there are also external threats including viruses, worms, phishing, and a host of other applets called *malware*.

Malware, or malicious software designed to interfere with the normal function of computer software, is pervasive. According to sources at ASIS International, the problem of malware is surging. The problem is attributed to hackers who sell off-the-shelf versions of malware to anyone willing to pay for it. Another concern is the way malware is spread. In the past, malicious software was principally spread via e-mail attachments. Today, it is most often spread by websites. IBM reported that the number of new malicious web links increased by more than 500% in the first half of 2009 compared to the first half of 2008.

Here is how malware works: A software *worm* has the ability to spread without the assistance from the user, whereas *viruses* require the user to actively assist the spread of the virus by such means as opening an e-mail attachment. *Phishing* leads a user to believe that he is visiting a legitimate Internet site and is tricked into entering private information such as credit card numbers or other identifiers. Today, websites are the principal vehicle by which malicious code is used to infect the computers of unsuspecting users.

Corruption of data is not always intentional. Corruption can result from defective programming or interruptions in the transmission of data. There are multiple ways to counteract both internal and external threats to integrity, including redundancy systems, algorithms, hash values, and error-correcting codes such as FEC. File hashing uses an algorithm to compute a single representative number after evaluating the number of bits in a file. This number is called the hash value. Each file has a recorded hash value; therefore, when the algorithm computes the hash value for a received file, if the two values are different, the file is deemed damaged. Similar to FEC, file hashing is used to ensure data integrity. This concept is critical to those responsible for information security. If the integrity of information cannot be confirmed, the information loses all its potential value.

Availability. The ease with which an individual can access information is referred to as its *availability*. Availability requires that accessing information is free from hindrance, and that once the information is accessed, it is in a usable form. Availability only applies to authorized users. Information security systems should be designed so that information is not easily accessible to any user—only the *authorized* user.

With increased availability or accessibility of information, the danger of unwanted users having access to information increases. This is especially true now that an organization's network can be accessed remotely from almost anywhere. One of the principal risks is that of authorized users accessing information through unsafe systems, such as public computers, or unsecured Internet connections. Though unintentional, the damage can be as great or greater than any intentional intrusion.

THE SECOND DIMENSION—PRIVACY

Privacy, as defined by Whitman and Mattord (2008), is the guarantee that "information that is collected, used, and stored by an organization is intended only for the purposes stated to the data owner at the time it was collected." In this sense, privacy does not refer to the protection from information being observed, only how it is utilized. This is where the second dimension (storage, processing, and transmission) of the security model becomes important.

Recently, some businesses have engaged in the practice of trading or selling the personal information of their customers. Personal information may be collected and stored in databases that are accessible in ways the individual did not authorize. Additionally, advances in phishing technology have now made it easier to create and access genuine marketing lists from companies that would not otherwise be willing to deliver such information. Unethical business practices have led many individuals to turn to the government to ensure their privacy is protected. Today, protecting the privacy of individuals and their personal information is of paramount concern. No information security management system is complete without adequately protecting individual privacy and personal information.

Identification. The initial process by which access to protected information is obtained is identification. Identification also sets the foundation for later authentication and authorization. Only when a system can distinguish one user from another, does it possess the characteristic of identification. Typically, a username or other ID (such as an e-mail address) is used for identification.

Authentication. Authentication is proof that a user is who she claims to be. Organizations such as SecurID and Authenx produce hardware devices that can authenticate a user's identity.

Authorization. Authorization occurs subsequent to authentication of identification. Once a user is properly identified, authorization is required to access secured information. Authorization can be granted to a person or to a computer and ensures that the user is allowed to observe, change, or delete information that he accesses.

There are two main elements essential to access control security:

- Discretionary access control—provides access to an individual who is a member of a group that has been granted the authority to use and view certain information. Only members of the group are granted access to those resources.
- Mandatory access control—grants access to information to specifically identified employees. It determines which individual can see, edit, alter, or even know about certain information.

One of the tools an organization can use to determine who has access to what data is a flowchart. A flowchart can identify employees who have access to information at different points by tracking data flow. This knowledge allows the organization to identify the points in the data flow where access control is needed. Additionally, once the data flow is understood, it becomes easier for an organization to make information accessible to employees within specific roles.

Accountability. Subsequent to the authentication of a user's identity and authorization to access information, there must be accountability for the user's access. A system possesses accountability when that system can track and attribute all actions to a known user or computer process.

THE THIRD DIMENSION—THE ROLE OF MANAGEMENT

Management's role in the information security model is to set policy, to provide education, and to ensure that the organization has the most effective technology for securing its information. A manager, according to Whitman and Mattord, "is a member of the organization assigned to marshal and administer resources, coordinate the completion of tasks, and handle the many roles necessary to complete the desired objectives." Among the many roles a manager may have within an organization is an informational role, an interpersonal role, a decisional role, or a combination of all three. The term *leader* should not be mistaken with manager, as the two have different methods by which they complete tasks.

- A leader encourages employees so that they are willing to achieve goals. A leader is someone who leads by example and exhibits personal characteristics that motivate others.
- A manager oversees the assets of the organization. A manager creates budgets, approves spending, and hires employees.

Although a manager may not always be in a leadership role, and nonmanagers are often in leadership roles, managers often function as both manager and leader.

MODERN INFORMATION SECURITY MANAGEMENT

Information security management functions similarly to the other types of management; it is the security team's goal to secure the organization's operations by means of protecting its information. This requires the active commitment of senior management.

However, some corporate managers give little thought to information security. Even though they have the authority to properly look after an organization's information assets, executives often delegate this responsibility. Individuals who do not have the authority should not be charged with information security. Therefore, it is a necessity that information security originates from the top of an organization (Moritz 2005). For some organizations it begins with the board of directors, the CEO, and the executive management team. If an organization lacks the active commitment of executives, a successful information security program cannot exist.

Typically, the delegation of information security duties breaks out as follows:

- Information security managers and professionals are responsible for protection from internal and external threats against an organization's information assets.
- IT managers and professionals are responsible for providing information technology sufficient to meet the company's needs and objectives.
- Nontechnical managers and professionals are responsible for conveying the company's policy and objectives and for distributing appropriate funds and assets to the other groups.

Regardless of its structure, or to whom information security management is delegated, one of the biggest vulnerabilities to a security system rests with the strength of the organization's management. That vulnerability is enforcement and compliance. Managers who also function as leaders should have the ability to motivate their employees to be compliant with established security protocols. Therefore, for the success of information security, it is essential that an organization hire only managers who have the ability to motivate their employees.

SUMMARY

Until recently, securing information was a relatively new concept; however, with globalization and the sharing of information outside an organization, it is becoming increasingly crucial for organizations to protect their information. Their survival may depend on it. In addition, unethical business practices by some, and the use of *malware,* (malicious software designed to corrupt information) is on the rise. For these reasons organizations must establish means and methods for protecting their information and information systems from unauthorized access, use, disclosure, disruption, modification, or destruction.

A successful information security management program must address physical security, operations security, communications security, and network security. One example of an information security model is the National Security Telecommunications and Information Systems Security Committee (NSTISSC) Security Model. The NSTISSC model was developed in 1991 by the Committee on National Security Systems (CNSS). The model is built on the concept that there are three dimensions of information security, each of which includes key components. To use this model, an organization must address the intersection of each dimensional component and understand the interconnectedness of the components. The first dimension includes such components as confidentiality, integrity, and availability. The second dimension includes storage, processing, and transmission. The third dimension includes policy, education, and technology.

Management is key to the success of any information security management program. Support for the program, as well as responsibility for the policies set and adherence to those policies, must originate from the top of the organization.

REFERENCES

Moritz, Ron. 2005. "Blending Corporate Governance with Information Security," *Larstan's Black Book on Corporate Security*. Potomac, MD: Larstan Publishing, Inc.

Nisbet, Jim. 2005. "Managing and Protecting Intellectual Property in a Shared Information Environment," *Larstan's Black Book on Corporate Security*. Potomac, MD: Larstan Publishing, Inc.

Whitman, Michael E. and Herbert J. Mattord. 2008. *Management of Information Security*, 2nd edition. Course Technology, Boston, MA: Thomson Learning, Inc.

13

Developing an Information Security Program

Eugene F. Ferraro, CPE, CFE, PCI
Courtney Fischer

CHAPTER 13

OBJECTIVES

The objectives of this chapter are as follows:

- describe the concept of an information security program and identify its key elements
- discuss the functions required to properly implement an information security program
- indentify the five steps to implement a content security strategy
- recognize the key elements of a quality and properly functioning information security program
- review the current best practices for securing an organization's intellectual property

INTRODUCTION

As discussed in the previous chapter, an *information security program*, as defined by Whitman and Mattord (2008), is "the structure and organization of the effort that strives to contain the risks to the information assets of the organization." Thus, the purpose of an information security program is not to eliminate all risk but to *contain* risk. This is a critical distinction. If the organization's expectation is to eliminate risk and risk can only be contained, then the organization's managers will be disappointed. For this reason, organizational expectations should be properly set and the responsibility for creating and maintaining an information security program should be established at the beginning of the development process.

The most difficult aspect of implementing an information security program is getting started. This is especially true if the organization is unclear about what it wants to protect and how to protect it. As a result, a logical starting point is asking the question, "What is it that is most critical to protect?" There may be more than one answer to this question, but each answer will provide a better idea of what is required to secure the organization's information. Once a starting place has been established, there is another important task to complete. It is essential for the organization to communicate with its customers to understand their needs. The program should be built around the assets that the organization desires to protect and the way its customers use those assets.

This chapter will examine the way in which an information security program is developed and the components that make it successful.

FUNCTIONS REQUIRED FOR IMPLEMENTATION

According to Whitman and Mattord (2008), there are several required functions that must be understood before an information security program can be successfully implemented.

- *Risk assessment*—risk assessments appraise the threat within a particular department or function. A risk assessment also helps identify the causes of threats and is used as a means to recommend processes that can decrease the identified threat.
- *Risk management*—risk management executes or monitors the processes used to decrease the level or risk to a company's information. Risk management is often used in conjunction with risk assessment.[1]

[1]The reader should be aware that the term *risk management* is also used to describe the activities around the management of an organization's casualty insurance. When used in this fashion, the *risk manager* is usually the person who administrates an organization's various insurance policies and serves as its authority relative to insurable risk.

- *Systems testing*—systems testing has two parts: First, it assesses the patches, or fixes, that have been used to secure known vulnerabilities. Second, it assesses the effectiveness of new systems and their compliance with organizational policy. These assessments are usually done in conjunction with incident response or risk management.
- *Policy*—policy is the function by which security throughout an organization is supported and sustained. The broad organizational policy must correspond with the information security policy. According to the ASIS International *Information Asset Protection Guideline* (ASIS 2007), "any successful program is based on a clear, practical, and communicated policy."
- *Legal assessment*—the function of the legal assessment is to be continually aware of laws, whether existing or anticipated, and the impact those laws may have on the organization's policies and processes. The legal assessment is conducted in conjunction with outside legal counsel and, if appropriate, law enforcement agencies. The legal assessment is typically removed from the information security and information technology departments.
- *Incident response*—there are three components of the incident response function. It manages how the first response to an incident, or disaster, is handled; it controls the intensity of actual incidents; and it carries out the preliminary response to possible incidents. Incident response should include employees from numerous departments across the organization.
- *Planning*—planning is the function that examines, produces, sustains, and supports the information security plan. Information security planning is often done using a project management style, as opposed to using strategic planning, which is often used for the organization. However, information security planning must work in conjunction with the policy processes in place throughout the organization.
- *Measurement*—the measurement function evaluates all parts of the information security program through the use of current processes. The data that results from this evaluation can be used by managers to make decisions about how well a security program is functioning.
- *Compliance*—the function of compliance is to ensure that known weaknesses are timely and accurately mended by system and network supervisors. Compliance, while it seems straightforward and positive, can also create a dilemma for organizations, especially those that are customer-service oriented. Many organizations find that a balance of maintaining good customer service and enforcing compliance is difficult to find.
- *Centralized authentication*—centralized authentication controls admission and withdrawal of authorization to the company network for all employees within an organization. This function is often performed by the help desk staff.
- *Systems security administration*—the systems security administration governs the setup of computer systems. Systems security is frequently prearranged based on the operating system the computer uses.
- *Training*—the function of training is to educate staff. For example, general staff are trained on general information security topics; information technology staff are trained in technical processes that are specific to information technology security; and information security staff are trained in technical and managerial areas of information security.
- *Network security administration*—while the systems security administration governs the setup of computer systems, the network security administration governs the setup of computer networks. Networks are often arranged into units based on the network area. Some security functions may have been initially assigned to these groups, which are removed from the information security function.
- *Vulnerability assessment*—a vulnerability assessment attempts to identify vulnerabilities so

that they can be mitigated before they can be exploited. This function is often contracted to a third party.

- *Content analysis*—content analysis is a process by which transmitted information can be monitored by an organization. The computer program that performs content analysis "learns" and can monitor and alert administrators when secured information is improperly transferred.

According to Jim Nisbet (2005), there are five steps to implementing a content security program:

1. Obtain agreement on what information should be monitored. The information can be confidential or protected. Additionally, the monitored information can be entire documents or pieces of information.
2. Determine the current security processes and procedures. Such efforts assist an organization in determining whether the current processes are adequate to protect the agreed-upon information.
3. Determine the type of information to be monitored. In order to accurately monitor the transmission of information, a company must know if all the information to be monitored is of one type. A different monitoring strategy may be needed if multiple types of information are to be monitored.
4. Identify the type of content analysis to be used to identify content. The types of content analysis may include a keyword search, a linguistic analysis, or a search for document types.
5. Execute a detection process that identifies where the monitoring device (or devices) should be positioned. To effectively monitor the transmission of information, the monitoring device should be positioned at the point where information moves outside of the company's protection.

Privacy issues may arise when an organization begins monitoring computer usage and transmission of information. These concerns can be resolved by implementing a policy regarding what information will be monitored, only monitoring that clearly identified information, and communicating the purpose of the monitoring. Such policies are critical for reducing the expectation of privacy by those who have electronic access to the organization's information as well as those who store or place information on the organization's systems.

The operative concept is the reduction of the reasonable expectation of privacy by those who use organizational information management systems. An organization's failure to properly set expectations creates unnecessary liability for it and its shareholders.

KEY ELEMENTS OF AN INFORMATION SECURITY PROGRAM

Whitman and Mattord (2008) identified 14 key elements of a security program along with the components of each element. These elements have significant overlap with the functions required to implement a security program already discussed.

1. *Policy*—policy includes program policy, issue-specific policy, and system-specific policy.
2. *Program management*—this includes central security programs and system-level programs.
3. *Risk management*—risk management includes risk assessment, risk mitigation, and uncertainty analysis.

4. *Life-cycle planning*—this includes the security plan, initiation phase, development/acquisition phase, implementation phase, and operations/maintenance phase.
5. *Personnel/user issues*—these issues include staffing and user administration.
6. Preparing for contingencies and disasters—this includes a business plan, identifying resources, developing scenarios, developing strategies, and testing and revising the plan.
7. *Computer security incident handling*—this includes incident detection, reaction, recovery, and follow-up.
8. *Awareness and training*—these include SETA (Security, Education, Training, and Awareness) plans, awareness projects, and policy and procedure training.
9. *Security considerations in computer support and operations*—these considerations include help desk integration, defending against social engineering, and improving system administration.
10. *Physical and environmental security*—these include guards, gates, locks and keys, and alarms.
11. *Identification and authentication*—these included identification, authentication, passwords, and advanced authentication.
12. *Logical access control*—access control includes access criteria and access control mechanisms.
13. *Audit trails*—audit trails include system logs, log review processes, and log consolidation and management.
14. Cryptography—cryptography includes public key infrastructure (PKI), virtual private network (VPN), key management, and key recovery.

BEST PRACTICES FOR SECURING INTELLECTUAL PROPERTY

Intellectual property can be defined as any intangible asset that consists of human knowledge and ideas. Some examples include patents, copyrights, trademarks, business processes, and proprietary software. Such assets are rarely identified on an organization's balance sheet. Because their value is often subjectively determined, an objective valuation is nearly impossible. Nevertheless, organizations, both public and private, value their intellectual property and often expend enormous resources protecting them.

Nisbet (2005) writes that "the value of this [the content security solution] rests in its ability to generate alarms with a high degree of accuracy." He identified seven best practices for securing intellectual property.

1. Define the organization's sensitive data. That data can be contained in or maintained by any of the following:
   - financial
   - human resources
   - research and development
   - legal
   - business strategy
2. Establish where sensitive data resides on the corporate network. That is, identify the intellectual property the organization owns and determine where it is being stored.
3. Document the means and methods by which the organization protects its intellectual property. Nisbet suggests these explicit steps be taken to document that the organization

is treating the content as valuable:
- spend money to protect sensitive data
- properly document the defined data
- create, implement, and enforce sensitive or proprietary information corporate policies

4. Implement a content security solution that will fingerprint, monitor, and immediately alert when content has left the network.
5. Use a detailed audit report, and begin researching transmission violations.
6. Prevent repeat offenses by immediately taking corrective measures, such as employee consultation and reassessment or realignment of corporate policies.
7. Require further investigation using in-depth forensics if habitual violations occur, such as: investigation to adequately prove violation and litigation Nisbet also identified methods of protecting intellectual property:
    a. Access control restriction—this is the basic type of access control in organizations. Access control lists are utilized to manage access to information. Access to information is granted to employees based on their respective department or on an individual level, such as when executives are allowed access because they are at a certain level within the organization.
    b. Physical partitioning—in this type of access control, protected information is located on a specific part of the network and access to the information is limited to a small number of physical locations. Additionally, the network on which the information is stored is either separated physically from the other networks, or the system does not allow connections between networks that house sensitive information.

SUMMARY

The protection of information assets is a time-consuming and tedious task. Most organizations do not do it well. However, in today's competitive business climate, information may be an organization's most valuable asset. A successful information security program not only contemplates the information requiring protection, but also the way its owners use it.

The initial process of defining and implementing an information security program can be quite a difficult endeavor for an organization. To ensure that your organization is implementing the appropriate security program, start by asking, "What is it that needs to be protected?" The answer to that question is the starting point for any information security program. There are multiple steps required for a successful information security program. The organization, as a whole, should be in agreement on what needs to be protected and the best way to achieve that protection.

REFERENCES

ASIS International. 2007. *Information Asset Protection Guideline,* Alexandria, VA: ASIS International, p. 8.

Nisbet, Jim. 2005. "Managing and Protecting Intellectual Property in a Shared Information Environment," *Larstan's Black Book on Corporate Security*. Potomac, MD: Larstan Publishing, Inc.

Whitman, Michael E. and Herbert J. Mattord. 2008. *Management of Information Security*, 2nd edition. Course Technology, Boston, MA: Thomson Learning, Inc.

14

Security Management Models and Practices

Eugene F. Ferraro, CPE, CFE, PCI
Courtney Fischer

CHAPTER 14

OBJECTIVES

The objectives of this chapter are as follows:

- identify the most common information security program models
- identify the two standards on which information security practices are based
- recognize the difference between *recommended practices* and the *gold standard*, and the limitations of each
- understand the principles required for program implementation and success

INTRODUCTION

To sufficiently implement a security program within an organization, a security model is needed. A *security model* is a generic blueprint that can be used to help an organization establish its own security model (Whitman and Mattord 2008). One way an organization can create a blueprint is by looking to other successful organizations and studying how they created their security model. This practice is called *benchmarking*. When creating a security model, developing one with multiple layers of security is recommended for the best protection. A key part of implementing a security program is to determine which pieces of any given model are relevant to the organization's specific needs or requirements. No single model will fit all organizations. The best approach is a customized model, using pieces of other models and combining them to best serve the organization's specific needs.

MODELS

Whitman and Mattord (2008) identify six security models from which an organization can draw to develop its own customized security model.

1. ISO/IEC 17799:2005. The ISO/IEC 17799:2005 security model is a widely popular and often cited security model. The goal of this model, as cited by Whitman and Mattord (2008), is to "establish guidelines and general principles for initiating, implementing, maintaining, and improving information security management in an organization." The model presents several suggested objectives for an organization regarding how best to implement, maintain, and manage a security program. Those objectives include:
    - security policy
    - organization of information security
    - asset management
    - human resources security
    - physical and environmental security
    - communications and operations management
    - access control
    - information systems acquisition, development, and maintenance
    - information security incident management
    - business continuity management
    - compliance

The objectives listed above should be enacted to meet the needs of an organization. To determine the organization's needs, a risk assessment should be conducted first. This specific model contains 133 different controls, many of which are not necessary for every organization. Customizing the information security program to fit the organization's needs helps to ensure that the organization has the most effective program.

2. ISO/IEC 27001:2005. Unlike the ISO/IEC 17799:2005, this model is not intended as an outline for executing a security program, but rather as an evaluation tool, which is its primary function. Organizations in need of certification are the primary users of this model. Whitman and Mattord (2008) identified six proposed uses for this model:
   - devise security requirements and purposes
   - guarantee that security risks are handled in a cost-effective manner
   - guarantee conformity to laws and regulations
   - provide a procedural outline for the execution and administration of controls to ensure that goals are achieved
   - define new information security management procedures
   - establish conditions of information security management activities
3. NIST security models. Information and documents regarding procedures for implementing security management programs can be found at the National Institute of Standards and Technology (NIST) website, www.csrc.nist.gov. The U.S. government has cited these documents as rationale for why organizations should not adopt the ISO/IEC 17799 standards. Specifically, NIST models have two distinct benefits that other sources of information lack:
   - the information is free and available to the public
   - the information has sustained a test of time and has been thoroughly assessed by the government and security professionals
4. RFC 2196 Site Security Handbook. The RFC 2196 was created by the Security Area Working Group, which is part of the Internet Engineering Task Force. The handbook identifies significant security concerns for an organization, a general description of the five basic tenets of security, and provides information regarding how to create and enact the program. The five tenets identified by the handbook are:
   - introduction to security
   - information regarding security policies
   - the structure of a security program
   - services and processes
   - incident management
5. COBIT. The Control Objectives for Information and related Technology is another information security model that offers suggestions regarding enactment of controls and explores the goals of an information security program. This model identifies 34 advanced goals that encompass 215 control goals. The control goals are divided into four categories as listed:
   - prepare and categorize
   - attain and execute
   - convey and maintain
   - examine and appraise
6. COSO. The Committee of Sponsoring Organizations of the Treadway Commission, like COBIT, is control-based. The COSO model was developed in 1985 by the private sector.

Unlike the models previously discussed, this model is focused specifically on fraud in financial reporting. The main goal of this model is to pinpoint the causes of fraud and to reduce the occurrence of fraud. This model also assists organizations in complying with the Sarbanes-Oxley Act of 2002 and other federal regulations.

PRACTICES

Information security has two standards on which practices are based: standards of care/due diligence and recommended security practices.

STANDARDS OF CARE/DUE DILIGENCE

Although the two terms (*standard of care* and *due diligence*) are frequently used synonymously, due diligence is a broader term that includes execution of the procedures and maintenance of protection. Whitman and Mattord (2008) define both terms and illustrate important differences between the two.

The *standard of care*, as defined by Whitman and Mattord (2008), exists "when organizations adopt minimum levels of security to anticipate a future legal defense, they may need to show that they have done what any *prudent* organization would do in similar circumstances." *Due diligence* is also "implementing controls at this minimum standard and maintaining them."

RECOMMENDED SECURITY PRACTICES

Security practices that go above and beyond the required level of protection of information are called *recommended security practices*. High-level security practices are deemed the top practices in the industry and are referred to as best security practices (BSPs). The accessibility of information and the requisite amount of protection is balanced when using BSPs.

The U.S. government operates a website, www.fasp.nist.gov that makes available the BSPs of many government agencies. Another website where organizations can find information on best practices is www.cert.org/archive/pdf/secureit_bestpractices.pdf. This source is operated by the Computer Emergency Response Team and Carnegie Mellon University. Microsoft also provides a website for access to its suggested best security practices: www.microsoft.com/privacy/safeinternet/security/best_practices/default.htm.

THE GOLD STANDARD

The *gold standard*, as defined by Whitman and Mattord (2008), is "a model level of performance that demonstrates industrial leadership, quality, and concern for the protection of information." Because this standard is the best in the industry, the actual execution of a gold standard for information security requires a large amount of resources—in financial requirements and manpower. For this reason, the gold standard is unattainable for many organizations; rather, it is the ideal that organizations strive to achieve.

SELECTING RECOMMENDED PRACTICES

There are many important things for an organization to consider before selecting a recommended practice. Some considerations include government rules and regulations. Even if an organization is not governed by federal rules and regulations, choosing a recommended practice can prove challenging. To ease this difficulty, there are questions an organization can ask, including:

1. Is the organization similar to the organization identified by the recommended practice?
2. Is the organization in a comparable market to those using the recommended practice?
3. Does the organization confront comparable issues to those using the recommended practice?
4. Is the structure of the organization analogous to those using the recommended practice?
5. Are the resources needed by the recommended practice comparable to those available to the organization?
6. Is the threat situation of the organization comparable to the one supposed by the recommended practice?

BENCHMARKING AND LIMITATIONS TO RECOMMENDED PRACTICES

Benchmarking, as previously defined, is the practice of examining other organizations and studying their practices and the results those practices produce. Benchmarking is an invaluable strategy for an organization seeking to implement an information security program. However, benchmarking can be challenging, as communication between organizations is not frequently practiced due to competing interests. The lack of communication makes it difficult for organizations to share models and policies. However, the practice is becoming more popular because there are now professional associations and societies that security professionals can join to facilitate communication of lessons learned, strategies tried, and challenges faced.

Another challenge to benchmarking is organizations are individualistic; no two are exactly the same. Therefore, practices of another organization that are similar to one's own may not meet one's exact needs. Using recommended practices also produces challenges for an organization. Mainly, recommended practices change with the times and security professionals must be aware of those changes and how those changes impact the practices currently used by the organization.

BASELINING

A *baseline,* as defined by Whitman and Mattord (2008), is the "value or profile of a performance metric against which changes in the performance metric can be usefully compared." Baselining, then, is the procedure by which an organization measures its standards against traditional standards. Referring to information security, baselining is the process by which an organization compares its security policies and procedures with the organization's potential operations.

PITFALLS TO AVOID

According to Stephen Foster, there are two key pitfalls to avoid when implementing an information security program: not confronting impractical beliefs and not addressing threats immediately (Foster 2005).

PROGRAM IMPLEMENTATION

Once an organization has chosen an information security model on which to base its program, the organization must consider how to effectively put the program into practice. Ron Moritz identified five over-arching principles necessary to implement a successful information security program (Moritz 2005). These principles are briefly described below.

CEO INVOLVEMENT

The chief executive officer (CEO) is the head of the organization and can serve as a leader and exemplify best practices for implementing an information security program. Additionally, the CEO, as someone with access and authority within the organization, should have a keen understanding of the security within the organization.

However, it is important that all employees be given access to the information they need. Access to information should be granted based on need, not on the employee's job title or level. Although the CEO of an organization is likely to need access to all information, many upper-level managers may not. It is up to the organization to decide who needs access to what information.

ORGANIZATIONAL UNDERSTANDING OF INFORMATION ASSETS

For an information security program to be successful, the organization and all its employees must be aware that it is everyone's responsibility to protect its information. The organization's policies and procedures should identify this fact.

To help an organization ensure that its policies and procedures have been clearly articulated to all employees, there are steps the organization can take (ASIS 2007). The first, as previously discussed, is the involvement of an organization's leaders in the information security program. The second step is to identify an individual or group of individuals responsible for overseeing the management of the program. Limiting the number of individuals responsible for this task will help eliminate oversights in security due to miscommunication. The third step is to ensure that all employees and individuals associated with the organization abide by the policies set forth by the organization. Finally, the organization should not expect employees to retain all of the information stipulated within the policies; therefore, continual training on those policies is needed.

INTEGRATING DATA STORAGE WITH THE SYSTEM LIFECYCLE

Data needs to be stored so that it is accessible to all necessary departments within the organization. Rather than thinking of data as belonging to a specific department, such as the HR or the finance department, the data should be analyzed for content then appropriately distributed to those who need access.

TESTING THE SYSTEMS

The systems that are implemented to ensure the security of the organization's information need to be tested to validate that they are actually performing desired functions. The systems can be tested using internal audits or by tracing actual security breaches to their sources. Tracing breaches helps an organization identify weaknesses in its security program and allows for corrective measures to prevent similar or future breaches.

COMPARATIVE ANALYSIS

Once the model is chosen, an organization should compare itself, and where its security policies and procedures fall, against other similar organizations' security policies and procedures. This is done to ensure that possible gaps in the system are filled and that the model is the best possible fit for the industry.

SUMMARY

Many models are available from which an organization can base its own information security program. However, program designers must be aware that no two organizations are identical. Each individual organization should customize its information security program to ensure that it is the best fit for the organization. These programs are vital for protecting information regarding its customers, employees, and vendors. Once the proper program has been chosen, the organization must adhere to the policies regarding information security, and the programs must be continuously tested. It is the entire organization's responsibility to maintain the integrity of its information security program. A wise organization will employ methods for reporting breaches of security and have efficient and effective means for investigating those breaches whenever and wherever they occur.

REFERENCES

ASIS International. 2007. *Information Asset Protection Guideline*, Alexandria, VA: ASIS International.

Foster, Stephen W. 2005. "Building a Secure Corporate Environment from the Ground Up," *Larstan's Black Book on Corporate Security*. Potomac, MD: Larstan Publishing, Inc.

Moritz, Ron. 2005. "Blending Corporate Governance with Information Security," *Larstan's Black Book on Corporate Security.* Potomac, MD: Larstan Publishing, Inc.

Whitman, Michael and Herbert Mattord. 2008. *Management of Information Security*, 2nd edition. Course Technology, Boston, MA: Thomson Learning, Inc.

Part 6

Physical Security Concepts

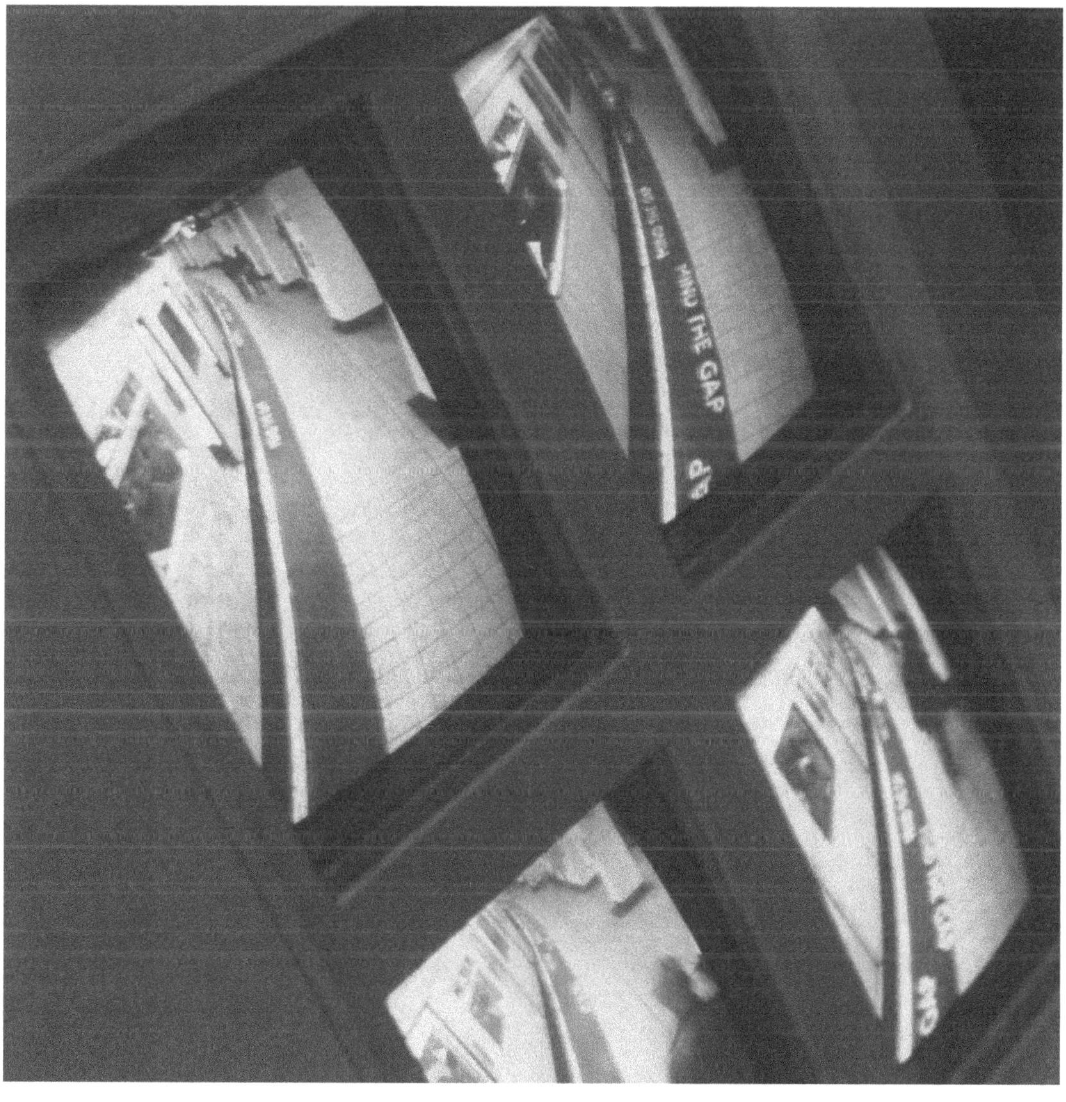

PART 6

15

Basic Risk Assessment for Security Management

Lauri A. Kingsbury

CHAPTER 15

OBJECTIVE

The objectives of this chapter are as follows:

- define *security* and the purpose of security management
- discuss how risk assessment forms an integral part of security management
- examine probability and criticality of loss as elements of a risk management program
- describe risk management methods

DEFINITIONS AND TERMS

Security "provides those means, active or passive, which serve to protect and preserve an environment, which allows for the conduct of activities within the organization or society without disruption" (Post, Kingsbury 1977). This "conduct of activities" involves personnel, property, and information. Within the field of security and loss prevention, the term *risk* may be defined as "expected results, or the probability that a loss has occurred or will occur . . . the possible occurrence of an undesirable event" (Broder 1984). Security management requires control of risks, which is attained through the implementation of physical and procedural methods that provide continuity. To accomplish this, factors that contribute to dangerous or harmful conditions must be identified before control measures are implemented. The method(s) must be matched to the identified risk.

There are two distinct categories of risk: dynamic and pure. *Dynamic risks* are connected to the cost of doing business. Decisions regarding whether to invest in product development, stock a particular brand of merchandise, or borrow money for a new manufacturing plant require evaluating dynamic risk factors. They have the potential for either profit or loss. The decision to accept a dynamic risk is founded on the relationship between current costs of acceptance and future return on investment.

Pure risks carry no possibility for return on investment or future benefits. The decision to accept this type of risk is based on how much an organization is able to lose. Pure risks include natural disasters; unfair competition; data theft and industrial espionage, including theft of trade secrets; traffic and aircraft accidents; criminal activity; and terrorism. Crime and loss prevention concentrates on the assessment and control of pure risk factors in order to remove, reduce, spread, transfer, or accept losses.

IDENTIFYING RISK FACTORS

Crimes are loosely grouped into two categories:

- Crimes against people—behavior in which physical harm is inflicted on a victim or the offender intends to harm the victim. Examples include assault, robbery, rape, and murder.
- Crimes against property—the target of these crimes is assets or possessions of an organization. Examples are burglary, vehicle theft, embezzlement and other financial crimes, larceny, computer hacking, and confidence games.

In general, law enforcement efforts to control crime are reactive. Police receive notification that a crime has occurred. Then, the event is investigated through witness interviews and evidence collection in an effort to identify the offender, prepare a case for prosecution, and, if possible, recover property. Prevention efforts are proactive in nature. This shifts the emphasis to forestalling a criminal act (or loss) before it occurs. Fred Hudson, MBE, director of the Home Office Crime Prevention Training Program in Staffordshire, England, developed the definition of *crime prevention* that has been adopted by law enforcement agencies and applied to private security practices worldwide:

> Crime prevention is the anticipation, recognition, and appraisal of a crime risk and the initiation of action to remove or reduce that risk.

This definition became the foundation for security surveys and is the basis of risk assessment (Kingsbury 1973). Note that within the security industry, the terms *survey* and *audit* are used interchangeably.

Three elements must be present, at the same time, for a crime or loss to occur:

1. a victim or target
2. the desire of an offender to commit the criminal act
3. an opportunity for the act

In Figure 15–1, these three elements are depicted as the sides of an equilateral triangle. If one of the elements is missing, the triangle collapses, and a crime or loss cannot occur.

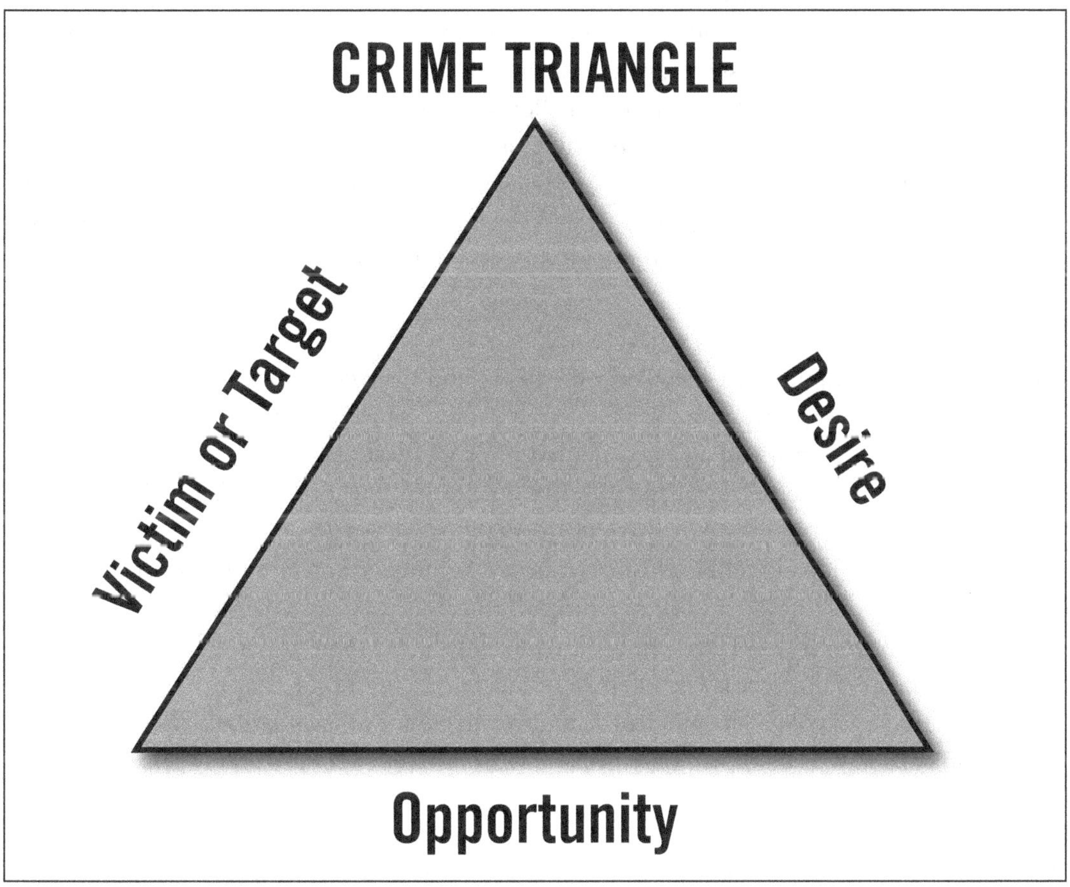

Figure 15–1. The Crime Triangle

Of the three factors in this triangle, opportunity is within direct control of a security program. Identification of risk factors requires the evaluation of potential opportunities for loss. Opportunity is site-specific. It is not enough to know that motor vehicles were stolen. The number of vehicles stolen from a given site, within a specified time period and frequency, and the location of the vehicles immediately prior to the theft must be examined. Another example is internal theft. Which items are missing—office and cleaning supplies, tools, employee briefcases, or purses? When were they last seen? Is there an area on the property where theft is more common than in other places?

Pure risks can be identified and managed in such a fashion as to reduce the losses associated with internal and external crimes by applying the definition of crime prevention to any facility, building, or location. To prevent a crime or loss from occurring, four steps must be taken (Kingsbury 1973):

1. Ask "Can this happen?" *Anticipate* that an event can occur.
2. Acknowledge that the event can happen; *recognize* the potential for a crime or loss.
3. Estimate the degree of risk associated with the recognized event; *appraise* the risk.
4. Take action to minimize the chances of the event occurring or reduce its severity; *initiate* protective measures.

Risk assessment examines number, frequency, and time of day of occurrence. It is applied to personal and property crimes to identify avenues of access and methods of committing the crime. It includes crimes committed by strangers and internal crime—fraud, theft, workplace violence, and so on. This underscores the importance of keeping company incident reports and periodically reviewing them as indicators of changing conditions. The results of this assessment are part of a physical security survey, which includes detailed, site-specific recommendations for reducing identified risks (see Chapter 16, The Physical Security Survey).

There are highly specialized types of risk assessments in the security field. One tests the vulnerability of information technology (IT) systems. Another examines access control hardware: alarms, access card systems, biometric readers, and closed-circuit video systems. Hardware assessment tests the function and response of the system. It does not analyze crime patterns.

LOSS PROBABILITY

After specific opportunities are identified, the likelihood for a crime or loss to occur must be assessed. The more opportunities for access to a person, object, or financial asset, the more likely access will be gained and a person injured or property stolen.

Loss probability is loosely grouped into five qualitative categories (Walsh 1986):

1. *Virtually certain*—If no changes are made, the loss will occur.
2. *Highly probable*—Likelihood of occurrence is much greater than that of nonoccurrence.
3. *Moderately probable*—The event is more likely to occur than to not occur.
4. *Improbable*—The event is less likely to occur than to not occur.
5. *Probability unknown*—Insufficient data is available for an evaluation. (Do not presume that probability is unknown only because facts have not yet been gathered.)

CRITICALITY OF LOSS

The impact of loss of personnel, materials, buildings and land, monetary assets, and information is the second aspect of risk assessment. The extent to which a company is affected has direct bearing on its continued financial health. *Criticality* is the level of severity; it is categorized in the same manner as probability (Walsh and Healy 1986):

1. *Fatal*—The loss would result in total abandonment or long-term discontinuance of the enterprise. Fire that destroys a building and its contents could be fatal to an organization.
2. *Very serious*—The loss would require major changes in investment policy and spending policies. Losing proprietary information is one example.
3. *Moderately serious*—The loss would have a noticeable impact on earnings and would require attention from senior executive management. Examples include losing key personnel or industrial facilities.
4. *Relatively unimportant*—The loss would be covered by existing monetary reserves. A small fire limited to a nonessential area and theft of office supplies are examples.
5. *Serious unknown*—This provisional rating should be replaced with a firm rating from one of the first four categories after sufficient data is collected.

There is a close relationship between probability and criticality. An improbable event can happen once with very serious or fatal consequences. Conversely, relatively unimportant events, such as theft of items that have low monetary value, can occur repeatedly over time with serious cumulative results. Criticality is not the sole factor when selecting security hardware and designing procedures.

MANAGING RISKS

Managing risks to limit exposure to loss requires:

- delineating problem areas precisely
- locating opportunities for crime or loss
- selecting appropriate countermeasures
- implementing and maintaining hardware and procedures

Opportunity to commit an act that results in a loss is site-specific; risk management methods are unique to every site and must be individually tailored to the needs of an organization, agency, or business. They must be practical, workable in a human context, and consistent within the framework of the organization's stated goals (Kingsbury 2003).

RISK MANAGEMENT METHODS

After the probability and criticality of a loss have been assessed, choose a method that most effectively reduces exposure to the specific identified risk (Small Business Administration 2008):

- Avoidance: Decide not to manufacture or sell a particular product; eliminate activities with high potential of harm, injury, or litigation. Avoiding a risk can have drawbacks, however. All human actions carry with them the potential for harm. If a product or service is necessary to the daily operation of a business, avoidance strategies may not be applicable.
- Reduction: Risks can be minimized to a level compatible with the daily operations of a business. Restrict the amount of cash on hand; store cash and negotiable instruments in a safe or vault. Choose a safer site; move objects from a dangerous or vulnerable location to one more easily controlled (a safe deposit box or appropriate storage building.) Install stronger physical security measures, such as impact-resistant glazing or door and window hardware. Implement procedural controls that restrict access to financial records or dangerous areas.

- Spreading: The risk is divided into smaller units to reduce the likelihood of one large loss. Store materials and records in several different locations; control access to each location. Use multiple backup systems for electrical power and data storage. Divide the duties of accounts receivable and accounts payable.
- Transference: The most common method of transferring risk is through the purchase of insurance. This shifts the risk to the insurance carrier. (Some insurance companies underwrite policies for crime loss.) Outsourcing also transfers risk. For example, a business that delivers items from its manufacturing facility might eliminate this service and hire an outside delivery firm. Another method of transferring risk is to pass on to customers the costs associated with internal and external theft.
- Acceptance: Many business owners accept (or retain) crime risks as part of doing business without examining the alternatives available to them. Not every risk can be reduced, spread, transferred, or avoided, but every effort should be made before the risk is accepted. The test for acceptability is the maximum dollar loss associated with the risk that a business is prepared to accept without reaching a "very serious" level on the criticality scale.

Losses due to crime and natural disaster can be reduced to acceptable levels by using more than one method: purchase crime insurance (transfer); implement procedural controls, such as entry logbooks and restricted access to sensitive areas where dangerous or valuable items are stored (spread); or install stronger door and window frames (reduce). Losses from floods, earthquakes, and storms can be controlled through installing hurricane shutters and impact-resistant glazing, moving assets and people to storm shelters, and installing on-site, stand-alone power generators. What is important for any risk reduction program is matching the method to the identified risk (Kingsbury 2006).

SUMMARY

There is no single method or particular security hardware system that protects people and property at a site. Layers of measures must be selected that meet the identified needs of a specific site within the framework of potential opportunities, probabilities, and level of criticality. The layers must be interrelated, interactive, and interdependent. This is not the same as fail-safe technology. The goal of a system of countermeasures that utilizes layers of security is to reduce opportunities for internal and external crimes to the lowest possible levels.

There are five primary reasons for losses (Kingsbury 2003):

1. Opportunity was not recognized.
2. Probability and criticality were improperly assessed.
3. Countermeasures were insufficient, improperly designed, or ineffective.
4. Implementation was partial or piecemeal.
5. There was a change in the level of vulnerability of an organization or business.

The examination of a site and the selection and application of physical security measures to reduce opportunity for loss, is discussed in Chapter 16.

REFERENCES

Broder, James F. 1984. *Risk analysis and the security survey.* Stoneham, MA: Butterworth.

Kingsbury, A. A. 1973. *Introduction to security and crime prevention surveys.* Springfield, IL: Charlcs C. Thomas.

Kingsbury, L. A. 2003. *Seguridad preventiva: métodos para prevenir y evitar robos.* Alexandria, VA: ASIS International.

Post, Richard S., and A. A. Kingsbury. 1977. *Security administration.* 3rd ed. Springfield, IL: Charles C. Thomas.

Small Business Administration. May 2008. *Small business insurance and risk management guide.* http://www.sba.gov/tools/resourcelibrary/publications/sev_pub_mplan.html.

Walsh, James, ed. 1986. *Protection of assets manual.* Santa Monica, CA: Merritt Publishing Company. (The *Protection of Assets Manual* is now edited and published by ASIS International, Alexandria, VA.

BIBLIOGRAPHY

ASIS International. General Guidelines Commission. *General security risk assessment.* Alexandria, VA: ASIS International, 2003.

ASIS International. *Protection of assets manual.* Alexandria, VA: ASIS International, 2005.

FEMA 426. Risk Management Series. *Reference manual: Providing protection to people and buildings.* Washington, DC: Federal Emergency Management Agency, 2003.

FEMA 429. *Insurance, finance, and regulation primer for terrorism risk management.* Washington, DC: Federal Emergency Management Agency, 2003.

Gitman, L. J., and C. McDaniel. *The future of business, millennium edition.* Cincinnati, OH: South-Western College Publishing, 2000.

Nickels, W. G., J. M. McHugh, and S. M. McHugh. *Understanding business.* 5th ed. Boston: Irwin McGraw-Hill, 1999.

Somerson, I. S. *The art and science of security risk assessment.* Washington, DC: ASIS International, 2009.

Vellani, K. H. *Strategic security management, a risk assessment guide for decision makers.* Woburn, MA: Elsevier/Butterworth-Heineman, 2007.

CHAPTER 16

16

The Physical Security Survey

Lauri A. Kingsbury

OBJECTIVES

The objectives of this chapter are as follows:

- understand the conceptual foundation of a security survey
- discuss risk evaluation and sources of data
- explain how to conduct a security survey, what to include in the report, and how to present findings and recommendations to the customer

DEFINITIONS AND FOUNDATION OF THE PHYSICAL SECURITY SURVEY

The purpose of security is to safeguard the assets of an organization—employees, physical facilities, inventory, equipment, and information—so that normal operations continue without interruption. To evaluate the effectiveness of physical security, practitioners conduct a *security survey*: "an on-site analysis and documentation of a location to determine the foreseeability of crime and the adequacy of the existing physical security in order to reduce or remove the physical opportunity for crime" (Kingsbury 2003). Security surveys are an extension and practical application of the definition of *crime prevention*: "the anticipation, recognition, and appraisal of a crime risk and the initiation of action to remove or reduce that risk" (Hudson 1963).

Loss prevention in the private sector closely resembles crime prevention in the public sector. The private sector examines the risk of loss to a corporation and initiates action to remove or reduce loss factors through the proper application of hardware and procedures.

Loss prevention practitioners who conduct security surveys are specialists. Managers and administrators do not have the time to acquire detailed knowledge of the hardware (locks, lights, glazing, alarms, access-control systems) and the techniques (policy and procedures) that reduce a company's vulnerability to criminal acts and internal losses. A written security survey provides management with a detailed analysis of the state of physical security at a specific site and a plan for improvement. It allows managers to make informed decisions about hardware and procedures related to loss prevention. It is the document that informs them of the identified risk factors and the methods for removing or reducing those risks to an acceptable level.

LAYERS OF PROTECTION

Controlling access is the primary means to reduce opportunity to commit internal and external crimes. Dishonest employees must have access to supplies, accounts, records and databases, cash, and negotiable instruments (checks, stock certificates) to steal from their employer. Thieves or vandals must gain entry to remove or deface property. Access control uses physical measures (locks, lights, alarm systems, glazing, walls, landscaping) and policy and procedures to reduce opportunity.

There are three ways to alter opportunity so that it can be removed or reduced to an acceptable level. Unauthorized personnel—whether invitees, employees, or intruders—can be *denied* access to people, property, and information. Their access can be *delayed*, and it can be *detected*. When all three methods are used at a site, in layers, the potential for loss or crime is greatly reduced.

Denying access is accomplished with physical impediments: landscaping, doors, and walls.

Procedural controls restrict and track entry into an area. Locks *delay* entry. If properly installed, the door and its lock may withstand attack long enough to discourage entry. *Detection* is the inner layer of access control. An alarm system signals that entry has been attempted or gained. (Card-access systems are an electronic version of metal keys. Neither the card nor the tracking software prevents entry.) Audit trails and visitor registers are procedural forms of detecting entry, but are only effective if used properly and consistently (Kingsbury 2003).

No organization should depend on only one method of access control. The more layers through which a potential offender (whether an employee or an outsider) must pass, the greater the probability that he will not attempt unauthorized entry. He is more likely to be discouraged by the difficulty of the attempt, or his entry will be detected. Installation of hardware and implementation of procedures at all three layers begins at the perimeter and is repeated throughout the site. The strategies of deny, delay, and detect should be present throughout a site.

APPLYING RISK ASSESSMENT TO PHYSICAL SECURITY SURVEYS

Security surveys are site-specific documents. Building design, function, and size, as well as corporate policy and procedures, can be duplicated throughout a chain business (the fast-food industry is one example). However, locations differ in demographic characteristics, geography, climate, economy, and crime rates. These factors influence recommendations made for physical security. Each site must be evaluated individually and recommendations formulated that meet the needs of that site. No two sites are exactly the same.

Risk analysis underlies the evaluation of a site's physical security. Security risk analysis requires:

- identification of potential risks or losses
- definition of the boundaries of these risks
- design or selection of countermeasures to reduce the potential for loss

Implementation of physical security employs this three-step process to achieve the goal of loss reduction and prevention.

SECURITY SURVEY STEPS

Every security survey, whether it examines 1,000 square feet of retail space, open parking areas, or hundreds of acres of a manufacturing or office complex, consists of six basic steps:

1. statistical analysis of losses and crimes that have occurred at the specific location, its immediate surroundings, and the community in general
2. on-site investigation and interviews with site users
3. collection of background and supplementary information
4. writing of the report
5. presentation of the formal report to management
6. follow-up meeting with management and a site examination

STATISTICAL ANALYSIS—IDENTIFYING POTENTIAL RISKS

Part of the identification process is a search for prior losses over the past two years. Internal documents that record losses, deficiencies, unusual occurrences, and so on, must be collected. Information necessary to identify potential risks comes from incident reports that describe external crimes (such as theft from or vandalism of cars in the parking lot) and internal crimes (such as missing inventory, embezzlement, workplace violence, and missing or stolen personal property of employees).

The surveyor must have a thorough knowledge of the patterns of criminal activity in the immediate area. This will enable the surveyor to determine what crimes may impact the client, who is likely to commit them, how likely they are to occur, and when they are likely to occur (National Crime Prevention Institute 1986). Identifying crime risks aids the surveyor in designing recommendations for a particular facility. A local police department or county sheriff's office can provide statistical information regarding local criminal activity.

The U.S. Department of Justice annually publishes information received from local law enforcement agencies. Federal sources provide information on reported criminal activity, victim profiles, and arrest and incarceration rates. This information is organized by geographic divisions (city, county, metropolitan area, state, nation), crime classification, offender characteristics, and victim demographics. It can be purchased as a book or CD from www.albany.edu/sourcebook. Data can also be downloaded from www.ojp.usdoj.gov/bjs.

Data can be compiled into the following categories to evaluate site-specific risk factors regarding the history of loss and criminal activity:

- target of attack: person or property?
- severity of loss: type of injury sustained or dollar value lost
- time of occurrence: time of day, day of week, daylight, darkness
- location: in an office, parking lot, loading dock, restroom, for example
- offender: outsider, visitor, employee, unknown

After these patterns are analyzed, measures can be designed to counteract undesired activities (Kingsbury 2003).

A statistical analysis of the performance of access-control systems and guard response can be gathered in a similar fashion. *The Design and Evaluation of Physical Protection Systems,* by Mary Lynn Garcia, provides a detailed method for this specialized analysis.

ON-SITE INVESTIGATION

An on-site investigation is an exhaustive examination of characteristics and details of physical layout, security hardware, procedures, and employee behavior. The surveyor examines the grounds and buildings for deficiencies and excesses, as well as for evidence that the owners or administrators are doing things correctly. This requires attention to detail, accurate recording of observations, and the ability to interpret how all these factors interrelate.

The site and its immediate surroundings are scrutinized to produce background information necessary to conduct interviews with the authorized (normal) users of a property. The surveyor must walk the site. Details essential to evaluation and recommendations cannot be seen from a car or an office window. It is important to take note of roadways, sidewalks, alleys, and all perimeters and their traffic patterns: volume, types, and frequency of traffic. Possible avenues of access and escape should be noted, as well as potential hiding places for people and stolen property and blind spots. A surveyor should walk slowly through each building to locate where tangible assets are and how employees use them. Assess the site in terms of deny, delay, and detect, and determine how hardware and procedures can be implemented to accomplish these strategies.

In addition to general characteristics of a site, a surveyor should record details about the type, construction, installation, position, and condition of the physical security hardware. This includes, but is not limited to:

- glazing materials—exterior and interior
- location and swing of doors
- location and type of exterior windows
- locks and card readers, including method of installation
- location and classification of safes and vaults
- components and performance of alarm systems
- components and field of view of camera systems

After the layout of a facility is understood, the surveyor should share findings with the owner or senior manager of the property. Physical layout and loss history provides the material necessary to conduct an in-depth interview. Emphasize to the property manager that her assistance is necessary to prepare a complete set of recommendations. Conducting an interview with the owner of a small business typically takes 90 minutes or less. Larger businesses may require more than one meeting. The owner or senior manager has knowledge of employee procedures, key control, alarm contractors, guard services, crime and loss activities (reported and unreported), and key personnel that yields information essential to a proper investigation and evaluation of a business. The surveyor should walk the site a second time with the owner or a senior administrator or manager. This gives the surveyor and the administrator an opportunity to discuss existing security conditions and what should be done to improve them.

Recommendations are not based solely on interviews with senior management. Informal discussions with the people who work at or visit a site yield valuable information about the ease or difficulty of daily tasks, constraints that current security measures place on performing their jobs, and other matters that may be unknown to management.

Make at least two physical inspections during hours of darkness to observe existing light levels and patterns of use. It is also important to visit the site during and after normal business hours to observe changes that may occur.

BACKGROUND AND SUPPLEMENTARY INFORMATION

Gather information that the owner and employees did not have or make available. Sources include, but are not limited to:

- alarm companies that installed the system(s)
- central monitoring stations for alarms
- law enforcement officers who patrol the area
- local zoning boards
- guard services at the business and adjoining businesses
- liquor control commissions (if applicable)
- state regulatory agencies for licensing (medical personnel, engineers, accountants, etc.)

These and other sources provide background information regarding socioeconomic factors that influence site use.

THE REPORT

Recommendations must be in writing so that an owner or senior manager has a document to refer to when budgeting for equipment or implementing a procedure. An administrator is unlikely to remember all the information a surveyor has discussed if the recommendations are described only during a meeting.

The document can be a checklist, a narrative report, or a hybrid. Each has its advantages and drawbacks. Checklists are convenient for the person who conducts the survey, but they do not adequately describe an individual facility, with its interplay of environmental and risk factors, nor does it supply the underlying reasons for the recommendations. An owner is more willing to invest time and money to improve existing conditions when she is given adequate explanations for doing so.

The report must be readable to be useful; it must be accurate and impartial in its observations and clear in its recommendations. The language must be clear so that the reader understands what to do and how to do it. Jargon should be avoided; technical terms should be defined so that a person unfamiliar with technical aspects of security hardware and procedures forms a clear picture of the recommendations and the reasons behind them.

The survey describes areas that require improvement, and it mentions areas where the current levels of protection are considered adequate. The value of implementation to the client should be clearly delineated. Security recommendations that correlate with the budget, site requirements, and business goals of an organization have a better chance of implementation and continued maintenance (Kingsbury 2003).

CONTENTS OF A PHYSICAL SECURITY SURVEY

The security survey report describes the site, its structures, business purposes, normal site users, and visitors. It lists the dates and times of inspection, including formal and informal interviews, and sources of all information. Many reports are divided by location into perimeter, exterior, and interior. Other logical divisions are by function: shipping and receiving, data processing, manufacturing, and research. The site and its functions dictate how the report is organized to easily convey recommendations. Whatever format is chosen, each section describes current conditions, then lists specific recommendations that will improve existing security.

Recommendations must be described in detail. It is not enough to describe existing lock hardware as adequate or insufficient. Record each observation in detail (e.g., a rim-mounted, single-cylinder, dead-bolt lock with a bolt throw of at least 1 in. (2.5 cm) into a reinforced wood frame 2 in. (5.1 cm) thick). Terms such as *remove, replace,* and *properly install* are found throughout the document. Details for screw length, wiring specifications, load capacity, angle of view, and type and thickness of glazing materials are necessary. Procedures require the same level of detail as hardware. They must be internally consistent with one another and not obstruct normal operations (Kingsbury 2003).

OUTLINE FOR A NARRATIVE SECURITY SURVEY

The following example provides an outline that was developed from practical experience writing hundreds of security surveys, as well as from a review of security literature. It is equally applicable to small and large businesses, nonprofit organizations, single structures, and sites with multiple buildings.

Example 1. Background Information

1. surveyor(s) names, titles, company
2. day(s), date(s), and time(s) of site inspections and interviews, including visits during hours of darkness and after normal business hours
3. each source of information, including, but not limited to: crime and incident data, persons interviews, database searches, reference books
4. detailed description of the site itself, including boundaries, size, zoning (residential, commercial, industrial, agricultural, etc.), names and addresses of property owners, purpose of the facility, hours of operation, number of employees
5. description of structures on the site including type of construction, year built, number of floors
6. description of surrounding area: population density, crime rate, income levels, housing types, zoning, natural boundaries, railroads, or limited-access highways

Site Description and Findings. Provide details for the following:

- perimeter
- landscaping
- parking lots, if applicable
- exterior of structure(s)
- lighting
- interior. If there is more than one room in a building, describe and make recommendations for each room based on its function, and for the building overall
- alarm systems
- closed-circuit camera systems
- card-access systems
- policy and procedures
- other functions specific to the site (research, shipping, manufacturing, etc.)
- general observations and conclusions in addition to the above listed specifics

Supplemental Information. This can be brochures, technical specification sheets, charts, tables, photographs, or drawings. A drawing or photograph does not stand on its own. You must include explanatory text.

Disclaimer. A disclaimer clearly states that there is no such thing as 100% security. The report is the professional opinion and recommendations of the surveyor, but selection, purchase, installation, implementation, and maintenance are the responsibility of the property owner or administrator.

Recommendations. The recommendations delineate one of three specific courses of action for hardware and procedures: maintain existing conditions in good repair, upgrade or improve existing conditions to meet current and projected needs, or install specified hardware/implement specified procedures. If the site examination/analysis finds an organization doing something right, the document should state this. There can be confusion about what can and cannot be done, and what is or is not effective. Organizations are more likely to adopt additional measures when told that the process of safeguarding the property has already begun (Kingsbury 2003).

Depending on the nature of the business, risk factors, and other variables, recommendations encompass one or more of the following areas:

- theft—internal, external
- information crimes, including theft of proprietary information and other data
- fraud, including embezzlement

- shipping and receiving
- burglary (breaking into a structure with the intent to permanently remove property)
- malicious destruction of property (vandalism), including data destruction or damage
- shoplifting, if applicable
- robbery (taking property from a person by force or threat of force)
- workplace violence

The surveyor develops and selects from a set of alternatives. When making recommendations,

> it is important to illustrate each recommendation in such a manner that only one conclusion may be drawn from each recommendation. Therefore, state one specific problem and identify one solution, unless the application of two or more security processes will be needed to modify or reduce the crime risk (Kingsbury 1973).

Follow-up. The physical security survey can reveal strengths and weaknesses of existing security. The surveyor must carefully inspect the site and develop a plan for layered security measures. Schedule a follow-up review for 6 to 12 months after submitting the report to the site owner. Examine every area of the site and note changes in conditions, including which recommendations are pending or have been implemented, and determine what, if any, difficulties prevented implementation.

SUMMARY

Physical surveys combine the investigative skills of a detective with the specialized knowledge of a physical security practitioner. They focus on a specific site in an effort to identify and, if necessary, improve existing security measures. Surveys are future-oriented. The detailed analysis and recommendations can lead to implementing changes before an organization suffers crippling losses.

The document becomes a tool to actively involve a business and its employees in the security of a facility. Active involvement in security decisions returns control of their environment to owners and to employees at all levels of the organization. In turn, this sense of control fosters commitment to the long-term security goals of a corporation (Kingsbury 2003).

REFERENCES

Hudson, Fred, MBE. 1963. Interview by Arthur A. Kingsbury. London: Home Office.

Kingsbury, A. A. 1973. *Introduction to security and crime prevention surveys*. Springfield, IL: Charles C. Thomas.

Kingsbury, L. A. 2003. *Seguridad preventiva, métodos para prevenir y evitar robos*. Alexandria, VA: ASIS International.

National Crime Prevention Institute. 1986. *Understanding crime prevention*. Newton, MA: Butterworth-Heinemann.

BIBLIOGRAPHY

Air Force Engineering and Services Center. *Protective Construction Design Manual*, ESL-TR-87-57. Tyndall Air Force Base, FL: Engineering and Services Laboratory, 1989.

American National Standards Institute: www.ansi.org

American Society for Testing and Materials. *Draft Standard Guide for Determining Design Criteria for the Physical Protection of a Facility*. Philadelphia, PA: ASTM, August 1996.

Atlas, Randall. *21st Century Security and CPTED*. Boca Raton, FL: CRC Press, 2008.

Checklists and hardware recommendations—Crime prevention units in state and local police and sheriff agencies.

Crowe, T. D. National Crime Prevention Institute. *Crime Prevention through Environmental Design, Applications of Architectural Design and Space Management Concepts*. 2nd ed. Woburn, MA: Butterworth-Heinemann, 2000.

Demkin, J., ed. *Security Planning and Design: A Guide for Architects and Building Design Professionals*. Washington, DC: American Institute of Architects. (undated)

FEMA 426. Risk Management Series. *Reference Manual: Providing Protection to People and Buildings*. Washington, DC: Federal Emergency Management Agency, 2003.

FEMA 427. *Primer for Design of Commercial Buildings to Mitigate Terrorist*

Attacks. Washington, DC: Federal Emergency Management Agency, 2003.

FEMA 429. *Insurance, Finance, and Regulation Primer for Terrorism Risk Management*. Washington, DC: Federal Emergency Management Agency, 2003.

FEMA E155. *Building Design for Homeland Security*. Washington, DC: Federal Emergency Management Agency, 2006.

Floyd, W. R. *Security Surveys, Guidelines for Evaluating Security*. Alexandria, VA: ASIS International, 2008.

Garcia, Mary Lynn. *The Design and Evaluation of Physical Protection Systems*, 2nd ed. Woburn, MA: Butterworth-Heinemann, 2007.

Iavarone, R. C. *1996 Dodge Unit Cost Data*. New York: McGraw-Hill, 1996.

Interagency Security Committee. *Security Design Criteria for New Federal Office Buildings and Modernization Projects*. Washington, DC: National Academies Press, 2003.

NFPA 730. *Guide for Premises Security*, 2006 ed. Quincy, MA: National Fire Protection Association, 2006.

NFPA 731. *Standard for the Installation of Electronic Premises Security Systems*, 2006 ed. Quincy, MA: National Fire Protection Association.

NIOSH. *Guidance for Protecting Building Environments from Airborne Chemical, Biological, or Radiological Attacks*. Washington, DC: Department of Health and Human Services (DHHS), Centers for Disease Control and Prevention, National Institute for Occupational Safety and Health, May 2002.

Post, R. S., and Arthur A. Kingsbury. *Security administration: An introduction*, 3rd ed. Springfield, IL: Charles C. Thomas, 1977.

Rea, M. S., ed. *Lighting Handbook Reference and Application*, 8th ed. New York: Illuminating Engineering Society of North America, 1971.

Sandia National Laboratories. Risk Assessment Methodology, http://www.sandia.gov/ram

Unified Facilities Criteria. Washington, DC: Department of Defense Minimum Antiterrorism Standards for Buildings. October 8, 2003.

U.S. Army Corps of Engineers. *Security Engineering Manual.* Missouri River, Omaha District, January 1990.

U.S. Department of Justice. *Protective Intelligence Threat Assessment Investigations*. Washington, DC: Office of Justice Programs, July 1998.

U.S. Department of Justice. *State Domestic Preparedness Equipment Program, Assessment and Strategy Development Toolkit.* Washington, DC: May 1999.

Vellani, K. H. *Strategic Security Management: A Risk Assessment Guide for Decision Makers.* Woburn, MA: Elsevier/Butterworth-Heineman, 2007.

17

Crime Prevention through Environmental Design

Mark C. Bach

CHAPTER 17

OBJECTIVES

The objectives of this chapter are as follows:

- explore the four basic tenets of crime prevention through environmental design (CPTED)
- understand how this concept has developed and evolved
- determine the three major factors in a site review process
- consider how the crime triangle can be useful in CPTED discussions
- review the four types of physical space and their uses
- consider examples of CPTED in action
- explore the future uses of CPTED

INTRODUCTION

Although crime prevention through environmental design (CPTED) is a relatively new concept, it has actually been in use for centuries. In Europe during the Middle Ages, moats and castles were used to defend key buildings. In the western United States, when Native Americans built dwellings on the sides of cliffs, they purposely designed entryways with a low overhead door lintel so intruders would have to stoop to enter and be vulnerable to defenders within.

Sometimes design elements used simply to enhance a site inadvertently function to prevent crime. A home might have a concrete pathway leading from the street to the front door. The purpose might have been to prevent the turf from wearing out, but this concrete walk clearly defines how a stranger should approach the residence. When someone does not follow this path, it raises suspicions and heightens surveillance. While a low white-picket fence along the side of a front yard might do little to physically stop a trespasser, it psychologically helps mark the territory and highlights a transgressor.

CPTED is now accepted as a legitimate crime prevention process.

> [O]fficers need to have a firm understanding of the contributions that building space, architectural design, lighting, and other such features of the environment have to criminality, especially vandalism, burglary, robbery, rape, and crimes involving automobiles. All too often, after a building is constructed and numerous crimes are committed on the premises, the owners must then address lighting and design problems or make the building and its surroundings safer by installing fences, gates, access control, closed-circuit television, and alarm systems (Peak and Glensor 1996, 105).

DEFINITION OF CPTED

In his 1977 book of the same name, C. Ray Jeffery coined the phrase *crime prevention through environmental design* to group the then-current efforts to use modern design theories and studies of criminal behavior to deter crime. He defined the phrase as "the proper design and effective use of the built environment that can lead to a reduction in the fear and incidence of crime, and an improvement in the quality of life."

HISTORY OF CPTED

While many people have had some impact on and contributed to the development of CPTED, six authors have consistently been acknowledged for their efforts. Through their writings and physical efforts, these people helped to advance and publicize today's current CPTED applications.

Jane Jacobs. Jacobs believed that the city designs of the post–World War II era were not working and sought ways to minimize crime in cities and revitalize inner cities. She felt "(i)n cities, liveliness and variety attract more liveliness; deadness and monotony repel life. And this is a principle vital not only to the ways cities behave socially, but also to the ways they behave economically" (Jacobs 1961, 99). She posited that streets of successful city neighborhoods required three items: a clear delineation between public and private spaces, encouragement of natural surveillance, and continual use of the neighborhood's sidewalks and streets (35). Jacobs recognized that the design of a city's neighborhood could have huge, long-lasting social impacts.

C. Ray Jeffery. Jeffery coined the term "cep-ted" (CPTED) and helped promote it as a process to prevent crimes from occurring in the first place. It has become accepted by architects and criminologists as a valid and successful crime prevention technique. Jeffery helped establish and publicize the original three tenets but was quick to note the contributions from others. He noted the unique distinctions between organized, mechanical, and natural methods for both surveillance and access control, noting "(t)he way we design our apartment buildings and housing complexes has everything to do with the crime pattern of the city" (Jeffery 1977, 193).

Oscar Newman. Newman was an architect involved in trying to make public housing a safe place for residents. In Dayton, Ohio, he focused his efforts in an area called Five Oaks, a deteriorating area between the old downtown and the booming new suburbs. His work helped reinforce the importance of territoriality and surveillance in creating safe environments. He created areas that, by their nature and design, the occupants could naturally defend. Newman coined the phrase *defensible space* as he designed these mini-neighborhoods. By dividing larger areas into smaller spaces, Newman felt he was helping the public to better distinguish strangers from residents. Newman wrote "(t)he larger the number of people who share a communal space, the more difficult it is for people to identify it as theirs or to feel they have a right to control or determine the activity taking place within it" (1996, 17).

James Wilson and George Kelling. In two short articles in *The Atlantic Monthly* in 1982, James Wilson and George Kelling helped revolutionize policing in America, ushering in the concept of community-based policing and the belief that public authorities needed to focus on basic, simple quality-of-life issues in their population. A major focus of these two articles, often collectively referred to as *Broken Windows*, was how society needed to concentrate its efforts on the first indications of a breakdown in a neighborhood, before the situation spiraled out of control and became hopeless.

The authors suggested that once a window was broken in a building and others saw it was not repaired, it became easier to break the remaining windows. When a building or neighborhood is not maintained, it becomes easier for other social controls to break down.

The same effect can be seen across the country with graffiti. If residents or business owners promptly cover up graffiti, at some point, artists go elsewhere. But if their markings stay up longer for others to see, this encourages more defacing at the site.

Kelling later wrote, "by enforcing rules, in the subway or on the street, (this) reduces the criminogenic conditions—the disorder and chaos—that both stimulate and provide cover for those dedicated to serious crime" (Kelling and Coles 1996, 248).

Timothy Crowe. Crowe focused on crime prevention methods and was instrumental in training a new generation of practitioners in CPTED concepts. His 1991 book helped place a wealth of observations and suggestions in the hands of individuals who recognized that preventing crime was much more helpful to society than solving crime after the fact. Crowe felt "(t)he design and management of parking lots, storefronts, parks, schools, and just about everything in your community have important connections to the problems of crime and the fear of crime" (1991, 27). Crowe recognized the value of and need to establish ongoing maintenance and activity management of a design to ensure that the space stayed safe and useful. He also grasped that the fear of crime was equally important, if not more so, than actual crime itself. If he could assist in promoting spaces that not only were safe but actually felt safe, the public would benefit.

GENERAL PRINCIPLES AND APPLICATIONS

CPTED focuses on using the following four major tenets as design elements to reduce the incidence of crime or unsafe behaviors. (Note that some actual applications and uses overlap.)

NATURAL SURVEILLANCE

It is critical that spaces allow for easy, continuous natural surveillance by all parties. If a window in a store offers a view of a commercial parking lot, many potential subjects—patrons and employees alike—can view this lot and any attempted crimes. An apartment dweller might be able to look out from her unit and see her children playing on some adjacent playground equipment. Criminals do not like to be seen, and the more opportunities a site presents for them to be observed, the more likely the criminals will go to less visible spots.

Surveillance can also include organized surveillance, such as police and private security making purposeful, visible patrols on scheduled routes or mechanical surveillance through closed-circuit television (CCTV) or enhanced lighting measures. Natural surveillance is more focused on intrinsic ways to enhance opportunities to see potential criminals.

Shops often place cashiers near entranceways. The cashiers serve as greeters as people come in and as gatekeepers as people leave. Just a simple welcoming greeting from that cashier can serve notice that a visitor will be watched and observed. Contrast that approach with a cashier stashed away at the back of the store. Now a visitor could wander through the front of the store without any contact or observation. Even after making a small purchase while at the back of the store, he has ample opportunity to gather other merchandise as he proceeds to the front of the store. Also, consider which cashier location is more likely to be seen by outsiders during a robbery attempt.

A home can be designed so that the front entrance is in constant view from the street, allowing neighbors and passing drivers to see and report suspicious actions. By eliminating tall shrubbery near windows, occupants are allowed to keep an eye on their exterior surroundings and passersby can see offenders trying to break in through these same windows. If a home's family room or kitchen (major focal points of most households) overlooks the front yard with strategic placement of windows, it is less likely that strangers can lurk nearby with impunity.

Natural surveillance is encouraged by strategic placement of windows to allow both internal and external views. A second-floor cafeteria in a large two-story furniture store/warehouse in Tempe, Arizona, was built with expansive glass windows looking out onto the parking lot. This design allows a large group of diners to casually observe not only their own vehicle but all the vehicles in the lot and report suspicious activity.

TERRITORIAL REINFORCEMENT

Allowing a property to project some territoriality or ownership gives notice to others that this is someone else's property and should be respected. Plus, if someone makes the effort to initially show ownership in her property, she is more likely to continue maintain those signs. Territoriality can be shown by using a similar color scheme for a set of apartments or through signage. A landscaped hedge can distinguish one property line from another, or different landscaping schemes can make the same distinction.

Businesses often use the same colors and signage to create continuity from one store to another. In fact, the interior floor plans and design often mimic each other so that customers instantly recognize that they are at "their" store.

Many of these efforts can be quite subtle yet distinctive. A store can use different flooring surfaces to establish intended major pathways and help direct customers to their desired destinations. As they leave the shop, a variety of surfaces can show them where the sidewalks end and traffic ways for cars begin (Figure 17–1).

Neighborhood residents can place markers at entrances to distinguish their specific section of town. In fact, some cities reinforce these efforts by using distinctive street signs to help label different neighborhoods. Likewise, cities often place entryway signs on major arterial streets, to tell visitors that they are now entering that city.

Figure 17–1. The exterior surfaces of this lot provide subtle clues to pedestrians and drivers as they approach the building.

NATURAL ACCESS CONTROL

> Before I built a wall I'd ask to know / what I was walling in or walling out.
> —Robert Frost (Lathem 1969, 34)

Access control involves a variety of elements that help to psychologically steer individuals to desired areas and limit the opportunity to commit crimes. Properly designed entrances and exits can direct people into and out of a property. Signage can be developed to reinforce these desired traffic flows. Lighting and fencing can be used to highlight the desired entrances. When first designing a property, conflicting shared uses—such as pedestrians and vehicles using the same entrances—should be avoided. Having a pedestrian pathway lead to a dead end should also be avoided, as well as elevated walkways that limit a person's escape paths.

Not everything has to be fenced or roped off. Landscaping, surface treatments, colors, and signage can all be used to indicate desired boundaries. Many cashier stands have a low-level wall or counter surrounding them. This wouldn't stop a determined robber but does provide notice to the general public that this is not a place for them to be in. A low hedge along a homeowner's property line will not stop a trespasser, but it does serve to deter the typical passerby.

Henry Cisneros, former secretary of the U.S. Department of Housing and Urban Development, noted:

> Outside spaces become more defensible if they are clearly demarcated (by fences, shrubbery, and so forth) for use by one household or a small number of households, and if they are easily observable by residents, neighbors, and passersby. It is hard to tell who is, and who is not, an intruder if too many people are entitled to use an outdoor space and no one feels responsible for its legitimate use. Defensibility is also helped by good lighting around possible means of entrance; removal of visual barriers such as high, solid fences and shrubs that create hiding places; and windows with good views of the space in question (Cisneros 1996, 17).

A shopping mall in downtown San Diego had a large grassy expanse along one of the main bordering arterial streets that served as an entryway into the site. Regrettably, this turf also became a gathering point for many homeless people and caused concern to shoppers and passersby, especially at night. Finally, the grass was replaced with a variety of mounding plants, some with thorns, to discourage any attempts to lie down in the vegetation. In addition, posts and chains were placed along the border to discourage entry into the garden. As an added bonus, the swooping set of chains did not offer a stable surface to rest or sit on. The homeless population dispersed to other more hospitable surroundings, and the tourists and customers reclaimed this space (see Figure 17–2).

Figure 17–2. Revised landscaping and some boundaries can discourage undesired loitering in an urban lawn area.

When discussing access control, instead of looking at natural designs, many people focus on traditional target-hardening methods, with more locks, gates, and alarms. These methods attempt to deny access to areas by creating physical devices or barriers. Target hardening uses either organized methods, such as more guards and fewer entrances, or mechanical methods such as locks, gates, and fences. Natural access control is focused more on spatial definitions and subtle, often subconscious, methods of controlling the public's access.

ACTIVITY MANAGEMENT/MAINTENANCE

Although it is not one of Jeffery's three original principles of CPTED, activity management/maintenance is advocated by many current CPTED proponents as an active guide for designers. Proper maintenance allows for the continuous use of space as originally planned and intended.

The "Broken Windows" articles emphasized the need for neighborhoods and property owners to actively maintain their property and repair the minor things affecting quality-of-life issues before major degradations occur. It does no good to install an expensive lock system throughout a property if there is no continuing maintenance to ensure these locks are used and functioning. Likewise, installing a CCTV system and then failing to staff sufficient personnel to monitor the cameras might be pointless.

If a landscape architect designs a splendid and lush garden, but the management/owner has no funding or intent to maintain such an effort, it will go to waste and become, at best, an eyesore of untended plants.

Activity management/maintenance refers to actively providing the resources and activities to perpetuate an intended lawful use. If a public park has a basketball court and the local government wishes to promote its active, positive use, the court striping needs to be maintained, the hoops and nets must remain functional, and the installation of lighting to promote additional hours of appropriate use should be considered. Tournaments could be held there to promote its location within the neighborhood, and signage should be in place to direct outsiders to the site. Within an apartment complex, a laundry room needs to be kept clean, with functional machines and sufficient space for the residents, or it will not be used for its intended purpose. Instead, it will be used for other unsupported and undefined purposes. In a shopping mall, if an area is designated a lounging and meeting spot, management needs to provide some seating space and good lighting to support those activities.

SITE REVIEW—ASSESSMENT

Whether in the initial architectural design stage, as a property is in use, or when it is undergoing renovations or modifications, it is critical to consider the three Ds of each unique space: designation, definition, and design. If there is a conflict between them, it will create confusion, chaos, and possibly crime. A site review helps to make sure that all steps are being taken to support the planned designated use. A site review will also help identify the risks inherent at the site and what steps have been taken to minimize them.

All space must have some obvious designation as it is first being designed and used. In a public park, space is set aside for parking, waste cans, and perhaps even an outdoor volleyball court. Physically, spaces are set aside for each of these uses or designations. A home also has specific set spaces designated for obvious, specific uses.

Each designated area must then have some definitions to show users these intended uses. A parking lot needs some obvious physical boundaries to prevent vehicles from parking on the well-tended grounds. A sand volleyball court needs the official boundaries of the court shown, as well as additional space for the players and observers. Signage might help further define the space and explain its purpose and expected use. Where waste should be deposited must be planned for and defined. Barrels or carts need signage, especially if recycling is part of the designation, and their visibility, stability, and accessibility must be ensured.

Finally, the actual design must be correct for the intended uses. A parking lot must be designed for heavy cars, frequent use, and weather extremes and generally must be flat and level, with sufficient entrances and exits. The trash cans must be near the users of a park, large enough to accommodate expected usage, and yet not noxiously close to people sitting on a park bench. A volleyball court must be sized for expected regulation use, have barriers to keep the sand inside the court area, and be far enough from other barriers to allow for errant shots without endangering other players or bystanders.

During a site review, consider these questions:

- What is the designated purpose of the site?
- What was the planner's original intention for the space?
- Has the original designation changed and who changed it?
- How well does the space support the current/intended uses?
- How is this space defined?
- Is it clear to all for whom the space is designed?
- Have necessary rules or regulations been posted to seek voluntary compliance?
- What are the borders for the space and how are they maintained?
- Does the current design support the current space utilization?
- Are there conflicts between the current designation, definition, and design?

THE CRIME TRIANGLE

It is helpful to use the crime triangle concept to focus our crime prevention efforts and decisions. The crime triangle, as introduced in Chapter 15 shows that in order to have a crime, three parts of the triangle must be in place: You must have a victim or target of the crime, an offender with the desire to commit the crime, and the opportunity/site to commit the crime. Without all three parts, no crime will occur. Refer to Figure 15–1 in Chapter 15. The CPTED philosophy allows us to actively work on all three parts. In a public parking lot, you can create lighting, cameras, and sight lines to promote surveillance and remove hidden areas that allow criminals to break into vehicles without detection. At the same time, you can remind potential victims to be aware of their surroundings, remove their valuables from sight, and lock their vehicles. Security efforts can become visible and obvious to discourage offenders from considering the area as a target.

Three groups can have an impact on the three parts of the crime triangle. The offender has handlers including peers, family, law enforcement personnel, and even parole/probation officers. The victim has guardians who help to look out for her, including family members, public service announcements, and prevention advocates. The opportunity/site has managers (owners, agents, and security officers) overseeing the proper use of the area, making sure that the area and its features are maintained and in use as designed.

TYPES OF SPACE

No matter where you go or what you do, any space can be defined as one of four possible types: *public*, *semipublic*, *semiprivate*, or *private*. How these spaces are designed and used helps determine how safe and secure they are.

As Figure 17–3 shows, a good design will allow for obviously separate, well-defined distinctions between the four categories. It becomes clear by use and demarcations where the public may go within these spaces. The closer a stranger approaches to a private space, the more surveillance he is subject to, and the more territoriality he should experience.

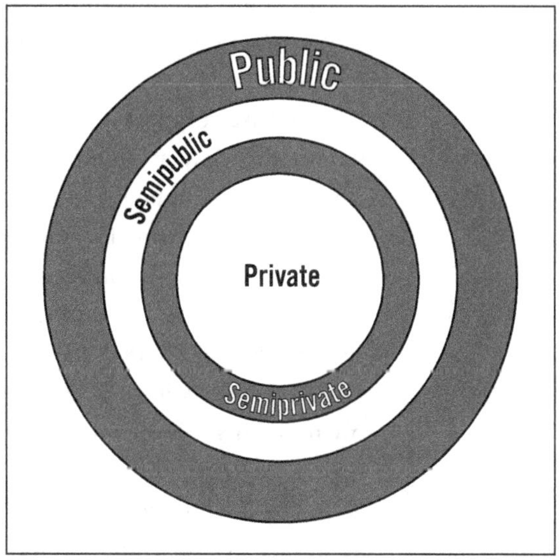

Figure 17–3. The four types of physical space help define appropriate behavior and defensible strategies.

On a residential street, the street and the adjacent sidewalks are generally public space open to all. If someone were to turn and walk up a sidewalk to a residence's front porch, she would be in a semipublic space. Without fencing or signage, the general public could be expected to be in this area, but generally only with some lawful purpose. A person on the front porch now has entered semiprivate space. The homeowner rightfully feels the space is predominantly his and a stranger must have a legitimate reason to be there. Of course, once somebody crosses the threshold and enters the house, she is in a private space and must be there by invitation.

The same principle applies to a local hardware store. The public is certainly welcomed to drive by and even encouraged to look at the display windows, as the exterior is public space. As a patron enters the store, he enters a semipublic space, welcomed to browse and shop during store hours. As he approaches a sectioned-off cashier area, he enters a semiprivate space where outsiders are not expected to enter or loiter. And of course, a locked display counter and an employee-only work area are private spaces where the general public is not expected to be.

When crimes occur, a review of the site of the crime and how it is defined often exposes conflicts between the intended use of the space and its present use. By reemphasizing the four basic space designations, owners can reclaim their space and take back their site's use.

THE FUTURE OF CPTED

As the CPTED concepts and success stories spread, CPTED will become more familiar to others and more widespread. The following three areas can continue to be impacted by CPTED.

REGULATIONS, CODES, AND LEGISLATIVE MANDATES

Through the use of building codes, municipal and state ordinances, and even federal regulations, it is fair to expect many CPTED concepts to be incorporated into regulatory compliance. As CPTED becomes more accepted and recognizable, it follows that end users and lawmakers might want to codify this acceptance. Already many cities have adopted CPTED guidelines, either specifically or indirectly, inserting them into revised building codes or design development requirements.

Policymakers recognize that it is far easier to mandate compliance for all than to seek voluntary compliance by some and later have to grandfather in exceptions. Since most buildings need to be identified with a visible numerical address, it does not take much effort to prescribe minimum size to that address or specific placement on a new building. Such a requirement not only emphasizes territoriality but makes a public safety response easier.

Rental codes or commercial business statutes will dictate the types of locks on entry doors, require the presence of fire and smoke alarms, and mandate the rekeying of locks with a change of tenants.

Because at some point a manager or small group of employees must leave or enter a business alone, often late at night, employee exits become an origin of crime and a point of oversight. The placement of employee service (or exit) doors, the existence of viewing panels in the doors, and the requirements for adequate lighting all can be codified to increase employee safety.

OSHA requires employers to provide a safe work environment, and many employer safety programs provide for employee training on CPTED principles to enlist their support in keeping their workplaces safe and crime-free.

Businesses with locations in multiple cities or states must pay particular attention to complying with variations in design requirements and may want to consider adopting the strictest regulation throughout all their sites. After all, if a viewing panel in a door is required in one state but is optional in another, the business runs a great risk and liability in not voluntarily accepting that stricter regulation in all its sites. If a tragedy occurs and it becomes known that the company provided a safer design in some stores but not in others, the negative exposure can be detrimental to business.

SITE AUDITS

Audits are commonly required by insurance companies or risk management directors within a large organization to ensure that common, everyday efforts are being made to safeguard not only the patrons of a business but its employees and, ultimately, its assets. A yearly audit may review the status of alarm systems, who knows the codes, and when their effectiveness was last tested. An audit may review the actual maintenance of the building and ensure that the lighting, signage, and landscaping are being sufficiently maintained to faithfully follow the CPTED concepts.

Rather than avoid an audit, a company should embrace this opportunity for an objective outsider's view of what is being done right and what might need improvement.

SECURITY PLANS

Through city ordinances, many types of businesses already are required to provide a detailed written plan establishing their site security plan. As these plans become more effective and publicized, it is reasonable to expect more cities to require them and to apply the standards to more business types. A security plan frequently provides details on staffing, entrances and exits, uses of the facility, and emergency policies.

As an example, a liquor-serving nightclub might need to indicate how many private security staffers are on site, how many employees are at the entrances to check identification of patrons, what areas are subject to CCTV surveillance, and from what parts of the building employees leave after the business closes.

A convenience store staying open 24 hours might need to dictate the minimum number of employees present, whether its CCTV system is remotely monitored, and how much window space must remain clear of signage to allow for natural surveillance of the interior.

Security plans frequently rely on the CPTED concepts to ensure public safety and minimize the need for police responses. The incorporation of CPTED principles helps a business focus its efforts, recognize design flaws in a building, and take steps to enhance the safety of all the users of the site.

SUMMARY

Crime prevention through environmental design (CPTED) concepts can form the framework for a comprehensive analysis of a site. CPTED has borrowed concepts and teachings from many disciplines and has been practiced, perhaps subconsciously, for centuries. It does not necessarily involve spending large sums of money or exerting great effort in the initial design of a building, and renovations frequently can be done with minimal additional effort. You can look at natural access control, territoriality, natural surveillance, and maintenance to verify how well a space is being used to design out crime and unwanted activities. CPTED will become more known in security avenues and will become the basis for the designs of future buildings or renovations.

REFERENCES

Cisneros, Henry. 1996. Defensible space: deterring crime and building community. *Cityscape: A Journal of Policy Development and Research*. Washington, DC: U.S. Department of Housing and Urban Development, Special December Issue.

Crowe, Timothy D. 1991. *Crime prevention through environmental design: applications of architectural design and space management concepts*. Stoneham, MA: Butterworth-Heinemann.

Jacobs, Jane. 1961. *The death and life of great American cities*. New York: Random House.

Jeffery, C. Ray. 1977. *Crime prevention through environmental design*. Beverly Hills, CA: Sage Publications.

Kelling, George L., and Catherine M. Coles. 1996. *Fixing broken windows*. New York: Martin Kessler Books.

Lathem, Edward Connery, ed. 1969. *The poetry of Robert Frost*. New York: Holt Reinhart and Winston.

Newman, Oscar. 1996. *Creating defensible space*. Washington, DC: U.S. Department of Housing and Urban Development.

Peak, Kenneth J., and Ronald W. Glensor. 1996. *Community policing and problem solving: strategies and practices*. Upper Saddle River, NJ: Prentice Hall.

BIBLIOGRAPHY

Brennan, Dean, and Al Zelinka. *SafeScape*. New York: APA Planners Press, 2001.

Schneider, Richard H., and Ted Kitchen. *Crime Prevention and the Built Environment*. New

York: Routledge, 2007.

Wekerle, Gerda R., and Carolyn Whitzman. *Safe Cities: Guidelines for Planning, Design and Management.* New York: Van Nostrand Reinhold, 1995.

Wilson, James Q., and George Kelling. "Broken Windows." *Atlantic Monthly,* March 1982.

18

Secure Facility Design

Kenneth G. Fauth, EdD, CPP

CHAPTER 18

OBJECTIVE

Security issues involved with facility design begin with its intended use, the land the facility will stand on or currently stands on, as well as what the facility will contain and whether its contents or occupants need to be protected. The objective of this chapter is to identify critical issues and areas that must be considered for a successful secure facility design.

INTRODUCTION

Understanding the purpose of the facility and its intended use are fundamental to developing a security design. For instance, a prison, an office building, and a manufacturing facility all have different requirements but have common elements that must be considered. First, consider the purpose or use of the facility. Second, understand what layout will give the facility the best protection from adjacent areas yet allow it to accomplish its objectives with ease. An analysis of the approaches to the facility for the public, employees, and commercial delivery must occur in order to accommodate the functional needs of the business. In general, since the terrorist attack on September 11 on the World Trade Center and Pentagon, security in general, and approaches to property in particular, has taken on a heightened role.

Any special requirements of business (hazardous materials, proprietary processes) must be considered. Traditional areas for visitors, employees, shipping and receiving, parking, and staging must be reviewed for inclusion in the security plan. Finally, the structure itself will impact security design. Is it a one-story, multistory, multibuilding, single-tenant, or multitenant facility? Is it an occupied building or is it used for storage? Clear design goals, access to the facility, and consistency of results are essential. Every facility has its own set of unique issues as well as those that are common to all properties.

Businesses, government complexes, university campuses, hospitals, airports, and many large commercial and industrial sites are multifunctional enterprises with many of the same services found in a city. Health clubs are active early and late. Dry cleaners and coffee shops are busy in the morning. Some sites have different specialty restaurants or retail stores in each segment.

How should security issues and goals be integrated into the fabric of a site so they are effective, as nonintrusive as possible, and of reasonable cost? The answer is the security investment must address realistic threats, meet them in a reasonable manner, and add value. The operation and function of the site, the threats facing it, its vulnerabilities, and the character of the organization all combine to determine the security design.

Site security designs must take into account the operation and function of the facility—whether it is an embassy or government building, a research lab, a general office building, a school, or a heavy or light manufacturing site. Operations and functions provide insight into the threats and vulnerabilities that need to be addressed. Realistic threats and vulnerabilities should be considered; sometimes the probability of a particular event occurring is so minimal (or there are no safeguards available to counter it) that realistic countermeasures are not appropriate. Finally, the character of the organization also plays a role.

This chapter creates an overview of the issues that should be addressed. Depending on the previously mentioned considerations, some protective measures will be appropriate while others will not. Some are relatively specific to one environment. This overview provides a starting point for the security design for any facility in the United States and some other relatively benign environments; however, enhancements will be needed for high-threat areas.

DESIGN GOALS

Determining the design goals for the security plan is the single most important task. Goals form the benchmark by which success is measured. In determining these goals, two themes seem to be most prevalent. First, security must provide for the authorized activities of the facility, complex, or enterprise to occur. Second, and a long-standing issue within the profession, is facilities should be designed in such a manner as to deter criminal activities or damage, detect those activities if they occur, delay the unauthorized intrusion or event, and, finally, cause a response to the event to occur. Addressing these two themes will help to formulate the design criteria necessary for a safe and secure facility.

Design goals need to take into account the principles of crime prevention through environmental design (CPTED). Incorporating these principles into any security design will enhance the overall protection level for a particular facility. In many locations throughout the United States, compliance with CPTED principles is a requirement of local building codes for commercial buildings. See Chapter 17 for a more in-depth discussion on CPTED.

Design goals should also derive from the particular activity that takes place in the facility. A production plant has very different goals than an office facility. Further, the proper training and awareness of employees impacts the way all security measures are viewed and therefore whether they are enhanced or undermined.

Security goals are achieved most effectively through a combination of management policies and procedures, CPTED applications, the physical environment, and physical security tools (barriers, locks, cameras, access controls, etc.). The mix of these will vary according to goals. Two examples of security goals may be securing high-value assets and securing the data in file servers connected to the Internet. The high-value assets would need to be physically secured, equipped with an alarm system and monitored, and a physical audit trail of accesses would have to be maintained. There is heavy emphasis here on physical security tools. The data in the file servers would best be protected by electronic security software, secure procedures, and the physical design of the data center where the computer hardware resides. The design goals presented within each of the following sections involve making decisions before completing the plans.

AUTHORIZED ACTIVITIES

An enterprise must define proper actions and activities about who can do what and when, who can take items or not, who should operate equipment, and so on. Once these parameters have been identified, the security professional can begin planning and establishing criteria that will meet these needs and accomplish the goal. Typically, there are various types of "authorized" activities. They are usually defined by an individual's job description and specific function within that classification as well as how the business operates overall.

DETER–DETECT–DELAY–RESPOND

The security industry safeguards and protects things, products, people, and activities by creating defensive rings or lines of defense around them. This redundancy, or the creation of hurdles to overcome, provides the means to implement the security principles of deterring, detecting, delaying, and responding to the event. Simply creating what may appear to be an insurmountable obstacle may cause a potential unauthorized intruder to go elsewhere or minimize damage if a man-made or natural disaster occurs. The detection of a breach or attempted breach (indicated by audio klaxons or flashing strobes, for example) provides warning and reaction time to activate a response; it also serves as an initial means of warding off a potential intruder. Physical or psychological barriers that must be penetrated to get to the items being protected creates delays and promotes a fear of being caught in the act. Finally, most security designs include the need for a human response—someone must actually go to the scene at some point to take charge. However, this last phase is more operational and procedural in the design process.

ACCESS TO THE SITE/FACILITY

Define the level of control that is appropriate for the site or facility. The property perimeter is the first line of defense for your site design. The property perimeter may be the building itself, as is the case in many cities, or it could be the grounds that surround the building. It can be an *open* environment, where little things (sidewalks, driveways) control the flow of people or vehicles, or it might be a *closed* environment, where iron fences stop or channel activity at the property line to central points.

Today, depending on the facility, or activities being conducted, physical barriers that will stop vehicles from penetrating a site are most common. The U.S. Department of State has designated a rating system for certain structurally designed vehicle-impact-rated barriers or anti-ram barriers. These ratings are defined as K-4, K-8, and K-12. A K-4 rating will stop a 15,000-lb (6,810-kg) gross-weight vehicle impacting a barrier from a perpendicular direction traveling at a nominal speed of 30 mph (50 km/h). A K-8 rating will stop a 15,000-lb (6,810-kg) gross-weight vehicle impacting a barrier from a perpendicular direction traveling at a nominal speed of 40 mph (65 km/h). A K-12 rating will stop a 15,000-lb (6,810-kg) gross-weight vehicle impacting a barrier from a perpendicular direction traveling at a nominal speed of 50 mph (80 km/h). Many government operations, as well as private businesses, are now incorporating these barriers into their facility design plans. The control can be one of authorization, where typical access onto the property is granted by either electronic means (i.e., card access) or by a security officer, or a combination of the two. For many sites, this real control is the perimeter of the structure itself, where doors and lobbies form the majority of the perimeter. *Authorization* is used in this chapter as the right of a person to be on site or in an area of the site at a particular time. Knowing who is on your site, and when someone is on your site, is critical to meeting the site's security and safety concerns.

A card-access control system, if not properly utilized, allows any holder of that card to open the door and allow as many into the facility as he chooses; in addition, someone can slip in before the door is totally shut (commonly referred to as "tailgating"). To eliminate these possibilities, biometric fingerprint or iris readers are being used in conjunction with interfacing revolving doors. The use of a security officer or uniformed guard, along with access lists or permission schedules, is also a time-honored method for reviewing credentials (ID cards, bills of lading, etc.) and granting access.

Access to a facility for pedestrians, vehicles, commercial vehicles, and rail and maritime transport must be planned. Normally, those needing access can be grouped into four general categories: employees, customers or visitors, contractors, and vendors. Each category must be defined in policy and procedure, along with its access levels and special freedom of movement (e.g., escort, no escort, limited access, etc.).

Another decision that should be considered early in the design process is what criteria should determine searches of property brought into or taken out of the facility. If packages, briefcases, or vehicles are to be searched, the company policy and procedures to accomplish this need to be fully developed and publicized and process room for this screening must be incorporated into the design.

Types of individuals or vehicles entering and leaving the facility, as well as work shifts and set schedules, can challenge a perimeter access-control system's physical limits if it has not been properly sized for traffic volume. In a professional office environment, employees tend to arrive over a two-hour period. This volume of traffic could be concentrated into a 10-minute period for a production environment with shiftwork. One goal that should be set is the designed maximum queuing times for the security perimeter. If swing doors are exclusively used, there is virtually no queuing. Any stricter control has a potential for creating queuing space. By setting a queuing space goal, the size of the lobbies and the security measures within can be properly designed.

THE PERIMETER

The perimeter is the defining point for what is under the control of the facility's site. The property line should identify the site and how it differs from public property, plus any restrictions on movement. A security challenge is present if the facility is actually located on and part of the property line. The challenge then is a limitation on space to place barriers between the perimeter and the public space.

The facility perimeter itself can be, and often is, at the property line, especially in an urban setting. When this is the case, the outer shell of the structure needs to be capable of stopping the free flow of movement into and out of the building. In cases where an open-air, atrium-type area has been created, the structure and occupants are more vulnerable to explosions. With a structure on the perimeter, openings to public areas become potential ways to breach the security of the building. The materials, products, and their placement in the outer walls; the doors and windows; the loading dock or freight movement locations; utilities access points; and the HVAC intake all present security issues that must be reviewed for potential impact. In addition, the roof and the underground access points cannot be neglected during a security evaluation and design of a structure.

BUFFER ZONES

The bombing of the Alfred P. Murrah Federal Building in Oklahoma City heighted the concern for creation of "buffer zones." Buffer zones are areas outside the perimeter of a facility that, in times of heightened security alerts, allow security practitioners to perform increased surveillance activities. Within the property perimeter, the establishment of "clear zones" allows for greater surveillance and provides additional reaction time to counter any developing threat.

Traditional barriers used in the physical security function may include steel fencing, boulders, chain-link fencing, barbed wire, and barbed tape or concertina wire. Signage creates a psychological barrier, as well as placing individuals on notice of any restrictions on access.

STANDOFF DISTANCES

The bombings of the U.S. Marine barracks in Beirut, Lebanon (1983); the World Trade Center in New York (1993); and the Murrah Federal Building in Oklahoma City (1995) have resulted in establishing certain minimal standards for federal buildings. The need to have a setback or "standoff" space between a building and vehicle traffic has become a requirement for U.S. government buildings. The use of a vehicle loaded with explosives has taken many lives, destroyed and damaged property, and caused operational delays for those affected. The minimum standoff distance ranges from 33 ft to 148 ft (10 m to 45 m) depending on the type of occupancy, construction, and applicable explosive weight that a vehicle bomb may present. Distances should, if possible, be greater, as an explosion, even with these minimal protections, will still do various degrees of damage.

CONSISTENCY OF SECURITY APPLICATIONS

Consistency of security applications can be understood with an example that uses validation as the goal. Validation involves checking credentials (typically an access card) against a current database to determine if the card is valid for a particular reader at a particular time. In addition, validation requires that only one person be allowed entrance with that credential. (This goal does not require authentication or package searches, which could be required, but is not in this example.) Although the need for access exists, it is imperative that only authorized and verified individuals and products get into the secured area of any facility.

In a lobby that is staffed, and where psychological control is a reasonable approach, optical turnstiles provide the fastest throughput of any validating barrier. (It would be difficult to justify more stringent control than psychological control for a lobby of a building constructed of glass walls. Such "hardening" of the lobby without also hardening the building perimeter does not appear to be justifiable.) Although some optical turnstiles are of the off-the-shelf variety and are unimposing, other custom designs are quite impressive. One site that aesthetically worked well was designed with red granite for the reception desk as well as the optical turnstiles. These barriers were considered attractive and imposing at the same time. This particular site also had other points of access and egress that were not staffed. Access-controlled revolving doors were used for locations that did not have a security officer present. The result was still validation. The goal was to provide a consistent perimeter barrier.

Loading docks in a validation environment should require a strict policy of "no regular traffic exit or entry" other than the receiving and shipping of products. In some applications, loading docks use a separate, adjacent door with "in" and "out" card readers that can be used to log the movement through the perimeter. Card readers are an excellent way to monitor such movement, provided they can also be monitored with a closed-circuit television (CCTV) arrangement. The loading dock staff must be part of the team in terms of controlling access or the system will be seriously flawed.

Why is it so important for the security design to be consistent? It turns out that people have a tendency to disregard procedures and rules that they do not see as valid. The typical independent American, encountering a way around the system, chooses the alternative path just for the sake of being different or for ease of movement. Others may consider the costs in capital and time a waste because of the lower degree of control at other points, all of which tend to make the system ineffective. Consistency of application can thwart these sometimes unconscious transgressions.

PARKING

Parking is an issue for all sites. In some urban locations, the issue is that there is limited or no parking for employees. In other urban locations, some parking is available but is not close to the facility, so shuttles and separate security measures for the parking lot or garage must be provided. In still other cases, aboveground parking structures or underground parking areas may have been designed into the complex. Whatever the arrangement, with the exception of public transit, the company also "owns" the liability for employees from the time that they park until they leave the facility. The goal should be to provide a safe work environment from the time a person arrives on company property until she leaves.

Parking lots surrounding a site have the advantage of proximity. They receive residual protection from this proximity, but the challenges to security still exist in many environments. Access control of the parking facility is a key issue. Limited and controlled access points are essential. Good lighting that reveals who is present and what is happening around the vehicles is another key to parking lot security. Painting the ceiling and walls of multilevel parking garages with white paint or a reflective coating increases the effects of lighting. Placement of emergency call-assist stations throughout parking areas is an important aid. The use of cameras and real-time video surveillance, if possible, is also ideal. Video can be used to reconstruct what happened after an event. Landscaping is an issue that must be considered in many locations, as well as seasonal issues (snow, freezing weather, heat, etc.). Good CPTED practices are a must, allowing for overlapping visibility and reductions of blind spaces. Identification of parking places—reserved versus open parking—creates additional security concerns, especially for executive protection.

Other design issues passively enhance security. One example is the direction of tree rows in relation to the facility. When trees are placed in rows that point away from the facility, the parking area is more visible to people in the building. The opposite is true of a concentric circle of trees set around buildings. When people in the building are able to see a crime in progress, the parking facilities seem safer to users than to would-be perpetrators, thus reducing the potential for crime to occur. Radial tree designs in parking areas allow perceived surveillance to enhance security.

PATHS AROUND THE SITE

The route to and from a facility usually defines what is an acceptable way to go from point A to point B. In other cases, it simply defines a place to take a leisurely stroll. In either case, the route should be clearly announced and purposefully designed. The walkway, while in most cases is landscaped, presents additional security and safety challenges ranging from would-be attackers to slips and falls. This is especially true in areas where the route has high vegetation, bushes, or trees, or is prone to blowing or drifting snow, ice, or standing water.

Landscape architecture is an art that can foster safety in an environment or detract from it. One simple rule makes the security difference: Design for open sight lines above 1.5 ft (0.5 m) and below 6.5 ft (2 m). This will put significant constraints on the foliage. For example, redwood trees of the California coast variety (*Sequoia sempervirens*) have low, dense branches. Unless they were imported as 20-ft (6-m) trees, it would not be feasible to cut off all of the branches below 6.5 ft (2 m) and expect the trees to survive. A tree that thrives despite this limitation is the crepe myrtle (*Lagerstroemia indica*), which can be easily pruned into a ball-shaped top sitting on a slender trunk. Slow growing shrubs are self-limiting and present problems. The early planting will be sparse enough for safety, but such shrubs will grow to large bushes within which a criminal can hide. Ensuring a sense of safety on paths with clear sight lines and good lighting is the most economical security measure available for transition areas.

LIGHTING

Security lighting varies from operational lighting in that the lighting level is generally lower than that required to perform operations. In general, lighting at entry and exit points, or at checkpoints to see identification, should be consistent with operational lighting. Garage and parking-lot lighting should be such that shadows are eliminated and faces can be recognized. As discussed earlier, ceilings and walls can be painted to reflect lighting.

Lighting just outside employee entrances should be sufficient to allow people who have just left a lighted building a good, clear view of their surroundings. Healthy eyes adjust rapidly, but not instantaneously, to a change in lighting. Higher levels of lighting near the entrance allow for this physiological transition. Although lighting is subject to many interpretations, the levels listed in Table 18-A provide good security in an urban environment. There may be reasons to raise these levels, but in most cases compromises on somewhat lower levels can be made during the design process.

Table 18-A. Security Lighting Requirements

Type	*Application*		*Illumination Width (Feet)*		*Minimum Illumination*	
	Lighting	*Area*	*Inside*	*Outside*	*Lux*[a]	*Location*
Boundary	Glare	Isolated	25	100	2 (0.2)[b]	Outer lighted edge and 30 ft out
	Controlled	Semi-isolated	10 10	70 70	4 (0.4) 2 (0.2) 4 (0.4)	At fence Outer lighted edge At fence
	Controlled	Non-isolated	20–30	30–40	4 (0.4) 5 (0.5)	Outer lighted edge Within
Sensitive inner area	Area	General	All	—	2–5 (0.2–0.5)[c]	Entire area
Entry point	Controlled	At structures Pedestrian Vehicular	50 25 50	— 25 50	10 (1) 20 (2) 10 (1)	Out from structure Entry pavement and sidewalk

[a] Horizontal plane at ground level (6 in. above grade) unless otherwise noted. Footcandles are shown in parentheses.

[b] Vertical plane, 3 feet above grade.

[c] Use the higher value for the more sensitive or confined areas.

Note: Initial field measurements will be multiplied by light-loss factors to determine the maintenance levels.

Source: U.S. Army Corp of Engineers, TM 5-811-1/AFJMAN 32-1080, p. 11-2.

LOBBY DESIGN

The lobby is the greeting spot and welcoming center for clients, customers, visitors, and, in many cases, staff and employees as well. A lobby is the central location where the opportunity exists to serve and screen those coming into the facility.

A policy and procedure should be in place to address the handling of clients, visitors, and delivery personnel. Lobby design should include space for clients and visitors to feel welcome, for a sign-in process, and, if practical, for temporary storage of flower arrangements and special delivery packages. (Ideally, however, these latter items should come through a delivery channel such as the loading dock, thus providing proper security screening of deliveries and internal delivery.) Since many nonemployees need to enter the lobby, it is appropriate to leave doors unlocked during normal business hours; when they are, a common security approach is to control access to the interior lobby doors that lead to the rest of the facility. Thus, the lobby is truly a semipublic space. Even when there is no physical barrier to entry from the lobby to the areas beyond, there is usually a perception of control at the lobby based on the reception station, which can be staffed by people who act as receptionists or by security officers who act more as gatekeepers.

If the design goal is *authorization*, a receptionist, a security officer, and/or a swinging-door access control into the rest of the space is appropriate. These systems and people can, with proper management and training, accomplish the desired result. If *validation* is the goal, a one-card-one-person system should be implemented. Optical or mechanical turnstiles can provide this control. The optical turnstiles are the fastest and least obtrusive method of validation available today. It is reasonable to expect to process 30 people per minute per lane with optical turnstiles. Because professional offices typically operate with relatively flexible schedules for the employees, estimating that most will show up for work over a two-hour period is reasonable. If, on the other hand, the lobby is for a set of shiftworkers expected to punch a time clock, the design needs to accommodate the high peaks of traffic flow associated with shiftwork.

Ideally, a lobby can be the location for a side room that can be utilized as a conference room. This facilitates meetings with visitors without bringing them into the secured areas of the site. This option is useful for organizations whose visitors, by the nature of the work going on inside, have access to potential trade-secret information. An example is a salesman for a product line that can also be used by a competitor. This salesman visits competitors' sites as well, and the best practice is to eliminate the opportunity for exposure to private information.

In-house telephone access is also useful in the lobby. This allows a visitor to call within the facility while restricting calling to outside numbers.

At least one unisex, key-controlled restroom should be available in the lobby. Restroom escorts are not practical, and the practice of letting a visitor enter the secured portion of the facility, unescorted, to roam the halls on the pretext of needing to use a restroom is not conducive to good security.

EMPLOYEE ENTRANCES

An entrance intended only for employee use requires different security criteria from that of the main lobby. The facility goals have a major bearing on the design of these doors. Package searches require staffing—there is no way to automate a briefcase search. All employee entrances need to be designed to accommodate security staffing if package searches are required.

If there is no requirement for package searches, the next goal to consider is the authorization-versus-validation decision. With authorization, access-controlled swinging doors operate well. This is the point at which to review the security tools for consistent results. The security process for the lobby should accomplish the same screening as that for employee entrances. If a video-audit trail of activity at this point is expected, the design should allow for a long enough walking path to support recording by a video system. The path should not be so wide that it is difficult for a single camera to cover.

Employees who work during hours when there are few people on site or when it is dark outside will appreciate being able to park relatively close by. Employee entrances and exits should provide this opportunity. One common and seemingly misguided approach is to close down all employee entrances except the one at the lobby after normal working hours. If staffing is an issue here because of the need for package searches, other means of allocating parking may offer a reasonable alternative, allowing those who work late to park close to the lobby. The case of employees working late and needing to get to their cars is best served by providing them with the shortest exterior path possible. The shortest path is usually the safest.

LOADING DOCKS

Loading dock or *shipping/receiving dock* is the common terminology applied to the designated area where product is coming into or going out of the facility. It is a security-sensitive area in that materials and access can be compromised at this point. The function should have both a policy and procedure to cover its operation. This area has many configurations. An office building loading dock can be as little as a doorway and a room for storage. More typical, however, is a roll-up door next to a swing door that opens onto a dock area that can have adjoining storage. One concept useful for some sites is provision for a trash compactor within the loading dock, allowing trash to be thrown into the compactor from within the building; once it is in the compactor, it is rendered useless. A typical theft opportunity for the cleaning service is eliminated because there is no way to stash computers or valuables near a dumpster to be retrieved after work hours.

Loading dock management needs to take ownership of this access/egress point and make sure that no normal building traffic has access to the loading dock as a means of entrance or departure. In addition, restricting traffic will keep the potential for theft and misdirected products to a minimum. An "in" and an "out" reader with lights or a bell that can be used for logging into or out of the building can be applied to a desk or wall. The manager of the area can make sure that anyone who does need to pass through the dock logs in or out. The readers also provide a tool for restricting this access point. The use of security camera coverage should be readily apparent in this area. In facility design, the things that cause losses at a loading dock are congestion brought on by lack of maneuvering space, confusion due to lack of storage, and lack of staffing. It is important to staff loading docks that have a lot of activity. High-value items should get special attention in terms of logging in and storage. If the high-value items are large, sufficient storage must be defined early in the design process.

Delivery personnel need limited and controlled access to this area. Provide a unisex restroom specifically for a loading dock for use only by dock/delivery personnel. Control is not an issue if a restroom is within the barrier of the loading dock.

Production and warehousing loading docks are potentially much bigger operations, with multiple roll-up doors. Management of areas like this is primarily procedural. One issue that seems to come up repeatedly is the request to leave the doors open for ventilation. This often results in confusion. First, the facility's mechanical engineering staff might say that the air conditioning works best if the doors are closed but the employees might say that they need to have the doors open for outside air. Second, incoming delivery drivers might see an open dock and bring their cargo load to that opening without guidance or direction. The facility design needs to take these two issues into account, clearly identify the specifics, and allow for chain or open-grid door closures that can be used for security while still allowing air to flow through. In addition, some means of traffic lights to notify drivers of dock space availability is also useful.

MULTITENANT BUILDINGS

Multitenant buildings represent many different security and safety concerns and issues. The simple division of overall structure security and safety versus the internal space that has been subdivided creates vulnerabilities due to various lines of responsibility and interpretations on how security issues are to be addressed.

In multitenant buildings, the conflict between saving lives and ensuring security becomes very apparent. If the building has a stairwell at each end of a rectangular floor plan but has its elevators near the center, the issue of controlling the floor will soon surface. In case of fire, elevators are designed to return to the ground floor. It is against building codes to restrict access to the stairwells. Securing doors so they cannot be readily opened with little effort is generally a well-established fire code requirement in most jurisdictions. The need for egress routes that can be readily opened creates security vulnerability. The use of "alarmed" panic hardware or delayed egress hardware, local enunciators, and/or security cameras provides some degree of security. Building codes prohibit dead-end corridors. A dead-end corridor is created when there is only one way to exit freely in more than 20 ft (6 m). The general rule is that there must always be two ways to exit from any corridor and from any large space. *Large* is defined in the code according to the type of building occupancy. A place of assembly, such as a cafeteria, is treated differently from an office space, even within the same building. Generally, the potential for more people drives the exiting requirements to higher levels. It is important not to design a security barrier into an egress path so as to create dead-end corridors. Another driving force is whether the structure is equipped with a sprinkler system or not.

In a multistory office building that has several tenants, or segregated occupancies within the same company, the elevator lobby design has a major impact on the availability of space for productive work areas rather than for passageways. Plan the building to locate one stairwell with the elevator bank. This will provide a means of egress from the elevator vestibule, will provide two means of egress from the floor, and will allow full locking of the doors from the elevator to an open-plan office area. Unfortunately, in many jurisdictions the doorway into the enclosed fire exit stairwells is required to remain unlocked during an evacuation period, thus exposing the internal workings of one tenant to another, who may utilize this opportunity to gain insight into another company's work area or to commit other criminal activities.

CURRENT ISSUES FOR SECURE FACILITY DESIGN

The expanding role of the U.S. Department of Homeland Security and its emphasis on Critical Infrastructure and Key Resources (CIKR)—protecting those basic elements needed for this country to survive—are driving many aspects of security design. Security principles remain the same; however, protection needs are being looked at from the standpoint of obtaining the best security results for the finances invested. The focus on survivability and sustainability places demands on physical security to harden areas of concern and reallocate resources to accommodate these requirements.

The business world is also struggling to maintain profitability. The economic recession has placed limitations on security designs, and spending for equipment and personnel have diminished greatly. The focus is on doing the basics, with expansion capabilities included in any designs. Doing more with less continues to be a big issue. The use of improved electronic security hardware and equipment takes center stage.

Smoking continues to be a security problem. Today, all federal buildings, airports, and most state buildings have been declared "smoke-free zones," and smoking is prohibited. In many states, it is becoming law that smoking is not allowed in office buildings or restaurants, at sporting events, or even in bars. Smoking is still allowed in outside designated areas.

A few years ago, when smoking was generally allowed in the business environment, there were few security-related issues concerning smoking. Mainly, there were fire issues. Today, smokers are a common cause of propped-open or forced doors, some of which are emergency-exit-only doors. People complain that they cannot get into the building because of the concentration of smokers near some entrances. A prudent business approach is to provide a reasonable exterior location for smokers, shielded from the weather, with places to sit. These should be away from the doors; some state laws require a distance of 20 ft (6 m) from any entrance. Another option is to provide an exhaust-only room inside the building. Many companies refuse to spend the money to provide either of these accommodations. Some are afraid that by creating a smoking room they are encouraging smoking, thereby increasing company liability. Others feel that by making any smoking accommodation relatively uncomfortable, they are discouraging smoking and that smokers will quit.

Unfortunately, violence in the workplace continues to be an issue for many employers. Pragmatic approaches are still the most cost-effective today, but there are areas and disciplines that should consider additional measures, some of which will have an impact on the facility design. Personnel departments should consider the location and design of areas where discharges and firings take place and where grievance meetings or hearings can be held. Measures such as screening for weapons and installing bullet-resistant glazing may be appropriate in such locations. "Panic rooms" that can be secured from the inside with turnbuckle locks and nonglass doors or side windows should be incorporated into designs for various floors and areas (i.e., storage rooms, conference rooms, and file rooms).

SUMMARY

The events of September 11, 2001, revealed vulnerabilities to acts of terrorism in the United States. The train bombings in Europe, gas attacks in Japanese subways, and bombings of embassies worldwide have led to new security design criteria and operating procedures. The location and operation of HVAC intake systems, parking, and access to a facility must be considered. To preclude large-scale damage, buffer zones and clear areas as well as parking must be restricted to greater distances from a facility. Lack of space in any city makes that unfeasible; therefore, considering other approaches to this challenge is important to any facility design. Vertical discrepancies and large-mass hardscape design elements can assist in locating vehicles farther from the critical areas of facilities and deflecting blasts. Restricted underground parking should be designed with street-level queuing space sufficient for managing garage traffic flow. Critical assets, such as data and communication centers, should be positioned within the core of a site, using the rest of the operation as a buffer. Large sites should be designed to allow for the addition of site-access controls, even when the initially designed security perimeter is a building and not the property line.

The public is demanding that safeguards be put in place but the costs remain difficult challenges to overcome. It is hoped that information gathering and other safeguards being instituted will preclude most incidents, eliminating the need to make wholesale changes to the architecture of the country.

Several means of safeguarding facilities can be employed, but each has ramifications. The following issues should be considered:

1. The security investment must add value.
2. The concept of deter, detect, delay, and respond needs to be incorporated into design thinking.
3. Security buffer zones and clear areas need to be a part of design criteria.
4. Controllable access processes need to be incorporated into designs. A conscious decision to operate the facility in open, authorization, validation, or authentication style must be made.
5. The visitor-processing system and throughput allowances must be addressed early in the design.
6. If package searches are to be a policy option, the design must accommodate this process.
7. A perimeter barrier that is consistent in result should be provided.
8. A working environment that is safe from the time a person arrives on company property is the primary goal.
9. Access control, reflectivity, patrols, and good lighting are the keys to parking security.
10. Radial tree designs in open parking areas allow perceived surveillance, enhancing security.
11. For planted areas, design should provide for open sight lines above 1.5 ft (0.5 m) and below 6.5 ft (2 m).
12. A secure lobby should consider the following items:
    - unlocked front door
    - locked interior door or turnstile control
    - storage of flowers, balloons, and parcels
    - house phones
    - secured conference rooms
    - restrooms
13. A trash compactor opening should be available within the loading dock.
14. A unisex restroom and a telephone should be convenient to a loading dock.
15. The building plan should accommodate one stairwell with the elevator bank.
16. Early coordination of secure zones and egress paths is essential to good security design.
17. Egress from an underground garage must be to the exterior of the site.
18. Proper functioning, review, and upgrading of electronic security equipment needs to occur.
19. Procedural management actions, when coupled with existing physical security safeguards, can provide an excellent defense to minimize workplace violence.
20. A reasonable exterior location for smokers should be provided.

BIBLIOGRAPHY

Bottom, N. R. Jr., and R. R. J. Gallati. *Industrial Espionage: Intelligence Techniques and Countermeasures*. Newton, MA: Butterworth-Heinemann, 1984.

Craighead, G. *High-Rise Security and Fire Life Safety*. 2nd ed. Newton, MA: Butterworth-Heinemann, 2003.

Crowe, T. D. *Crime Prevention Through Environmental Design*. 2nd ed. Newton, MA: Butterworth-Heinemann, 2000.

Demkin, J. A. *Security Planning and Design: A Guide for Architects and Building Design Professionals*. New York: John Wiley & Sons, 2003.

Fennelly, L. J. *Handbook of Loss Prevention and Crime Prevention*. 4th ed. Elsevier/Newton, MA: Butterworth-Heinemann, 2004.

Garcia, M. L. *The Design and Evaluation of Physical Protection Systems*. 2nd ed. Newton, MA: Elsevier/Butterworth-Heinemann, 2007.

Patterson, D. G. *Implementing Physical Protection Systems: A Practical Guide*. Alexandria, VA: ASIS International. 2005.

Post, R. S. *Determining Security Needs*. Des Plaines, IL: Nickerson Collins, 1973.

Purpura, P. P. *Security and Loss Prevention: An Introduction*. 5th ed. Newton, MA: Elsevier/Butterworth-Heinemann, 2007.

U.S. Government Printing Office. *National Infrastructure Protection Plan: Partnering to Enhance Protection and Resiliency*. Washington, DC: U.S. Department of Homeland Security, 2009.

———. *Physical Security*. U.S. Army Field Manual No. 3-19.30. Washington, DC: 8 January 2001.

———. *Physical Security Program*. U.S. Department of Defense, DoD 5200.08-R. Washington, DC: U.S. Government Printing Office, 27 May 2009.

———. *Reference Manual to Mitigate Potential Terrorist Attacks Against Buildings*. Washington, DC: U.S. Federal Emergency Management Administration (FEMA), FEMA-426, December 2003.

———. *TM 5-811-1/AFJMAN 32-1080, Electrical Power Supply and Distribution*. Washington, DC: Joint Departments of the Army and Air Force, 28 February 1995.

———. *Unified Facilities Criteria (UFC)*. DoD Minimum Antiterrorism Standards for Buildings. UFC 4-010-01. Washington, DC: 8 October 2003.

19

Barriers and Locking Systems

Kenneth G. Fauth, EdD, CPP

CHAPTER 19

OBJECTIVE

The objective of this chapter is to explore the issues surrounding the appropriate selection of various portal-related physical security hardware systems. The benefits and drawbacks of the following systems are considered:

- gates, gate areas, and bollards
- keys and key control
- locking hardware cautions
- building perimeter openings
- interior doors
- cipher or combination locks

INTRODUCTION

Physical security often is described as a series of steps that is repeated in concentric circles. The first step is to notify and, therefore, define the boundary. In notifying and defining the boundary, physical security also acts as a deterrent by creating what is or may be a perception that this is a "difficult fortress" to try to breach. The second is to detect the crossing of the boundary. And the third is to delay further ingress. This process is often repeated numerous times the closer one gets to the center of the circle or to where key assets are located.

One example is a site with a fenced perimeter (fencing notifies and defines boundaries and creates a deterrent). Inside the fence line, a photoelectric beam, fence sensors, or buried sensors can detect the entry across the fence. Inside this initial fence line, some delay can be applied. A few examples are another fence, difficult landscaping and various types of plants, vertical discrepancies, or water. Another possibility is a strong physical response (uniformed security guards/personnel and patrol vehicles), as might be used at a nuclear or military site. This first set of define–deter–detect–delay tactics is typically followed by another at the perimeter of a structure within the space. The structure itself is the definition of the line; alarm systems, lighting, and security personnel are the detection; and delay is provided by obstacles such as the walls, the portals in the walls, and the security personnel. Any critical assets that exceed the value of the general space within the area probably will have a third, fourth, or larger set of these rings.

In each case, a ring of protection needs one or more portals that allow access and egress to those who have duties to perform within the space. The portal is analogous to a switch for electricity in that it should be completely closed to traffic when there is no need for access and fully open when someone or something needs to pass through. Although this concept appears to be obvious, putting it into practice is not always easy. In most jurisdictions, fire codes require a second means of egress to be available to depart from the inside, thus creating an additional vulnerability. The structure of this chapter is based on the types of barriers and the portals for those barriers, starting with the site perimeter and working toward the center, where the most critical assets should be located. Thus, a property line is the logical starting place.

The boundary of a property can be defined by a fence, a line of trees, walls, or any other obvious means. Signage at this point should define the line that should not be crossed so as to provide legal notification of trespass. Signage also deters nondeliberate intruders, thus lowering the load on the response team and process. Often, a planted berm is used for aesthetic reasons. This berm effectively limits the ability to drive onto the site except through defined paths, which may be gated or just monitored and observed. In today's world of terrorism, vehicles and individuals who may be carrying explosives need to be kept at a distance, outside the boundary of the property. However, for the purposes of this chapter, discussion will be limited to portal-related issues—those that apply to the openings in the barrier.

GATES, GATE AREAS, AND BOLLARDS

Traditional measures at a perimeter or boundary continue to be fences and gates to limit foot traffic, bicycles, and, in some cases, light vehicles (when steel posts are employed). The threat scenario for a site greatly influences the design of the perimeter protection. For a threat that assumes no physical force, the gate can be a staffed kiosk, with no physical barrier at all, or it can be automated—controlled via a card reader or intercom connection to a staffed post. There are many other options.

It often is easy to *tailgate* (follow another authorized user through an automated gate before it closes). This is a result of the way the gate controls and the gate area are designed. In many cases, a timer is the only trigger for closing the gate, and the timer is set for a slow, large vehicle. It is entirely possible to use sensors that close the gate once the vehicle has passed through the portal, thus reducing the time needed for the open cycle. Another method that can assist in limiting tailgating is to use separate gates for ingress and egress. This is especially useful where there is an island between the two directions of travel. This configuration makes it harder for the person who is tailgating to traverse to the other side of the road. Tire shredders buried in the pavement are a reasonable means of restricting the direction of travel. Also, shorter gates close faster. Therefore, if the gate that opens for a vehicle is designed to be used only in one direction of travel, significantly better security is achieved by combining these factors.

Safety is a key element in automated-gate design, and the challenges to security that a safe gate operation brings cannot always be easily solved. It is possible to design a *sallyport* of two gates, requiring one to close before the other opens, but this is typically reserved for high-security applications or cargo departures.

One note of interest is the commonly applied technique of using a *buried cable exit loop sensor*. In a San Francisco case, a steel overhead garage gate was used at a high-value condominium complex. The door would open when a vehicle preparing to exit approached and ran over the exit loop buried cable sensor. The bottom of the gate, when closed, was positioned above the concrete surface by about 2 in. (5.1 cm) and a garbage enclosure was accessible from the street side. Thieves were taking garbage can lids, flattening them, and sliding them shuffleboard-style under the door. When their aim was accurate enough to hit the corner of the vehicle loop, the door would open and the thieves had access to all of the expensive vehicles within. This vulnerability needs to be kept in mind when designing the exit system for any opening that employs these loops.

Many business and public government buildings have physical devices to stop a moving vehicle before it breaches the boundary or perimeter. This has now become commonplace in the aftermath of various car bombings. The use of vehicles to ram perimeter defenses, locate explosive payloads next to structures, and detonate them results in greater structural damage than if they are detonated at a distance from the structure. The need to stop vehicles at the perimeter or boundary is paramount to good security.

The U.S. Department of State has designated certain ratings of structurally designed vehicle-impact barriers or anti-ram barriers for stopping certain types of vehicles (see chapter 18 for details on these ratings). These bollards come in a variety of designs; some are stationary, while others move up and down or are electrically or hydraulically operated.

Anti-ram barriers are used where there is a threat of force. Such devices have been applied at many locations, including the White House, airports, government structures, corporate headquarters, and some commercial office buildings. A typical application would be stationary bollards on possible vehicular avenues of approach to the structures. Often a combination of an automated or staffed gate or barrier arm at a card-reader checkpoint, followed by the vehicle barrier/bollard, is used. Once cleared, the vehicle would be allowed to proceed past the initial checkpoint, to continue through a sharp turn, and then to pass over the lowered vehicle barrier/bollard. Any inappropriate activity would keep the vehicle barrier/bollard in the raised position. As soon as the vehicle passed over the vehicle barrier/bollard, the barrier/bollard would be raised again, pending the next transaction.

KEYS AND KEY CONTROL

Key locks remain the most popular method for securing buildings today. Key locks are used even when electronic methods have been incorporated into the same access design to allow entrance through various portals. The primary reason is cost.

Key designs over the years have progressed to a point where they range from a basic "skeleton key" design to high-security designs. With proper application, these various designs can provide momentary privacy or a high level of entry delay time. However, the major point from a security standpoint is that anyone who possesses the key may use it to enter.

Four critical questions arise when using a key system:

1. What is the importance of what you are trying to protect? This will determine the level of security needed in the keying system.
2. Who will have access to the key, and will that person (or persons) safeguard the key against improper use?
3. Is there an ongoing administrative record of where all keys are at all times, including all keys on hand, assigned, and disposed of?
4. Are policies and procedures in place for describing proper use and replacement of keys or the rekeying of the entire facility?

LOCKING HARDWARE PRECAUTIONS

Before we discuss applications of locking systems on doors, several precautions to safe and secure operation should be understood. The same rules apply to all door-locking hardware for doors in the exit plan for the building. Always use a mechanical means of egress. This design will not leave an employee or fire fighter trapped in a building, whether the emergency is a fire, an active shooter, or some other situation.

A common misapplication of hardware is the installation of a dead bolt to the bottom of a lobby double door. This approach is often suggested by the building architect, with little concern for panic hardware. Basic life-safety design requires free egress from all locations within a structure. If a lobby door is part of a building exit plan, the best approach is to apply mechanical egress, which allows the door to be opened simply by putting a hand on the panic bar or door lever even if the access-control system is locked on the outside, broken, or out of power. Yet the building designer often drives the door design to an awkward state in which local inspectors approve the operation based on exceptions in the code. All too often these exceptions are misread, misapplied, and misunderstood, thus resulting in costly modifications after installation, as well as potential safety issues.

When access control is applied to doors with no mechanical exit equipment, a passive infrared sensor may be used to unlock the door as someone approaches it from the inside. (A passive infrared device detects a person in its field of view and triggers an event.) This system is being replaced by sensors within the door handle or a floor mat that releases on touch. In this case, a dead-bolt or magnetic lock is usually applied to the header of the door, with a card or biometric reader on the outside. The problem here is that no passive infrared or other device is approved within building codes for use in a life-safety egress application. A second difficulty is that the device is often wired into an access-control system "request-to-exit" switch (input), to disengage the lock, instead of directly to the lock. If the access-control system is improperly programmed or broken, people cannot get out of or through the doorway. In this environment, responsible designers have analyzed the failure modes of all parts of the overall system that allow egress from a building. Any broken or missing component should lead to an unlocked door.

DESIGNING FOR ELECTRONIC LOCKING SYSTEMS

During the facility design process, the electronic lock system is often incorporated into the fire alarm system design so that, when a fire alarm is triggered, doors are automatically unlocked. The fallacy here is not all emergencies are fires. Gas leaks and workplace violence also take place within buildings. There is no guarantee that a fire alarm will be activated in every emergency, and free egress is necessary for many types of emergencies.

For these reasons, fail-safe (open) and fail-secure (locked) locking hardware must be understood during the design process. Fail-safe locks are those that require power to operate and without power are left in the open or unsecured position. An example is a magnetic lock that holds a door secured until a command from an authorizing access system has been given (i.e., use of an access card, biometric, etc.). Fail-secure locks require power to unlock as well as a proper authorizing device. If the rule of mechanical egress has been followed, there is almost never a reason to use fail-safe locks (exceptions might be highly sensitive and unoccupied areas). Problems with fail-safe equipment include the failure of security when there is no power. While most security applications have temporary battery backup, the batteries usually last for only 2 to 4 hours. Often, there is no generator to back up the main power. Fail-secure locks provide security even if the access-control system is not functioning. Most access-control doors should be designed with fail-secure lock sets but with emergency means of mechanical egress.

One area where fail-safe (open) locks are required is in high-rise building stairwells. From the floor side, the lock must permit mechanical egress. From the stairwell side, if the door is to be locked, the hardware must be fail-safe (open) because there is a potential for smoke accumulation in the stairwell, and people may need to reenter the floor to get away from stairwell smoke. There is no way to use an electric strike for a fail-safe, fire-rated door because a fail-safe strike has no latch-holding potential when it is unlocked. An electric lock set, on the other hand, latches whether it is locked or not.

A few basic lock-set functions must be understood for good design:

- A *passage set* is a pair of handles and a latch; there is no key or locking function.
- A *privacy set* is a pair of handles with a button or turn handle on the inside of the door and some unlocking capability on the outside. Typically, the unlocking function works through a

small hole on the handle into which a rod is inserted to unlock the door. Neither the passage set nor the privacy set should ever be used or considered as a security locking device.

- *Office-function lock sets* operate with a button or turn handle on the inside or edge of the door that locks or unlocks the exterior handle. A key cylinder is used to gain access when the door is locked. The inside handle always provides unlocked mechanical egress. The office-function lock set, while once popular, should not be used or considered a security locking device as the system can unknowingly be compromised by pushing the disabling mechanism.
- *Classroom-function lock sets* use a key to operate the locked-unlocked function of the exterior handle. A key cylinder provides access when the door is locked. The inside handle is always unlocked.
- *Storeroom function lock sets* use a key to unlock the door from the outside and the handle from the inside. When the key is removed from the lock set, it is locked from the outside.

Typically, only storeroom-function lock sets are electrified because only they offer secure access control. A classroom or office function left in the unlocked mode would not alert the system and would allow for unrecorded access or egress to occur. Electric strikes used instead of electrified lock sets are often misapplied with office or classroom lock sets, leaving the system vulnerable. If a door must remain unlocked for specific periods, the access-control system application program should be used to schedule locking and relocking; no mechanical means of achieving this state should be employed. Whenever possible, for access-controlled doors, use storeroom-function lock sets.

CYLINDRICAL VERSUS MORTISE LOCKS

Two styles of lock sets are available to the facility designer. One is called *cylindrical* and the other *mortise*.

- *Cylindrical lock sets* are so named because they are installed by drilling one cylinder through the side of the door and another cylinder into the edge of the door so that the two intersect. This is a relatively simple and inexpensive installation, and the cylindrical lock sets are less expensive than mortise lock sets.
- *Mortise lock sets* are installed by mortising out a rectangular opening in the edge of the door. When done professionally by door hangers or a well-equipped lock installer, mortise installation is not difficult. Manually doing this with a drill, hammer, and chisel is exacting. There is a big difference in the level of security offered by cylinder lock sets and mortise lock sets.

Recently, the use of a "bumper key" has gained popularity among the criminal population to thwart the pins utilized in most cylindrical locks. The technique called *bumping* is used on locks that contain pin tumblers, which includes most locks used on houses and buildings. The "bumping" process causes the pins to fall in place (in preparation to opening) when a specially designed key is placed in any lock of this type and gently tapped or "bumped." High-grade security locks are less vulnerable to this technique, and some are not affected at all (i.e., wafer locks, rotating disk locks). To be effective, the bumping key must be able to fit into the lock's keyway. Most lock manufacturers have taken steps to address this technique when designing some of the newer locks.

A deadlocking latch is a lock-set latch in the edge of a door that, when pushed in by door closure, inhibits the actual security latch from being pushed into the door. This is the feature that keeps a door secure when using a credit-card or similar type of penetration to gain access. The deadlocking latch for a cylindrical lock set is a small, half-round piece that sits next to the latch. It activates when the latch goes into the door-jamb hole and the half-round piece is blocked from following. This, in turn, causes a blockage within the lock cylinder and holds the latch secure until a key is inserted to unlock the mechanism. However, it is subject to the door itself being forced, pulled, or pushed in too far, causing the deadlocking latch to fall into the strike hole and rendering the security void. With mortise lock sets, the deadlocking latch is positioned above or below the latch. Pushing or pulling on the door cannot "pop" the deadlock latch into the strike hole. The problem of the registration on the cylindrical lock set's deadlocking latch is especially difficult when used with electric strikes. The slight play in the keeper of the electric strike allows the deadlocking latch to be slipped into the hole in the strike, placing the lock set into a latch that can be easily pried open with a wire, knife, or credit card. Whenever possible, use mortise lock sets, especially with electric strikes. Neither is an issue with the use of magnetic locks.

Electric strikes were once thought of as the most appropriate electrified locking function for securing a door. A clean and long-lasting installation is still done with electrified mortise lock sets. However, UL-certified magnetic locks are often the choice for access-controlled doors. Magnetic locks are chosen over electrified mortise lock sets because they are faster to install and require less maintenance. In order to install an electrified mortise lock set, a hole must be drilled horizontally across the door from the lock set to a wired hinge or mortised wire transfer; these door cords are vulnerable and present ongoing maintenance problems. Preparing the doors appropriately for electrified mortise lock sets is easy when specified for new construction. Retrofitting a door with this hole is more difficult if the door is a fire-rated door. In some cases, new doors must be installed.

Doors have a tendency to rack if pulled on repeatedly. (A racked door is one that is not flat.) This problem occurs when the hinge side of the door is held by two or three hinges or pivots, but the latching side is held only at the top or sometimes at the bottom. Tempered glass doors rack very little because the glass has a strong tendency to return to its "normal" position. Wood doors rack to some extent because of force and warp if exposed to moisture. Storefront doors (those that have a metal frame, typically aluminum, with large glass inserts) tend to rack more than other doors because the frame stays bent once it is forced into the racked position. To limit or stop racking, doors must either be latched in the center where the door handles are or, in the case of double doors, held at the top and the bottom of the latch side, typically by the top and bottom rods of a piece of "panic" hardware. This is especially true when they are used in conjunction with magnetic locks.

Some access-control retrofit situations involve taking the bottom rod off a piece of panic hardware and applying an electric strike, which interacts with the remaining top rod of the hardware, to the header. This provides a worst-case situation for racking, and makes the door somewhat unsecure. In other cases, the vertical-rod panic hardware is removed, with magnetic locks applied at the head of the door, and various means of unlocking the magnetic lock have been used. In addition to the life-safety problems created by the potential lack of mechanical egress, the application of magnetic locks on storefront doors leads to racking, which, in turn, leads to unreliable closing and locking.

Where access-controlled double doors are required for the exterior of a building, the best solution is electrified panic hardware coupled with magnetic locks. By their very nature, these locks are fail-secure from the outside and allow mechanical egress from the inside. They work well as lobby doors that are scheduled to be unlocked for a period of time, and they act as push-and-pull doors during the open period.

Another approach used to maintain a security level is the use of access-controlled revolving doors. These installations are expensive compared to standard revolving doors, but the newest ones detect if a compartment contains more than one individual, can stop, and can go in reverse. Some access-controlled revolving doors also are equipped with voice activation messages.

A cost-effective and reliable lobby operation uses two single doors to replace double doors. Seldom is there a need for double-door width. Typically, double doors are designed for aesthetic reasons when a well-designed entry with two single doors is just as effective. Electric strikes can be used with single doors with rim-style panic hardware, keeping the installation and maintenance costs down and reliability high. Rim-style panic hardware is designed with a latch at the rough vertical center of the door where the panic bar is located, as opposed to vertical-rod panic hardware, which latches at the top and bottom of the door.

BUILDING PERIMETER OPENINGS

The perimeter of any commercial industrial building typically has the following:

1. a main entry
2. employee entries
3. emergency exits
4. a loading dock
5. a main power entrance room
6. utilities and HVAC
7. roof access

The main entry usually is open for visitors, with a receptionist in the lobby. Most multitenant buildings have very little control of visitors at the lobby entrance, whereas most single-tenant buildings use the lobby for controlling visitors. This generalization and those that follow refer to U.S. locations and areas of the world whose threat scenarios are similar. In many other locales, control is much more stringent. All these portals should be augmented with electronic video coverage as well as alarm sensors.

As discussed in the previous chapter, the operation of a lobby is usually scheduled to allow free access and egress during normal operation of the perimeter doors. After business hours, these doors are often locked from the outside while allowing free egress from the inside, appropriate for a building in which there are after-hour activities. Inside, a reception function may take place. Locking the interior doors leading from the lobby into the rest of the building is a widely accepted approach. All access-controlled doors must have a door closer. This allows for reasonable control of visitor access.

The lobby is a key point to establish a "visitor management system" to identify those who are not regularly credentialed to enter the building, verify their identity, and in many cases have "nondisclosure agreements (NDAs)" signed. This is also an excellent place to provide a safety orientation if special processes or conditions are present in the building.

When more stringent control and an audit trail of activity are needed, mechanical or optical turnstiles can be installed in the lobby. These are used to provide a validating barrier of one-card, one-person access. Biometric readers are coming into wider use within this entry area. Swinging doors are not validating barriers because they permit virtually no control of tailgating. Turnstiles of either type are designed to allow only one person to enter per access reading. By turning on the anti-passback function in the access system, it is possible to keep track of the people who are on site and get a good audit trail of their activity. Anti-passback is a software function that limits the use of an access-authenticating process (i.e., card, biometric) to one use on one side of a barrier. The access-authenticating process must be used on the other side of the barrier before it can be used on the first side again. This application was first developed for parking lots to control the number of cars into the lots. Anti-passback works well with validating barriers such as optical turnstiles or "mantraps" but poorly with swinging doors. Tailgating is so easy with swinging doors that people tailgate often, sometimes even inadvertently. Then, when they come to the same perimeter again, the system "thinks" that they are on the other side and refuses to permit entrance. This is important in high-security applications and in critical safety areas where it is necessary to account for the presence of each individual. Some chip-fabrication plants (fabs) use validating barriers primarily for safety reasons.

Employee entrances are those points of access to and egress from a facility that do not have reception resources. These points should have a barrier consistent with the front-lobby design. If validation is used in the lobby, the employee entrances should also use validation. Another approach to a validating barrier is full-height revolving doors or turnstiles. These are good physical barriers and can operate without a security officer present, which is not the case with optical or low turnstiles. Employee entrances must permit access to many without the need for staffing, so revolving doors are a good option for most sites that have a goal of validation.

Without validation, employee entrances are typically unstaffed access-controlled doors. The designer or planner should expect some tailgating and security breaches at these doors. They are out of view of a security officer, and social conventions call for people to hold the door open for one another. A security design that conflicts with social mores is doomed to failure. In another scenario, a person might leave a facility and head to the right, allowing the next person out to close the door. Another person, standing outside at the left of the door, can catch the door before it closes and walk into the facility. The person exiting may not even be aware of the tailgate entry. Swinging doors provide little security when access is to be controlled.

The number of emergency exits should be sufficient to make a building safe and consistent with building and fire codes. Not every exit needs to be a normal access and egress point, as employee entrances are. Therefore, different controls are appropriate. Alarmed doors with local annunciators/sounders, in which both a central monitoring point is alerted and a sound is heard at the door, are appropriate for many emergency exits. The use of battery-operated annunciators is allowable, but each one needs to be on a regular testing and replacement schedule. It is critical that these portals be monitored—otherwise, the facility will be vulnerable if the controls fail.

Loading docks typically are served by roll-up doors. These can be access controlled by integrating the access-control output with the door controller. Often, there is no need to open the door from outside, but access control is applied to the OPEN button on the inside. Often this button is a key switch, providing for additional local control and operation by authorized users. Roll-up doors are not legal exits, so a pedestrian door next to the larger opening is usually part of the dock design.

The security issues that surround loading docks are typically more operational than physical. When the roll-up doors are open, they create a wide opening for entry. It is common security operational practice to prohibit general access or egress through these openings, reserving the portal for freight use only. To control this use, the responsible individual(s) who oversee the loading dock must enforce the rules of conduct for the space. Operating the dock remotely with cameras and intercoms usually results in weak security. A responsible human presence at the dock when it is open is the best possible practice.

The main power entrance room is only noted here because it is often unusual. This is probably the only room that has access from the outside with no connecting door into the building itself. This is the best design approach because the keys to this room are controlled by the power company. If there is an internal door from this room into the building, it is important to keep it locked on both sides. This is for safety reasons as much as it is for security.

The building's other main utility feeds (water lines, gas lines, data lines) are normally buried and come into the property from buried vaults in the street. On a campus where several structures are located, these feeds—including some centralized HVAC items—traverse the property via a buried series of tunnels. These feeds are vulnerable to attack by either explosive damage or by tampering/hacking into the lines to destroy, divert, or compromise resources or data. No matter what the goal, these openings need to be secured and in many instances monitored. Access points are normally protected with manhole covers or other bolted-down protective metal grates. One-way bolts, welded grates, or specially designed "security manhole covers" should be incorporated into the design.

Heating, ventilation, and air conditioning (HVAC) present other issues as they provide a means for airborne pollutants that may contain toxic chemicals to be inhaled by occupants. The attacks by groups and individuals using sarin gas (on the Japanese subways) and anthrax revealed the need to protect the intake points as well as to segregate or contain the ventilation systems for buildings. The lowest edge of the outdoor air intakes should be placed at the highest feasible level above the ground or above any nearby accessible level (i.e., adjacent retaining walls, loading docks, handrails). The goal is to minimize public accessibility, with higher being better.

The roof of a building is often overlooked from a security standpoint. Depending on the building and whether it is freestanding or adjacent to other buildings of its height, the roof can be an access point into the structure. In some jurisdictions codes require free access to and from the roof to prevent someone from being trapped there and/or to allow for an escape route. The security of any roof opening at a minimum should be a sensor point that is monitored by a security intrusion system. This provides knowledge that the point was breached, whether in an authorized manner or an unauthorized one. In most applications, the roof access is secured from within the building, often by means of a maintenance padlock. In other cases, the door may be secured with a card reader.

INTERIOR DOORS

An interior door in a structure has the potential to be a rated assembly, which means that the door itself, its frame, and all hardware are tested to withstand a fire for a specified period of time. Exterior doors are seldom fire rated. They are usually fire-egress doors, but those pose a completely different set of issues. The criteria applied depend on the location of the door. Stairwells are typically rated for 2 hours of protection from fire intrusion, whereas internal hallways that are part of the exit path are usually 1-hour corridors. Wherever a door penetrates a rated wall, the door assembly must be rated.

The primary hardware criterion for fire rating is that the door be self-closing and self-latching. This implies that each door leaf must have a door closer; that the door may not be held open by a mechanical device, including a fusible link (a soft metal chain link that melts when exposed to fire); and that whenever the door closes, it must latch and close securely. Note that there is no requirement for locking, just latching. As part of the security needs of a building, security doors are often rated.

Security doors must latch securely as well. When a rated double-door assembly is used, it typically uses a coordinator to make the correct door close first, automatic flush bolts in the inactive leaf, and a lock set in the active leaf whose latch penetrates the inactive leaf. The "inactive" leaf is the leaf of the door that does not have a lock set in it. It is not truly inactive, just less used and therefore less active than the "active" leaf. Of course, door closers should be on both leaves. The automatic flush bolts project down into the floor and up into the frame on the basis of the pressure of the active leaf pushing on the tabs that project from the inactive leaf. A great deal of friction is created in this process, and doors having this hardware are a constant maintenance problem because they tend to not latch properly. The best idea is to use single doors to the greatest extent possible. Single doors are many times more reliable than double doors. Security doors need to be solid core or metal doors, have protected hinges, and be properly installed within the door jamb.

The most common interior door is a card-access (or other authentication means—biometrics, keypad), free-exit, single door. This door normally includes a request-to-exit input on an access-control system, to shunt the alarm for the door when a person exits using the unlocked handle from the inside. The handle outside the space is locked except when an access card or other authentication means is presented to an associated reader. This lock set should be fail-secure so that if the system is broken or out of power, the lock operates just as any typical lock set would. The inside handle is always mechanically free.

Mantraps (sallyports) can be used for perimeter-access or interior-access portals. While they provide better security, the length of time it takes the user to move through both interlocked doors and possibly be subjected to biometric screening in between the two doors is greater. Therefore, mantraps are usually used for higher-security areas within a structure, where a small percentage of the population needs access. This keeps the time users must spend on this special security process down to a minimum.

Mantraps make poor egress points. The best way to implement a mantrap is to have alternative routes for egress; then it is possible to truly interlock the doors of the mantrap. By making the exterior door fail-safe and the interior door fail-secure, it is also possible to put an emergency door release within the mantrap that cuts all power to the locks. If this release is pulled, the outer door unlocks and the inner door locks. This leaves the interior protected space secure and allows the person who pulled the release to exit the mantrap toward the public side.

CIPHER OR COMBINATION LOCKS

Access-control systems are more expensive than some other devices that provide similar protection but with fewer audit and authenticating processes. When there is a need for greater than normal security, but not ongoing monitoring or higher level authenticating, sometimes a combination lock or keypad is installed in place of the former keyed lock set. Although this may seem to solve the key-control problem, it does not provide the level of security that a keyed lock does. It is easy to watch surreptitiously while someone enters a code. Sometimes the combination or number sequence is written near the door. In case after case, the code is just blurted out across the room or over the telephone, or the combination is the building's address, 911, or another code just as easy to figure out. It is impossible to know when a code has been compromised. If a key is lost because of its tangible nature, it is reasonable to assume that it is not in the possession of the owner. A code does not provide that assurance. Thus, combination or keypad locks provide a lower level of security than keyed lock sets.

Combination or keypad locks are useful when tight security is not essential. One example might be the interior doors of a lobby for a site without access control. The front lobby door can employ a classroom-type lock set to the unlocked state during the day and to the locked state after business hours, when visitors are not expected. Within the lobby, an employee can come out to greet a visitor and, using the combination or keypad, lead the visitor to the rest of the site. Such doors are intended for operational control, not high-perimeter security, and the application is cost-effective. There is often little use for an audit trail of the activity of such interior lobby doors. If the site had an access-control system, using card readers in the lobby would probably be reasonable as well. Video monitoring can augment security at this location to ensure a record of who is in the area and what activity occurs.

Combination or keypad locks can be battery operated, allowing multiple codes and an audit trail. Many lock manufacturers use an infrared data transfer to program the codes and to collect the lock history. Although this history is available, it is certainly not convenient to collect if there are many such locks on site.

SUMMARY

For effective security, several rules should be observed:

- Always use a mechanical means for egress.
- Gates and bollards provide perimeter protection from vehicle ramming.
- Remember not to treat all emergencies as you would fires.
- Use of keys requires correct application, and key control is critical.
- Design access-control doors to implement fail-secure lock sets, whenever feasible.
- Whenever possible for access-controlled doors, use storeroom-function lock sets.
- Whenever possible, use mortise lock sets, especially with electric strikes.
- Whenever possible, use electrified mortise lock sets for access-controlled doors.
- Incorporate a door closer into the design of all access-controlled doors.
- Remember that a security design in conflict with social mores is doomed to failure.
- Use single doors to the greatest extent possible.
- HVAC system intakes need to be placed at higher elevations in building designs.
- The roof provides a means of access into the structure.
- Combination or keypad locks provide a lower level of security than keyed lock sets.

BIBLIOGRAPHY

Craighead, G. *High-Rise Security and Fire Life Safety.* 2nd ed. Newton, MA: Butterworth-Heinemann, 2003.

Demkin, J. A. *Security Planning and Design: A Guide for Architects and Building Design Professionals.* New York, NY: John Wiley & Sons, 2003.

Fennelly, L. J. *Handbook of Loss Prevention and Crime Prevention.* 4th ed. Newton, MA: Elsevier/Butterworth-Heinemann, 2004.

Garcia, M. L. *The Design and Evaluation of Physical Protection Systems.* 2nd ed. Newton, MA: Elsevier/Butterworth-Heinemann, 2007.

Khairallah, M. *Physical Security Systems Handbook: The Design and Implementation of Electronic Security Systems.* Newton, MA: Elsevier/Butterworth-Heinemann, 2006.

Patterson, D. G. *Implementing Physical Protection Systems: A Practical Guide.* ASIS International, 2005.

Post, R. S. *Determining Security Needs.* Des Plaines, IL: Nickerson Collins, 1973.

Purpura, P. P. *Security and Loss Prevention: An Introduction.* 5th ed. Newton, MA: Elsevier/Butterworth-Heinemann, 2007.

U.S. Government Printing Office. *Guidance for Protecting Building Environments from Airborne Chemical, Biological, or Radiological Attacks.* Washington, DC: Department of Health and Human Services (DHHS), DHHS (NIOSH) Publication No. 2002-139, May 2002.

———. *Physical Security.* U.S. Army Field Manual No. 3-19.30. Washington, DC: 8 January 2001.

———. *Physical Security Program.* U.S. Department of Defense, DoD 5200.08-R. Washington, DC: U.S. Government Printing Office, 27 May 2009.

———. *Reference Manual to Mitigate Potential Terrorist Attacks Against Buildings.* Washington, DC: U.S. Federal Emergency Management Administration (FEMA), FEMA-426, December 2003.

———. *Unified Facilities Criteria (UFC).* DoD Minimum Antiterrorism Standards for Buildings, UFC 4-010-01. Washingon, DC: 8 October 2003.

20

Secured Storage Areas

Kenneth G. Fauth, EdD, CPP

OBJECTIVES

The objectives of this chapter are as follows:

- consider the protection afforded by safes against fire, burglary, or both
- explore the benefits and limitations of vaults—small and large, modular, and those that address special needs
- examine alternative systems for short-term and long-term storage

INTRODUCTION

What is your view of a secure storage compartment? Do you envision a small hidden compartment that is secret and therefore safe from others? Or do you picture a large commercial bank cash vault that may contain hundreds of millions of dollars in paper currency that need to be accessed by many people? Each of these views could be considered an extreme case, but neither is outside the appropriate range of discussion. Some assets have very little volume and extremely high value. One small-volume example is a set of hardware and software keys stored in a computer system used to authenticate legal software. The cash vault example is typical of a large-volume asset. Basically, size and contents determine what type of container, vessel, or area you are going to use to protect the contents from known hazards such as fire, theft, destruction, and compromise.

The common standards for protection of these varying assets are based on practicality and common sense. For instance, fire is one important hazard. Typically, the assumption is that fires usually start accidentally. Safes can carry a rating or listing for fire that determines how long paper items stored within can withstand the fire. Underwriters Laboratories, Inc. (UL) is one of the better known rating agencies that have created a set of ratings, standards, or listings describing the protection offered by a particular design. With the advent of computer media and the more delicate nature of the magnetic media on which data is stored, new categories for media safes have been created. The ratings are based on exposure to fire at conditions that approximate a hot building fire, along with physical trauma to the safe that might occur when a building collapses. Fire rating is only one of the protection standards or listings.

Security standards or listings are based on a physical, forced attack using standard sets of tools, with different ratings for different circumstances and penetration times. The design features that create burglary resistance are almost the opposite of those that create fire resistance. Since the protection needed to guard against fire is so different from that needed for a physical, forced attack, the two types of safes have different designs. Therefore, one cannot expect a fire safe to provide much physical protection in a burglary. Similarly, a burglary-resistant safe provides little protection against fires. This is why it is not uncommon to find one type of safe physically located within the other. However, multipurpose containers or safes that will meet both fire and security protection needs are available.

A container is generally described as a reusable receptacle that can accommodate smaller objects or items. A safe is similar in that it is considered something into which you would reach to place and retrieve items, but it offers a higher degree of safety than a container. A vault, on the other hand, is a walk-in space that is protected according to similar considerations. Containers can be small metal cash boxes, file cabinets, or vertical wall-type cabinets. Safes are generally small to medium in size, ranging from the size of a shoe box to the size of a washing machine. Vaults start at the size of closets and can be large enough to house many assets—and even the people handling these assets. Some military vaults house control centers that are key sites designed to survive a thermonuclear war.

FIRE AND SECURITY RATINGS, STANDARDS, AND LISTINGS

Throughout the world, in different countries or areas, organizations or governmental bodies have been created to test and provide a certification, rating, standard, or listing regarding the capabilities of certain products to withstand certain actions taken against them. In the United States, great emphasis is placed on ratings, standards, or listings provided by Underwriters Laboratories, Inc. (UL). UL is an independent organization that tests products produced by different manufacturers to ascertain if they meet certain thresholds. The American Society for Testing and Materials (ASTM) International and Insurance Services Office, Inc. (ISO) are other rating systems and services used for establishing resistance ratings. For our purposes in this chapter, UL references will be used.

For fire, these ratings are based on standard time-temperature curves for typical types of fires. UL Classification 350 is for records, UL Classification 150 is for media, and UL Classification 125 is for computer data media and hardware. Thus, the lowest rating for records is 350-1, for 1 hr, and the highest record rating is a 350-4, for 4 hrs. The lowest rating for media is 150-1, for 1 hr, and the highest rating for media is 150-4, for 4 hrs. The lowest rating for computer data media and hardware is 125-1, for 1 hr, and the highest rating for computer data media is 150-3, for 3 hrs. All three UL classifications (350, 150, and 125) are Fahrenheit temperatures and the container rating/listing means that the internal compartment where objects are stored will not exceed these temperatures.

The comparisons and exact characteristics for each rating are available from UL and all companies that manufacture safes. These ratings must be considered within the context of the application. A site with a high fire loading that may react atypically to a fire must use a higher-than-normal rating. (The National Fire Protection Association (NFPA) 232 is another fire-rating system for safes.)

To measure protection against physical, forced entry, UL utilizes ratings or listings such as TL-15, indicating that the safe is resistant to an attack with a specified set of tools for 15 mins. Thus, a TL-30-rated safe is more resistant to an attack than a TL-15-rated safe. Ratings for a key-locked safe that is resistant to burglary can range all the way up to TRTL-60, which conveys resistance to torches, tools, and explosives for 60 mins. It is important to review the cost benefit of the many options available. To do this, a good safe company will normally show how its products fit into the range of ratings and how its costs apply to each category. Lock-manipulation ratings are assigned to the locking mechanisms themselves.

For vaults, there is a UL Classification 608-1988 Burglary Resistant Vault Doors and Modular Panels with Class M (15 mins of resistance); Class 1 (30 mins)—vault wall having 12-in.-thick (30.48 cm) poured, reinforced concrete; Class 2 (60 mins)—vault wall having 18-in.-thick (45.72 cm) poured, reinforced concrete; and Class 3 (2 hrs)—vault wall having 27-in.-thick (68.58 cm) poured, reinforced concrete.

The ANSI/UL 768-1987 covers combination locks. The ANSI/UL 887-1994 defines the standards for Delayed Action Time Locks. The ANSI/UL 680-1991 defines Emergency Vault Ventilators and Vault Ventilating Ports. The UL 140 defines the Relocking Device Standard.

What creates the different ratings/listings for containers is the combination of things they include: what materials the walls, top, bottom, and hinges are constructed of; the type of insulation; the type of locking mechanism(s) used; what occurs when the container is dropped from a height; the integrity of the locking mechanism; and the assembly of the product. Modifications to the listed or rated product after purchase beyond the manufacturer's guidelines may void all ratings and listings. For instance, drilling holes for cables or alarm wires can void the rating on fire safes. The heat-penetration testing is based on insulation that restricts the spread of heat. Placing holes or conduits through the insulation normally lowers the resistance to heat transfer, rendering the safe worthless for its intended rated purpose.

CONTAINERS

As previously mentioned, a container is basically a reusable receptacle into which items or objects may be placed to provide for their storage away from general view. Containers can differ greatly in purpose, as some are simply used to help organize items and provide for efficient storage, while others are designed to provide additional protection. This additional protection may be in the form of security safeguards, fire protection safeguards, or both. In many cases, containers may be made of wood or light metal and have "privacy locks" (to avoid an accidental opening)—think of a locked desk drawer or a file cabinet. Other containers, such as a U.S. General Services Administration (GSA)–required file cabinet to hold classified material, may need to meet the same requirements as a high-security safe.

An issue with containers is that they are generally considered just a place to put items, where they remain out of sight until needed. Unless they are upgraded to provide for fire protection and/or protection from physical, forced entry, they will not be very secure. This is a concern, as employees often see them as "out of sight, out of mind" and place materials in them that should really be placed in a safe or vault to achieve the needed protection level.

SAFES

Safes are an excellent type of container that can protect objects from damage or theft but, to provide protection, they must be of the proper type and rating and be used correctly.

Safes that are used for protection against forced, physical attack or burglary should carry a UL label rated or listed for an appropriate type of attack and time period. What is appropriate? This is a judgment call that should be based on common sense or may be specified by insurance or contract (i.e., for recipients of U.S. classified documents). If the safe is in a room that is covered by an intrusion alarm system and the alarm response time is known, the safe rating should be based on these conditions. The decision will be different if the safe is not in an alarmed area but can be viewed through the front window of a store and patrols are common. In many cases, the decision is likely to be based on the judgment of the purchaser of the safe about cost and hassle and the time and effort required by a perpetrator to break into the safe. In the absence of insurance or contractual requirements, the entire process is ultimately based on the security paradigm: "Define the perimeter; detect the intrusion; delay access to the asset; and have a response."

Safes provide security only if they are managed appropriately. Here are a few examples of the findings of security surveys over the years that bring to mind the need for reasonable management:

- The safe is closed, but the dial on the combination lock that is used to secure the door has not been spun, or the handle has not been turned to lock it. This safe can be opened without a combination as easily as an unlocked door.
- The safe is located near a desk within which the combination is noted on a piece of paper or on the last page of the desk calendar.
- The combination is given over a desktop, portable, or cellular telephone.
- The combination has not been changed since former employees, who know the combination, left.
- The special-use safe has a cylinder and crank on top so that money can be dropped in without opening the safe, but a stack of cash is left just sitting on top in the cylinder.

The actual location and installation of a safe is critical to its overall protection. Safes can be installed in a structure. This is commonly seen in small-merchant locations, where the safe is installed before the slab is poured. A lip on the back of the safe precludes removal of the safe from the structure. Some safes are manufactured with wheels so they can be moved. These typically rely on the physical mass of the safe to lower the potential for removal of the entire safe. Obviously, if the safe is stolen and moved to a private location, the penetration time is not a limiting factor for the criminal. Portable safes are almost a classic oxymoron.

One case pointed out the perils of safes that are not securely affixed to the structure. An electronic manufacturer that used gold in the process of making electronic parts stored the gold in a 3,000-lb (1,361-kg) safe in a building whose parking facility was located right next to the building. The safe was on the interior of the alarmed building but backed up to an exterior wall. One night, thieves who knew the location of the safe came with a truck. Using a large forklift, they punched a hole through the exterior wall, dragged the safe out, and put it on the truck. The stolen forklift was left at the scene. The thieves were never caught. Unlimited time was available to crack the safe at some remote site.

CASH-HANDLING SAFES

Retail and financial institutions are utilizing different variations of safes to not only store money but also disburse it in various amounts to preauthorized cashiers and users. Use of "smart safes" has become very popular, as they not only provide physical protection but their computer software now reconciles deposits or disbursements. The various manufacturers of these specialized safes have ferreted out a business market for them. Their use has improved efficiency for many businesses, contributing to cash-handling integrity and providing for more flexibility in staffing.

DOCUMENT SAFES

Document safes are special safes designed to protect documents from being destroyed or rendered unusable. The main purpose is protection of the document from external environment actions (such as fire, light, and fumes) that would react with the paper or ink in a negative manner. Fire-rated safes for paper and media carry UL ratings similar to those for the burglary-resistant safes discussed earlier. They can protect such items as corporate incorporation records and board of directors' actions that are critical to establishing the legitimacy of the business and yet may be needed on a day-to-day basis. At a personal level, safeguarded items may include negotiable bonds, wills, special one-of-a-kind pictures, or heirlooms.

COMPUTER MEDIA SAFES

Electronically stored records, such as data cartridges, flash drives, and CDs, are particularly vulnerable to heat and humidity, making insulated storage and fire protection a necessity. Of course, these types of media also need theft protection. Computer media safes are designed for the storage of computer software and electronic records or media conveyance hardware (hard drives, data storage files, CDs, etc.). Computer media data are easily damaged and should not be exposed to temperatures over 125 F, humidity over 85%, dust, magnetic fields, accidental damage, and electrostatic charges. In addition, computer media safes provide protection from unauthorized access that can lead to data compromise, destruction, or corruption of the files or programs. As such, these types of safes require special attention from a security point of view.

AUTOMATIC TELLER MACHINES (ATMS)

Automatic teller machines (ATMs) have become a way of life, with many located within buildings for the convenience of employees and visitors. While the financial institution usually accepts responsibility for them, there are also conditions that need to be met for the installation to take place. Although these machines are alarmed, thieves have not bothered to disconnect the alarms or the 120-v power when they have used trucks and winches to remove the whole machine. One ATM may contain hundreds of thousands of dollars when it is full. Careful placement of ATMs has made this type of attack less successful than in the past.

VAULTS

Vaults are essentially large safes in the form of reinforced rooms constructed of concrete. Vaults normally have a steel door equipped with various styles of locking mechanisms. The size of vaults may vary. The U.S. Federal Reserve vault is so large inside that an automated warehouse delivery system "puts and pulls" various bins from high-rise storage racks using a computer program.

In an armored car service main office, the vault is surrounded by a large secured work area controlled by mantraps for the workers. Truck bays with several doors are used to load and unload trucks of cash. A truck delivery starts with all doors closed and the truck showing up at the entrance. The outer door is open as long as the other doors are closed. This is an interlock system that never leaves a clear path to the exterior. Only when the truck is parked inside, with the exterior door closed, is either of the interior doors opened. Two interior doors in series are common, providing for the receiving team an interposing chamber between the drivers and bank personnel. Cash is put into this interposing area, and the appropriate doors opened and closed, in such a way that the drivers and bank personnel are never actually in the same space at the same time.

One critical element of large-vault operation is the exit plan. In a fire it is important to get people to safety; however, one cannot just put exit doors on a vault. This is why a common element in the design of large vaults is two vault doors. Typically these are within a secured working area as just described. It is important to control exiting from this area. If the controls were weak, an employee could be in collusion with thieves dressed as fire fighters waiting outside the building. When the employee pulled a fire alarm and people started exiting, the fake firefighters could rush in and take over the vault. Therefore, good fire detection and suppression allow time to use interlocks and special-circumstance controls on egress paths without risk to life in a true emergency.

SPECIAL VAULTS

Archives often are kept in vault systems. State and federal archives are commonly kept in vaults where temperature and humidity are carefully controlled to ensure preservation. Archives require both security and environmental controls at the openings. Sometimes these are in operational conflict. One solution has been to create smaller openings to accommodate people traffic and possibly a small cart. Larger items or large volumes of material can be moved through interlocked environmental chambers that keep the appropriate controls in place.

Archive documents are used for historical studies. This activity should be carried out within a controlled environment adjacent to the archive vault. Areas for restoration, cleaning, cataloguing, and study all need to be environmentally controlled and secured.

In rare cases, high-value documents are displayed in cases that are lowered into custom vaults at night. During the day, the display case is on a public level fitted with special glass to screen out harmful solar rays. If a potential problem is sensed by the operators of the archives, the display can be quickly lowered and secured.

The size of the vault normally creates additional issues such as lighting and ventilation for occupants who come and go. There are two major components for a vault: the walls and the door entryway.

Walls. Normally, buildings such as banks are constructed around their vault. The vault is located in a basement or ground floor due to the weight of its walls. When constructed of poured reinforced concrete, the walls can range from a UL Classification Class M to a 1, 2, or 3. Poured reinforced concrete has been the most cost-effective construction material. However, modular walls can provide greater flexibility and may be more cost-effective, depending on the situation.

As mentioned, an alternative to poured concrete vaults are the modular-style vaults produced by different manufacturers. One type of modular vault wall is made of proprietary (Diebold) designed high-density composite concrete mix materials, steel fibers, and other substances; this wall has resistance similar to that of poured reinforced concrete but is lighter and thinner. For example, a UL Class 1 rating requires 12 in. (30.5 cm) of reinforced concrete versus 2.5 in. (6.4 cm) of the proprietary material. This represents a big difference in volume displacement as well as weight. A second example is a modular vault wall made of steel plate having strength capabilities similar to concrete.

The relative thickness or portability of modular vault systems is making these increasingly popular. They are either lighter in weight or thinner than concrete vaults and often can be placed in existing structures. They have all of the attributes of large, poured-in-place, reinforced concrete vaults, although the ratings for steel plate are limited in some cases. Steel plate is more subject to a burning attack than concrete. It is relatively easy to punch a series of holes into the side of a steel-plate vault with burning bars made of pipes fed with oxygen and lit with acetylene torches. The process, however, is time-consuming, and for this reason, smoke detectors are a common part of any vault's security system. Since some of these modular vault walls are not rated as highly, they are not sufficient for all applications; however, they are excellent for many.

Doors. A typical vault door is made of steel and is very heavy. Manufacturers have made a great effort to design these heavy doors with a hinge design that allows for easy opening and closing. Some doors are constructed with heat-resistant ceramic materials, while others use concrete filling between steel plates.

LOCKING DEVICES

Locks come in many different forms and sizes to perform a myriad of tasks, from simply keeping a door from opening due to uneven floors or air currents to preventing access unless the lock and materials around it are destroyed. There are many types of locking devices, even locks that lock other locks. Electronic locking mechanisms are coming to the forefront and are regularly combined with manual mechanisms to increase the level of security.

Security depends on keeping unauthorized individuals out of the protected area and away from valuable objects or items. In order to do this, the integrity of the locking system needs to be maintained. Whether it is accessed via a key, a series of numbers, or an access card, the locking operation requires care and attention. With the introduction of biometrics access systems, the level of protection can be raised. How the locks interface with the latches or locking bars that actually hold the door or drawers in place is critical. Again, testing and classifications, listings, or ratings from testing agencies are paramount to achieving reliability and confidence in the locking system.

LOCKS FOR SAFES, WALL STORAGE UNITS, AND FILE CABINETS

Many manufacturers provide a basic lock set on safes, wall storage units, and file cabinets. In most cases, unless otherwise specified, they are of basic quality that provides little or only a "privacy" level of security (for example, they may prevent the unit from accidentally falling open). Again, depending on the degree of protection needed or required, locks can usually be specified to meet certain standards. These standards are established by various organizations including UL and, for many related government activities, by the U.S. General Services Administration (GSA). Multiple U.S. manufacturers provide locks and locking mechanisms that also include "high-level security" capabilities. One type of lock that is considered to provide "high-level security" has diagonal pins and requires a diagonally cut key. Padlocks can have specially designed shells constructed of hardened steel. Combination locks have specially constructed discs and lead plates to block the use of x rays to reveal settings. Newer electronic locks require the entering of a special sequence of numbers. The use of biometrics, especially a designated fingerprint, is gaining in popularity and application. The time-honored version of a timing mechanism that only unlocks after a set time period remains a good secondary level of security in any locking mechanism configuration.

All of these features can be utilized together to form multiple usage groupings. In some cases two locks must be activated by two different individuals before access is allowed. One method that is required in certain applications is a "relocker" (UL 140). Relocking devices are intended to relock the bolt mechanism or door of a vault, safe, or chest in the event that the combination or other lock is subjected to attack. This serves as a backup to the basic lock and creates additional obstacles that an unauthorized intruder must overcome to gain entry.

VAULT LOCKS

The types of locks described above for safes also apply to vaults. One additional component often found in conjunction with a vault lock is a time lock. A vault door is set with a clock timer that will not unlock (create a clear pathway for the locking bar or bars to slide through) until the clock reaches a set time. Normally, the clock would be set for the next business morning, when multiple employees would be present to conduct transactions. Unfortunately, the setting of this mechanism also limits access before that time arrives. Any person locked inside the vault is unable to get out until the vault is next opened. Thus, ventilation systems were required in vaults for life-safety purposes.

SAFE AND VAULT PROTECTION SYSTEMS

Safes are small objects, and the most effective method of protecting them is locating them within a protected area. This can be accomplished by using typical burglar alarm devices and possibly closed-circuit television (CCTV) for alarm qualification. Contacts can be added to safe doors if care is taken not to void any of the ratings. Weight sensors below the safe can be used. Securing the safe to the floor or anchoring it in concrete is also advised. Security officers may be warranted for patrols, and definitely for response to alarms.

Vaults should be procedurally protected and guarded, when appropriate. Specific to vaults is the use of heat detection, smoke detection, vibration detection, and motion detection. The first three are good ways of detecting the typical methods of breaking into a vault. Motion detection is used mainly to detect a person who hides in a vault, waiting for closing. Once the vault is closed, this person can set up a delivery or possibly a takeover for the morning.

Cash handling is one area in which actual CCTV monitoring of hand movements on a work surface can be a reasonable and cost-effective process. Video surveillance of these transactions is crucial to good security. Video records of this activity can easily be accomplished through the use of digital video recording (DVR) technology.

Good lighting is critical; halogen lighting has proven to be effective because of its accurate color rendition and its brightness. Use a relatively dark work surface, and position the lighting and cameras so that they complement the image and do not create glare. Recent developments in camera technology, however, have made lighting almost a nonissue.

Ultimately, safes and vaults are no different from other stored assets. The critical issues are consistent policy and procedures, definition of boundaries, detection of boundary penetrations, and delay of access to the assets.

SUMMARY

Selecting the appropriate safe, vault, or storage system requires careful analysis. Typically, a container has fire protection or burglary protection as its main focus. The relative risks present, the size of the operation, the number of people who have access to the area, the location, and whether the system is to be installed in a preexisting structure or included in the building plan are crucial to determining the system or combination of systems ideally suited to a particular facility or use.

BIBLIOGRAPHY

Demkin, J. A. *Security Planning and Design: A Guide for Architects and Building Design Professionals*. John Wiley & Sons, 2003.

Fennelly, L.J. *Handbook of Loss Prevention and Crime Prevention*. Newton, MA: Butterworth-Heinemann, 1982.

Garcia, M. L. *The Design and Evaluation of Physical Protection Systems*. 2nd ed. Elsevier/Butterworth-Heinemann, 2007.

Khairallah, M. *Physical Security Systems Handbook: The Design and Implementation of Electronic Security Systems*. Elsevier/Butterworth-Heinemann, 2006.

NFPA 232: Standard for the Protection of Records. Quincy, MA: National Fire Protection Association, 2007.

Patterson, D. G. *Implementing Physical Protection Systems: A Practical Guide*. ASIS International, 2005.

Post, R. S. *Determining Security Needs*. Des Plaines, IL: Nickerson-Collins, 1973.

Purpura, P. P. *Security and Loss Prevention: An Introduction*. 5th ed. Newton, MA: Elsevier/Butterworth-Heinemann, 2007.

Roper, C., and B. Phillips. *The Complete Book of Locks and Locksmithing*. 3rd ed. McGraw-Hill Publishing, December 1990.

Tobias, M., *Locks, Safes and Security: An International Police Reference*. 2nd ed. Charles C. Thomas, 2000.

Underwriters Laboratories Inc. Online Certification Directory. http://database.ul.com/cgi-bin/XYV/template/LISEXT/1FRAME/fireressrch.html (accessed August 27, 2009).

U.S. Government Printing Office. *Federal Specifications 71 III E Category Listings: Miscellaneous Furniture—Security Files, Safes, Vaults*. Washington, DC: General Services Administration (GSA), 2008.

U.S. Government Printing Office. *Physical Security Handbook 440-2-H*. U.S. Geological Survey Manual. Washington, DC: Department of the Interior, August 2005.

U.S. Government Printing Office. *Physical Security Program*. U.S. Department of Defense, DoD 5200.08-R, Washington, DC: U.S. Government Printing Office, 27 May 2009.

Part 7

Critical Incident Management

CHAPTER 21

21

Developing an Emergency Operations Plan

Michael J. Fagel, PhD, CEM

OBJECTIVE

The objectives of this chapter are as follows:

- define an emergency operations plan, its purpose, and its functions
- understand the steps in the emergency planning process
- explore the hazard profile as a starting point to an emergency operations plan
- learn the key elements that should be included in a jurisdiction profile
- create effective instructions for implementing an emergency operations plan

INTRODUCTION

An emergency operations plan (EOP) is a jurisdiction's public document that assigns responsibility to organizations and individuals for carrying out specific actions, at projected times and places, in emergencies that exceed the capability or routine responsibility of any one agency. This chapter sets forth lines of authority and organizational relationships, and shows how all actions will be coordinated. In addition, an EOP:

- describes how people and property will be protected in emergencies and disasters
- identifies personnel, equipment, facilities, supplies, and other resources available within the jurisdiction, or by agreement with other jurisdictions, for use during response and recovery operations
- identifies steps to address mitigation concerns during response and recovery activities

PURPOSE OF AN EOP

Planning a response to emergencies and disasters is typically the responsibility of state and local governments. The elected leadership in each jurisdiction is legally responsible for ensuring that necessary and appropriate actions are taken to protect people and property from the consequences of emergencies and disasters.

When a disaster threatens or strikes, citizens expect their elected leaders to take immediate action to deal with the situation. The government is expected to marshal its resources, channel the efforts of voluntary agencies and private enterprises in the community, and solicit assistance from outside the jurisdiction, if necessary. The development of a comprehensive, all-hazard EOP will help ensure that all government response activities are undertaken efficiently and effectively.

THE PLANNING PROCESS

In today's system of emergency management, local government must act to attend to the public's emergency needs. Depending on the size and nature of the emergency, state and federal assistance may be provided; however, local governments should not assume this type of assistance will be available.

Therefore, the local EOP should focus on the functions that are essential for protecting the public before and after a disaster. Minimally, these functions include warning, emergency public information (EPI), evacuation, and providing shelter.

Emergency planning is not a one-time event. It is a continual cycle consisting of planning, training, exercising, and revising that takes place throughout the four phases of the emergency management cycle (mitigation, preparedness, response, and recovery).

The planning process does have a single purpose: the development and maintenance strategy for addressing critical needs in an emergency—to protect life and property. Although the emergency planning process is cyclical, it does have a definite starting point, as follows:

1. Hazard analysis. Emergency planning begins by analyzing the hazards facing the jurisdiction. Hazard analysis is the process by which hazards that threaten the community are identified, researched, and ranked according to the risks they pose and areas and infrastructure that are vulnerable to damage from an event involving those hazards. The outcome of this initial step is a written hazard analysis that quantifies the overall risks to the community from each hazard.
2. EOP development. The next step in the emergency planning process is EOP development. The outcome of this step is a completed plan, which is ready to be trained, exercised, and revised, based on lessons learned from the exercise.
3. EOP testing. The third step in the planning process is testing the plan through training and exercising. Exercises of varying types and complexity allow for evaluation to see what in the plan is unclear and what does not work. The outcome of this step is a series of lessons learned about weaknesses in the plan. These weaknesses can then be addressed in the final step—plan maintenance and revision.
4. EOP maintenance and revision. This step in the planning process can be completed based on needs and resources, which may have changed since the development of the original EOP. After the EOP is developed, step 3 can be repeated in a continual cycle to keep the plan up to date. If the community becomes subject to a new hazard, however, the planning team will need to revisit steps 1 and 2.

PERFORMING A HAZARD ANALYSIS

A hazard analysis determines:

- what can occur in the community
- how often the damage is likely to occur
- the damage it is likely to cause
- how likely it is to effect the community
- how vulnerable the community is to the hazard

There are four steps in the hazard and analysis process:

1. Identify hazards and profile each hazard.
2. Develop a community and jurisdiction profile.
3. Determine vulnerability.
4. Create and apply scenarios.

There are many potential sources of hazard information. A starting point may be to check local newspapers. To get a more complete picture of the types of hazards that a community has faced historically, it may be necessary to check sources such as the State Department of Agriculture, Bureau of Labor Statistics, or other similar agencies; the National Weather Service;

local historical societies; or long-time residents. If your community has an existing hazard analysis, start by reviewing it with an eye toward what has changed in the area since the hazard analysis was completed.

During the hazard analysis process, it is important to keep in mind that the hazards a community faces may change over time because of new mitigation measures, the opening or closing of facilities, local development activities, or terrorist threats that were not considered before the terrorist attack of September 11, 2001. There may be other long-term changes to investigate as well. Changes in average temperature or rainfall and snowfall amounts may be harder to track but will play an important role in a complete hazard analysis.

IDENTIFY AND PROFILE THE HAZARD

A hazard profile should consider four factors:

1. magnitude
2. duration
3. seasonal pattern
4. speed of onset

A hazard profile should address each hazard's magnitude—or size. How strong a hazard is and the areas that it could affect can dramatically change response plans. For example, a storm that drops 2 in. (5 cm) of rain very quickly over a small area requires a much different response than a nor'easter that drops 20 in. (51 cm) of rain over a four-state area.

It is also important to consider a hazard's frequency, including whether a seasonal pattern exists. In some parts of the country, thunderstorms are a near daily occurrence. On the other hand, hurricanes are a seasonal occurrence, which may or may not present a high risk to your area.

Consider each hazard's duration, or how long the hazard is expected to last. For example, the duration of even the most severe thunderstorm is much shorter than that of a hurricane.

Finally, consider the speed of onset of the hazard. This is important for determining the available time for issuing a warning, and is also critical to the response. The amount of damage and loss of life that an extreme hazard could cause can be mitigated if emergency personnel and the public have time to take protective action.

A profile should be completed for each hazard to which the community is vulnerable, but it is important to keep in mind that some hazards pose such a limited threat that additional analysis may not be necessary. However, low-risk hazards that have a high potential for damage if they occur should not be ignored. These low-risk hazards may not be a planning priority, but they should be planned for nonetheless. (See the hazard profile worksheet shown in Figure 21–1.)

CREATE A COMMUNITY AND JURISDICTION PROFILE

After identifying the hazard, it is necessary to combine hazard-specific information with a profile of the community to determine the community's vulnerability to or risk of damage from the hazard. Because different communities have different profiles, vulnerabilities to the same hazard will vary. Table 21–A summarizes key factors that are included in the community profile. After gathering information about the community, develop the community's jurisdiction profile by plotting vulnerable areas on a jurisdiction map. Table 21–B shows the use of community factors in a jurisdiction profile.

Hazard:
Potential Magnitude (Percentage of the community that can be affected): ❏ *Catastrophic:* More than 50% ❏ *Critical:* 25 to 50% ❏ *Limited:* 10 to 25% ❏ *Negligible:* Less than 10%
Frequency of Occurrence: ❏ *Highly likely:* Near 100% probability in the next year ❏ *Likely:* Between 10 and 100% probability in the next year, or at least one chance in the next 10 years ❏ *Possible:* Between 1 and 10% probability in the next year, or at least one chance in the next 100 years ❏ *Unlikely:* Less than 1% probability in the next 100 years
Seasonal Pattern:
Areas Likely to Be Affected Most:
Potential Speed of Onset (Probable amount of warning time): ❏ Minimal or no warning ❏ 6 to 12 hours warning ❏ 12 to 24 hours warning ❏ More than 24 hours warning
Existing Warning Systems:
Does a Vulnerability Analysis Exist? ❏ Yes ❏ No

Figure 21–1. Hazard Profile Worksheet

Table 21–A. Key Community Factors

Geography	*Property*	*Infrastructure*	*Demographics*	*Response Organizations*
• Major geographic features • Typical weather patterns	• Numbers • Types • Ages • Building codes • Critical facilities • Potential secondary hazards	• Utilities, construction, and layout access • Communication systems layout, features, and backup • Road systems • Air and water support	• Population size, distribution, and concentrations • Number of people in vulnerable zones • Special populations • Animal populations	• Locations • Points of contact • Facilities • Services • Resources

Table 21–B. Use of Community Factors in the Jurisdiction Profile

Types of Information	*Used In*
Geographic	Predicting risk factors and the impact of potential hazards and secondary hazards
Property	Projecting consequences of potential hazards to the local area Identifying available resources
Infrastructure	Identifying points of vulnerability Preparing evacuation routes and emergency communication Projecting response and recovery requirements
Demographic	Projecting consequences of disasters on the population Disseminating warnings and public information Planning evacuation and mass care
Response Organizations	Identifying response capabilities

DETERMINE THE COMMUNITY'S VULNERABILITY TO RISK

After compiling the jurisdiction profile, the next step is to quantify the community's risk by merging the information. Risk is the predicted impact that a hazard would have on the people, services, and specific facilities in the community. Quantifying risk enables jurisdictions to focus on those hazards that pose the highest threat to life, property, and the environment. Quantifying risk involves:

- identifying the elements of the community (populations, facilities, and equipment) that are potentially at risk from a specific hazard
- developing response priorities (Risk to life is always the highest priority.)
- assigning severity ratings based on the potential impact to life, essential facilities, and critical infrastructure
- compiling risk data into the community risk profiles that show the areas of the community that are at highest risk from the hazard

In analyzing risk, it is helpful to develop response priorities, using the following hierarchy for setting priorities:

- Priority 1: Life safety, including hazard areas, high-risk populations, and potential search and rescue situations.
- Priority 2: Essential facilities. Response personnel cannot respond if their own facilities are affected.
- Priority 3: Critical infrastructure, including utilities, communication, and transportation systems that are essential to life safety and would adversely affect response efforts if they were disrupted.

Next, assign each hazard a severity rating, or risk index, that will predict, to the degree possible, the damage that can be expected in the community as a result of that hazard. This rating quantifies the expected impact of a specific hazard on people, essential facilities, property, and response assets. Table 21–C provides an example of severity ratings that may be used. Develop a risk index for each hazard and assign a value to each characteristic, using the following values:

- 1 = catastrophic
- 2 = critical
- 3 = limited
- 4 = negligible

Ratings should be assigned for each of the following types of hazard data:

- magnitude
- frequency of occurrence
- speed of onset
- community impact
- special characteristics

The value of all factors can then be averaged to determine the overall risk level for each hazard. The result of this process is a prioritized list of hazards that pose the greatest threat to the community. As a result, a community planning team can plan for each hazard for which the risk index exceeds a predetermined threshold.

CREATE AND APPLY SCENARIOS

The final step in the hazard analysis process is to create and apply scenarios for the highest-risk hazards. Scenarios should be realistic and based on the community's hazard and risk data. To create a scenario, brainstorm to track the development of a specific type of emergency. Each scenario should describe:

- the initial notification that the event is occurring or is about to occur
- the potential overall impact on the community
- the potential overall impact of the event on specific community sectors
- the potential consequences, such as casualties, damages, and loss of services
- the actions and resources that would be needed to deal with the situation

Creating scenarios helps to identify situations that may exist in a disaster. These situations should be used to help ensure that a community is prepared if a hazard event occurs. Figure 21–2 can be used as a guide to determine the key factors that affect a particular community.

Table 21–C. Hazard Severity Ratings	
Severity	***Expected Impact***
Catastrophic	Multiple deaths Complete shutdown of critical facilities for 30 days or more More than 50% of property severely damaged
Critical	Injuries and/or illnesses result in permanent disability Complete shutdown of critical facilities for at least two weeks More than 25% of property severely damaged
Limited	Injuries and/or illnesses do not result in permanent disability Complete shutdown of critical facilities for more than one week More than 10% of property severely damaged
Negligible	Injuries and/or illnesses treatable with first aid Minor quality of life loss Shutdown of critical facilities for 24 hours or less Less than 10% of property severely damaged

COMPONENTS OF AN EMERGENCY OPERATIONS PLAN

An EOP describes actions to be taken in response to natural, man-made, or technological hazards. It also details tasks to be performed by specific organizational elements at projected times and places based on established objectives, assumptions, and assessment of capabilities.

An EOP should be comprehensive. It should cover all aspects of emergency preparedness and response, and it should address mitigation concerns. It also should address all hazards and, thus, be flexible enough to use in all emergencies—even unforeseen events. Finally, the EOP should be risk-based. It should include hazard-specific information based on the risks that were established in the hazard analysis. The EOP is written to provide an overview of the jurisdiction's response policies. It also should provide a general understanding of the jurisdiction's approach to emergency response for all involved agencies and organizations. An EOP consists of three parts:

1. the basic plan
2. functional annexes that address the performance of a particular broad task
3. hazard-specific appendices that provide additional response

Although the basic plan provides the general approach to emergency response, it does not stand alone. Rather, it forms the basis for the remainder of the plan. Also, each part of the plan may have addenda in the form of standard operating procedures (SOPs), maps, charts, checklists, tables, forms, and so on. These addenda may be included as attachments or be incorporated by reference.

THE BASIC PLAN

Although there is no standard format, for the sake of compatibility with other jurisdictions and levels of government, it is recommended that the basic plan include the following components:

- introduction
- purpose statement
- situation and assumptions

Make notes in the space next to each category of key factors to assess the jurisdiction hazards of a particular community
GEOGRAPHY Major geographic features: Typical weather patterns:
PROPERTY Numbers: Types: Ages: Building codes: Critical facilities: Potential secondary hazards:
INFRASTRUCTURE Utilities, construction, and layout access: Communication system layout, features, and backup: Road systems: Air and water support:
DEMOGRAPHICS Population size, distribution, and concentrations: Numbers of people in vulnerable zones: Special populations: Animal populations:
RESPONSE ORGANIZATIONS Locations: Points of contact: Facilities: Services: Resources:

Figure 21–2. Jurisdiction Hazard Assessment Form

- concept of operations
- organization and assignment of responsibility
- administration and logistics
- plan development and maintenance
- authorities and references

The *introduction* consists of five elements:

1. *The promulgation document* is signed by the senior elected official within the jurisdiction. The document affirms the official's support for the emergency management agency and for the planning process. It gives organizations the authority and responsibility to perform their tasks. It also mentions the tasked organizations' responsibility to prepare and maintain implementing instructions, gives notice of necessary EOP revisions, and commits to the training necessary to support the EOP.
2. The *signature page* is signed by all partner organizations and demonstrates their commitment to EOP implementation.
3. *Dated title page and record of changes* includes the date, description, and parts affected by changes to the EOP.
4. *Record of distribution* lists EOP recipients and facilitates and provides evidence of EOP distribution.
5. *Table of contents* lists the sections outlined in the document.

The *purpose statement* should include a broad statement about what the EOP is meant to do. It should also include a synopsis of the EOP, annexes, and appendices. The purpose statement need not be complex, but it should include enough information to establish the direction for the remainder of the plan.

The *situation and assumptions* section characterizes the planning environment, making it clear to the community why emergency planning is necessary. It draws from the hazard analysis to narrow the scope of the EOP and includes:

- hazards addressed by the plan
- relative probability and impact
- areas likely to be affected
- vulnerable critical facilities
- population distribution
- special populations
- interjurisdictional relationships
- maps

The *assumptions statement* delineates what was assumed to be true when the EOP was developed. In addition, the assumptions statement shows the limits of the EOP, limiting liability for the jurisdiction. It may be helpful to all communities to list in the plan even obvious assumptions, such as:

- Identified hazards will occur.
- Individuals and organizations are familiar with the EOP.
- Individuals and organizations will execute their assigned responsibilities.
- Assistance may be needed, and if so, will be available.
- Executing the EOP will save lives and reduce damage.

The fourth component of the basic plan is the *concept of operations*, which explains the community's overall approach to emergency response (i.e., what, when, and by whom). The concept of operations includes:

- the division of local, state, and federal responsibilities
- when the EOP will be activated and when it will be deactivated—and more importantly, by whom
- alert levels and the basic actions that accompany each level
- the general sequence of actions before, during, and after the event
- forms necessary to request assistance of various types

The *organization and assignment of responsibilities* section lists, by organizations and positions, the general areas of responsibility assigned. It also identifies shared responsibilities and specifies which organization has the primary responsibility for a given function and which have supportive roles. In other words, the organization and assignment of responsibilities section specifies reporting relationships and lines of authority for an emergency response.

The sixth component of the basic plan is the *administration and logistics* section. This section provides resource management policies and policies for augmenting response staff with public employees and volunteers, as well as a statement that addresses liability issues. It also includes the assumed resource needs for high-risk hazards—resources that are available within the community and resources that may be available through mutual-aid agreements. It is important to note, however, that the community should not rely on mutual-aid agreements because neighboring jurisdictions may be faced with the same emergency or disaster.

The next component is *plan development and maintenance*. The responsibility for the coordination of the development and revision of the basic plan, annexes, appendices, and implementing instructions must be assigned to the appropriate people. Therefore, this section:

- describes the planning process
- identifies the planning participants
- assigns planning responsibilities
- describes the revision cycle (training, exercising, review of lessons learned, etc.)

The final section of the basic plan is the *authorities and references* section. This section cites the legal basis for emergency operations and activities, including:

- laws, statutes, and ordinances
- executive orders
- regulations
- formal agreements
- predelegation of emergency authorities
- pertinent reference materials, including related plans for other levels of government

ANNEXES

An *annex* delineates how the community will carry out broad functions in any emergency, such as issuing warnings or managing resources. Early in the planning process, it is important to determine the functions that will be included in the basic plan as annexes. When making this decision, it is important to consider the organization of the state government and that of the jurisdiction, the capabilities of the jurisdiction's emergency services agencies, and the established concept of operations.

During this process, it is important to keep in mind the hazard analysis information developed for the community. What the community's planning team knows about its vulnerability is key to developing meaningful functional annexes. Because communities vary so widely, there is no single list of functional annexes that is right for everyone. There are, however, eight core functions that the Federal Emergency Management Agency (FEMA) recommends be addressed as annexes in every EOP:

- The *Direction and Control Annex* allows a jurisdiction to analyze the situation and decide the best response, direct the response teams, coordinate efforts with other jurisdictions, and make the best use of available resources.
- The *Communications Annex* provides a detailed focus on the total communications system and how it will be used.
- The *Warning Annex* describes the warning systems in place and the responsibilities and procedures for using them.
- The *Emergency Public Information (EPI) Annex* provides the procedures for giving the public accurate, timely, and useful information and instructions throughout the emergency period. It is important to note that while a Warning Annex focuses on the procedures that the government uses to alert those at risk, an EPI Annex deals with developing messages and accurate information and monitoring how the information is received. Because the warning system is one means for an EPI organization to get information out, an EPI Annex must address coordination with those responsible for the warning system.
- The *Evacuation Annex* describes the provisions that have been made to ensure the safe and orderly evacuation of people threatened by hazards that the jurisdiction faces.
- The *Mass Care Annex* deals with the actions taken to protect evacuees and other disaster victims from the effects of the disaster, including providing temporary shelter, food, medical care, clothing, and other essential needs.
- The *Health and Medical Annex* describes policies and procedures for mobilizing and managing health and medical services under emergency or disaster conditions.
- The *Resource Management Annex* describes the means, organization, and process by which a jurisdiction will find, obtain, and allocate resources to satisfy needs that are generated by an emergency or disaster.

In addition to these annexes, the planning team may want to consider annexes that make sense for the community. For example, if the community has a nuclear power plant, the planning team should add an annex on radiological protection. Additional functional annexes may be added based on state law or jurisdictional requirements. Examples of annexes that may be added include:

- Damage Assessment
- Search and Rescue
- Emergency Services
- Aviation Operations

HAZARD-SPECIFIC APPENDICES

An *appendix* is a supplement to an annex that adds information about how to carry out the function in the face of a specific hazard. Therefore, every annex may have several appendices, each addressing a particular hazard. The hazard-specific appendices that the planning team deems appropriate depend on the community's hazard analysis. The decision about whether to develop an appendix rests solely with the planning team.

Unlike annexes, hazard-specific appendices are not attached to the basic plan, but are linked to each functional annex. Topics addressed in hazard-specific appendices include:

- special planning requirements
- priorities identified through the hazard analysis
- unique characteristics of the hazard that require special attention
- special regulatory considerations

Table 21–D suggests appendix topics for each functional annex.

Table 21-D. Topics for Functional Annexes

Annex	*Appendix Topics*
Direction and Control	• Response actions keyed to specific time periods or phases • Urban Search and Rescue (US&R) inspection • Protective gear for responders • Detection equipment and techniques • Laboratory analysis services • Containment and cleanup teams
Communications	• Provisions made to ensure the effects of a specific hazard do not prevent or impede the ability of responders to communicate with each other during a response operation
Warning	• Hazard-specific public warning protocols • Required or recommended notifications of state and federal officials
Emergency Public Information	• Information the public will need to know about the specific hazard (evacuation routes and shelters, in-place protective actions, etc.) • The means by which information to the public will be conveyed
Evacuation	• Evacuation options and timing • Evacuation routes • Transportation resources to support mass evacuation
Mass Care	• Shelter locations out of the hazard's vulnerable areas • Protection of shelter occupants • Food and water stocks to support extended shelter stays • Capability to decontaminate people exposed to hazardous materials
Health and Medical Services	• Unique health consequences and treatment options for people exposed to the hazard • Environmental monitoring and/or decontamination requirements
Resource Management	• Provisions for purchasing, stockpiling, or otherwise obtaining special protective gear, supplies, and equipment needed by responders and disaster victims

IMPLEMENTING INSTRUCTIONS

Like the basic plan, each annex or appendix may use implementing instructions in the form of:

- standard operating procedures (SOPs)
- job aids
- checklists
- information cards
- record-keeping and combination forms
- maps
- charts, tables, and forms

Implementing instructions may be included as attachments or by reference, and the planning team may use them as needed to clarify the contents of the plan, annex, or appendix. For example, the evacuation annex may be made clearer by attaching maps with evacuation routes marked. Because these routes may change depending on the location of the hazard, maps may also be included in the hazard-specific appendices to the evacuation annex. Similarly, the locations of shelters may be marked on maps supporting the mass care annex.

SOPs provide response protocols for carrying out specific responsibilities. They describe who, what, when, where, and how. SOPs are appropriate for:

- complex tasks requiring step-by-step instructions
- tasks for which standards must be specified
- tasks for which documentation of performance protocols is required as a protection against liability

When developing SOPs:

- Develop a task list.
- Determine who, what, when, where, and how. Remember "who" includes who performs the activity, to whom he reports, and with whom he coordinates.
- Identify the steps for each task.
- Identify the standards for task completion.
- Test the procedures.

It is important to keep the SOPs updated through the review and revision process.

Another type of implementing instructions is job aids. A *job aid* is a written procedure that is intended to be used on the job while the task is being done. SOPs may be presented as job aids. Job aids are also appropriate for:

- complex tasks
- critical tasks that could result in serious consequences
- tasks that are done infrequently
- procedures or personnel that change often

Job aids are also useful when conformity is needed among workers or across locations. Job aids should specify:

- the task title
- the purpose of the task

- when to do the task
- materials needed to perform the task
- how to perform each step of the task
- the desired results
- standards to which the task must be performed
- how to ensure that the work is done properly

A job aid may include graphics, flow charts, if/then decision tables, and dos and don'ts. Because job aids are used in the midst of completing a task, they must be clear to be effective. They should use action verbs and everyday language. Important information should be highlighted and warnings placed before steps to which they apply. Formatting is also important when creating job aids. Numbering steps and using space, boxes, or lines to separate steps allows users to find their place easily after looking away. Job aids may not be useful for all tasks, especially simple tasks that are performed regularly or must be accomplished quickly and from memory. If a task cannot be completed while referring to a job aid, then a job aid is not appropriate.

Checklists are also useful implementing instructions. They provide a list of tasks, steps, features, contents, or other items to be checked off as completed. They often take the form of boxes to be checked off but can be developed in any form, including rating scales. Checklists are particularly useful for tasks made up of multiple steps that must be completed in sequence or for when it might be necessary to document the completion of steps. Checklists may be less useful when observations must be recorded, when calculations or evaluations must be made, or when detailed instructions are required to complete the task.

Information cards provide information that is needed on the job in a convenient, often graphic, form. Examples include:

- reference forms
- diagrams, labeled illustrations, charts, or tables
- information summarized in matrix form

Things that might be presented in the form of information cards include:

- call-down rosters
- contact lists
- resource lists
- organizational charts
- task matrices
- equipment diagrams

Common record-keeping and combination forms used as implementing instructions include record-keeping forms on which calculations, observations, or other information (damage assessment) can be recorded, and combination forms that serve multiple functions, such as checklists with recordkeeping sections.

Maps may be used as implementing instructions to highlight geographic boundaries and features, jurisdictional boundaries, locations of key facilities, and transportation or evacuation routes. It is important to note that, when using a map as an implementing instruction to show a particular feature, extraneous details are often eliminated.

CREATING EFFECTIVE IMPLEMENTING INSTRUCTIONS

To be effective, implementing instructions must be appropriate for both the audience and the intended use. They must also be:

- Complete, in that they cover all of the components or steps.
- Clear, concise, and easy to use. They should avoid jargon and ambiguity, be organized in a logical manner, and include instructions that identify the purpose and applicability of the particular implementing instruction.
- Sufficiently detailed, in that they give all necessary information.
- Up to date. The latest revision should be included.
- Sufficient in scope. They must cover each function fully.
- Identified in the EOP so that their existence is recorded. Implementing instructions should be incorporated by reference in the basic plan, annex, or appendix to which they refer.

Implementing instructions are used by all emergency agency personnel who respond to disasters, whatever their function. They are developed at the specific agency level because agency personnel will be using them, and therefore, will know if they are effective. Implementing instructions should support a particular agency's roles and responsibilities as described in the basic plan. For this reason, only some of the types of implementing instructions described here will be useful to a particular agency, depending on its function in a response.

When developing any type of implementing instruction, first consider the job title or position and the tasks that go along with that position. Then decide what type of implementing instructions would be most useful for those tasks.

SUMMARY

The end product of emergency planning is the EOP, a document describing how citizens and property will be protected in a disaster or emergency. There are four steps in the emergency planning process:

1. the hazard analysis or process by which hazards in the community are identified and ranked according to the risks that they pose to the community
2. EOP development, including the basic plan, functional annexes, hazard-specific appendices, and implementing instructions
3. testing the plan through training and exercises to determine weaknesses and strengths in the plan
4. maintenance and revision based on needs and resources that may have changed since the development of the original EOP

Completing each of these steps thoroughly is the key to developing an effective EOP that requires fewer and less-significant changes following training and exercising. Exercising, which is a focused practice activity, is designed to place all participants in a simulated situation, requiring them to function in the capacity that would be expected of them in a real event. The purpose of exercise is to promote preparedness by testing policies, plans, and training personnel.

22

The Critical Incident Planning Process

Michael J. Fagel, PhD, CEM

CHAPTER 22

INTRODUCTION

Given the creativity of those committed to carrying out acts of terrorism, it is more important than ever that communities are equipped to respond to terrorist events. Planners must consider a broad range of incidents, including assaults on infrastructure and electronic information systems that could result in consequences affecting life, health, and safety of the population.

State and local governments have the primary responsibility of planning for and managing the consequences of a critical incident using available resources in the hours before federal assistance can arrive.

THE PLANNING PROCESS

In Chapter 21, Developing an Emergency Operations Plan, we learned that every basic emergency plan should include annexes. An *annex* delineates how the community will carry out broad functions in an emergency, such as issuing warnings or managing resources. Linked to these annexes are *hazard-specific appendices* that add information about how to carry out those function in the face of a specific hazard. The process of developing a critical incident annex to an emergency operations plan (EOP), which addresses such an event as terrorism, for example, is similar to that used to develop other EOPs. The planning process must begin before an emergency and prior to any planned special event that could be subject to an attack.

Traditionally, the planning process has consisted of six phases:

1. initiation
2. concept development
3. plan development
4. plan review
5. development of supporting plans, procedures, and materials
6. validation of plans using tabletop, functional, and full-scale exercises

However, given the intense and multifaceted nature of a critical incident such as a terrorist attack, a seventh phase in recommended: Thorough coordination of plans, internally and externally.

It is important to carefully compare plans for various response functions within an agency and revise the plans, if necessary, to remove any discrepancies. This helps prevent disconnects between vital functions that support one another and helps ensure that each function does what is expected in a timely basis. Similarly, the various departments and agencies within a local jurisdiction should also compare plans, focusing on issues of consistency and coordination. This review ensures that each organization does what the other expects, when it is expected.

Such reviews are especially important in planning for response to a terrorist incident because local jurisdictions are likely to be aided by neighboring communities, counties, and possibly states during the response to the incident.

TERRORISM HAZARDS

The terrorism annex to an EOP should identify and discuss the nature of the terrorist hazard.

The hazard may be a weapon of mass destruction (WMD), including conventional explosives, secondary devices, and combined hazards; or it may be another means of attack, including low-tech devices, attacks on infrastructure, and cyberterrorism.

WEAPONS OF MASS DESTRUCTION

A *weapon of mass destruction (WMD)* is any weapon that is designed or intended to cause death or serious bodily injury through the release, dissemination, or impact of toxic or poisonous chemicals, disease organisms, radiation or radioactivity, or explosion or fire. Two important considerations distinguish these hazards from other types of terrorist tools. First, in the case of chemical, biological, and radioactive agents, their presence may not be immediately obvious, making it difficult to determine when and where they have been released, who has been exposed, and what danger is present for first responders and medical technicians. Second, although there is a sizable body of research on battlefield exposures to WMD agents, there is limited scientific understanding of how these agents affect civilian populations.

Chemical Agents. Chemical agents are intended to kill, seriously injure, or incapacitate people through physiological effects. A terrorist incident involving a chemical agent will demand immediate reactions from emergency responders—fire, police, hazardous materials (HazMat) teams, emergency medical services (EMS), and emergency room staff—who will need adequate training and equipment.

Hazardous chemicals, including industrial chemicals and agents, can be introduced via aerosol devices (munitions, sprayers, or aerosol generators), breaking containers, or covert dissemination. Such an attack might involve the release of a chemical warfare agent, such as a nerve or blister agent or an industrial chemical, which may have serious consequences. Table 22–A lists some indicators of the possible use of chemical agents.

Table 22–A. Indicators of Use of Chemical Agents
Stated Threat to Release a Chemical Agent
Unusual Occurrence of Dead or Dying Animals • For example, lack of insects or dead birds
Unexplained Casualties • Multiple victims • Surge of similar 911 calls • Serious illnesses • Nausea, disorientation, difficulty breathing, or convulsions • Definite casualty patterns
Unusual Liquid, Spray, Vapor, or Powder • Droplets, oily film • Unexplained odor • Low-lying clouds/fog unrelated to weather
Suspicious Devices, Packages, or Letters • Unusual metal debris • Abandoned spray devices • Unexplained munitions

Early in an investigation, it may not be obvious whether an outbreak was caused by an infectious agent or a hazardous chemical; however, most chemical attacks will be localized, and their effects will be evident within a few minutes. There are persistent and nonpersistent chemical agents. Persistent agents remain in the affected area for hours, days, or weeks. Nonpersistent agents have high evaporation rates, are lighter than air, and disperse rapidly, thereby losing their ability to cause casualties after 10 to 15 minutes. They may be more persistent in small, unventilated areas.

Biological Agents. Recognition of a biological hazard can occur through several methods, including identification of a credible threat, discovery of bioterrorism evidence, diagnosis, and detection. When people are exposed to a pathogen, such as anthrax or smallpox, they may not know that they have been exposed, and those who are infected, or subsequently become infected, may not feel sick for some time. This delay between exposure and onset of illness—the incubation period—is a characteristic of infectious diseases. The incubation period may range from several hours to a few weeks, depending on the exposure and the pathogen.

Unlike acute incidents involving explosives or some hazardous chemicals, the initial detection and response to a biological attack on civilians is likely to be made by direct patient-care providers and the public health community.

Terrorists could also employ a biological agent that would affect agricultural commodities over a large area (wheat rust or a virus affecting livestock), potentially devastating the local or even national economy. The response to agricultural bioterrorism should also be considered during the planning process. Responders should be familiar with the characteristics of biological agents. Table 22–B lists some indicators of the possible use of biological agents.

Nuclear/Radiological Agents. The difficulty of responding to a nuclear or radiological incident is compounded by the nature of radiation itself. In an explosion, the fact that radioactive material was involved may or may not be obvious, depending on the nature of the explosive device used.

The presence of a radiation hazard is difficult to ascertain, unless the responders have the proper detection equipment and have been trained to use it properly. Although many detection devices exist, most are designed to detect specific types and levels of radiation and may not be appropriate for measuring or ruling out the presence of radiological hazards.

The scenarios constituting an intentional nuclear/radiological emergency include the following:

- Use of an improvised nuclear device includes any explosive device designed to cause a nuclear yield. Depending on the type of trigger device used, either uranium or plutonium isotopes can fuel these devices. Although "weapons-grade" material increases the efficiency of a given device, material of less than weapons-grade can still be used.
- Use of a radiological dispersal device (RDD) includes any explosive device utilized to spread radioactive material upon detonation. Any improvised explosive device (IED) could be used by placing it in close proximity to radioactive material.
- Use of a simple RDD spreads radiological material without the use of an explosive. Any nuclear material (including medical isotopes or waste) can be used in this manner.

Conventional Explosives and Secondary Devices. A conventional explosive device, or improvised bomb, which may be used to cause massive local destruction or to disperse a chemical, biological, or

Table 22–B. Indicators of Use of Biological Agents
Stated Threat to Release a Biological Agent
Unusual Occurrence of Dead or Dying Animals
Unusual Casualties • Unusual illness for region/area • Definite pattern inconsistent with natural disease
Unusual Liquid, Spray, Vapor, or Powder • Spraying; suspicious devices, packages, or letters

radiological agent, is the easiest of weapons to use and to obtain. The components are readily available, as are detailed instructions for constructing such a device.

Improvised explosive devices are categorized as being explosive or incendiary, employing high- or low-filler explosive materials to explode and/or cause fires. Explosions and fires also can be caused by projectiles and missiles, including aircraft used against high-profile targets, such as buildings, as was the case for the September 11, 2001 terrorist attack.

Combined Hazards. WMD agents can be combined to achieve a synergistic effect—greater in total effect than the sum of their individual effects. They may be combined to achieve both immediate and delayed consequences. Mixed infections or toxic exposures may occur, thereby complicating or delaying diagnosis. Casualties of multiple agents may exist; casualties may also suffer from multiple effects, such as trauma and burns from an explosion, which may exacerbate the likelihood of agent contamination.

Attacks may be planned and executed to take advantage of the reduced effectiveness of protective measures produced by an initial WMD agent. Again, the potential exists for multiple incidents in single or multiple municipalities.

OTHER TERRORISM HAZARDS

Planners need to consider the possibility of unusual or unique types of terrorist attacks that were not previously considered likely. For example, prior to the attacks of September 11, the use of multiple commercial airliners with full fuel loads as explosive incendiary devices in well-coordinated attacks on public and private targets was not considered a likely terrorist scenario.

Although it is not realistically possible to plan for and prevent every conceivable type of terrorist attack, planners should anticipate that future terrorism attempts could range from simple, isolated attacks to complex, sophisticated, highly coordinated acts of destruction using multiple agents aimed at one or multiple targets. Therefore, the plans developed for terrorist incidents must be broad in scope, yet flexible enough to deal with the unexpected.

These considerations are particularly important in planning to handle the consequences of attacks using low-tech devices and delivery, assaults on public infrastructure, and cyber terrorism. In these cases, the training and experience of the responders may be more important than detailed procedures.

Low-Tech Devices and Delivery. Planning for the possibility of terrorist attacks must consider that explosives can be delivered by a variety of methods. Most explosive and incendiary devices used by terrorists would be expected to fall outside the definition of a WMD. Small explosive devices can be left in packages or bags in public areas for later detonation at a time and place when and where the terrorist feels that the maximum damage can be done.

The relatively small size of these explosive devices can also be brought onto planes, trains, ships, or buses, within checked luggage or carry-ons. Although present airline security procedures minimize the possibility of explosives being brought on board planes, planners will need to consider the level of security presently employed on ships, trains, and buses within their jurisdictions.

Larger quantities of explosive materials can be delivered to their intended target area by means of car or truck bombs. Planners need to consider the possible need to restrict or prohibit vehicular traffic within certain distances of key facilities identified as potential terrorist targets. Planners may also need to consider the possible use of concrete barriers to prevent forced entry of vehicles into restricted areas.

Infrastructure Attacks. Potential attacks on elements of the nation's infrastructure require protective considerations. Infrastructure protection involves risk management actions taken to prevent destruction of or incapacitating damage to networks and systems that serve society.

Infrastructure protection often is focused more on security, deterrence, and law enforcement than on emergency-consequence management preparedness and response. Nevertheless, planners must develop contingencies and plans in the event critical infrastructures are brought down as the result of a terrorist incident.

Cyberterrorism. Cyberterrorism involves that malicious use of electronic information technology to commit or threaten to commit acts dangerous to human life, or against a nation's critical infrastructures, to intimidate or coerce a government or civilian population to further political objectives. As with other infrastructure guidance, most cyber protection guidance focuses on security measures to protect computer systems against intruders, denial of service attacks, and other forms of attack, rather than addressing issues related to contingency and consequence management planning.

THE SITUATION OVERVIEW

The situation section of a terrorism annex to the EOP should discuss what constitutes a potential or actual terrorist incident. It needs to present a clear, concise, and accurate overview of potential events and discuss a general concept of operation for response. Any information already included in the EOP need not be duplicated in the terrorism annex, but should be referenced. The situation overview should include as much information as possible that is unique to terrorism response actions, including the suggested elements listed in Table 22–C.

A state or local emergency manager or planner should consider the possibility of unusual or unique types of terrorist attacks in addition to those that have occurred in the past. Creative thinking about possible scenarios and response needs is critical. Plans developed for terrorist incidents must be comprehensive in scope, yet flexible enough to deal with the unexpected. The key to successful emergency response involves smooth coordination among multiple agencies and officials from various jurisdictions regarding all aspects of the response.

Because critical incidents such as terrorism require interacting with a wide range of organizations and individuals within organizations, up-to-date directories of the points of contact

Table 22–C. Elements Unique to Terrorism Response Actions

Suggested Emergency Operations Plan Elements	
Maps	• Use detailed, current maps or charts. • Include demographic information. • Use natural and man-made boundaries and structures to identify risk areas. • Annotate evacuation routes and alternatives. • Annotate in-place sheltering locations. • Use geographic information and analytical tools, as appropriate.
Environment	• Determine response routes and times. • Include bodies of water with dams or levees (these could become contaminated). • Specify special weather and climate features that could alter the effects of a WMD.
Population	• Identify those most susceptible to WMD effects or otherwise hindered or unable to care for themselves. • Identify areas where large concentrations of the population might be located, such as sports arenas and major transportation centers. • List areas that may include retirement communities. • Note locations of correctional facilities. • Note locations of hospitals, medical centers, schools, day care centers, or any other locations where multiple evacuees may require assistance. • Identify non-English-speaking populations.
Regional	• Identify multijurisdictional perimeters and boundaries. • Identify potentially overlapping areas for response. • Identify rural, urban, suburban, and city mutual-risk areas. • Identify mutual-aid resources from adjoining municipalities. • Identify terrorism-specific or unique characteristics, such as interchanges, choke points, traffic lights, traffic schemes and patterns, access roads, tunnels, bridges, railroad crossings, and overpasses or cloverleafs.
Resources	• Identify mutual-aid resources. • Identify terrorism-specific resources.

must be maintained in the course of the planning process. It is important that these directories are kept current to reflect changes in personnel and telephone numbers.

While assistance from federal agencies will be needed in the event of a terrorist incident, planning by state and local jurisdictions should take into account the difficulty that can be experienced when incorporating federal resources into the initial local response. Coordination among state, local, and federal officials should take place well in advance of any events that could be targeted so that all response organizations clearly understand their responsibilities and how they will be integrated.

POTENTIAL TARGETS

In determining the risk areas within a jurisdiction, the vulnerabilities of potential targets should be identified, and the targets themselves should be prepared to respond to a WMD incident. In-depth vulnerability assessments are needed to determe a response to such an incident.

Areas of vulnerability may be determined by population, accessibility, criticality and economic impact. It may be beneficial to coordinate vulnerable areas with the Federal Bureau of

Investigation (FBI). Various criteria may be used in determining the vulnerability of facilities to terrorist attacks. In evaluating the vulnerability of facilities, state and local planners need to consider the existing security measures in place and the need, if any, to upgrade security. Table 22–D shows some vulnerable areas that may be potential targets. These areas should be considered during the planning process.

INITIAL WARNING

While specific critical incident events may vary, the emergency response and protocol followed should remain consistent. When an overt WMD incident has occurred, the initial call for help will likely come through the local 911 center. This caller may or may not identify the incident as a terrorist event, but may state only that there was an explosion, a major accident, or a mass casualty event. Information relayed through the dispatcher, prior to the arrival of first responders on scene, as well as the initial assessment, will provide the basic information necessary to begin responding to the incident.

With increased awareness and training about terrorist incidents, first responders should recognize when a terrorist incident has occurred. The information provided in this chapter applies where it becomes obvious or strongly suspected that an incident has been intentionally perpetrated to harm people, compromise the public's safety and well-being, disrupt essential government services, or damage the area's economy or environment.

Be aware of the likely occurrence of false warnings. Since these cannot be ignored, they must be investigated, resulting in wasted resources and psychological stress. Develop procedures and training to deal with false warnings as well.

INITIAL DETECTION

The initial detection of a WMD terrorist attack will likely occur at the local level by either first responders or private entities (hospitals, corporations, etc.). Consequently, first responders, the business community—public and private—should be trained to identify hazardous agents and to take appropriate actions. State and local health departments, as well as local emergency first responders, will be relied upon to identify unusual symptom occurrence, and any additional cases of symptoms as the effects spread throughout the community and beyond.

The detection of a terrorism incident involving covert biological agents, as well as some chemical agents, will most likely occur through the recognition of similar symptoms or syndromes by clinicians in hospital or clinical settings. Detection of biological agents could occur days or weeks after exposed individuals have left the site of the release. Detection will occur at public health facilities receiving unusual numbers of patients, the majority of whom will self transport. Similarly, a biological attack aimed at agricultural assets might first be detected by veterinarians or agricultural inspectors.

First responders must be protected from the hazard prior to treating victims. Planning for response to terrorist acts must include provisions for appropriate personal protective equipment (PPE) for emergency responders, specifically first responders. This equipment should include protective clothing and respirators, with high efficiency particulate air (HEPA) filters. Detection equipment for chemical, biological, nuclear, or explosive materials will assist in identifying the nature of a potential hazard.

Determine the availability of protective and detection equipment within your jurisdiction, determine if additional resources would be needed to adequately protect first responders, and identify the funding needs to upgrade resources, if needed.

Table 22–D. Areas of Vulnerability to a Critical Incident	
Motor Vehicle Traffic	• Determine which roads, tunnels, or bridges carry large volumes of traffic. • Identify points of congestion that could impede response or place citizens in a vulnerable area. • Note time of day and day of week that this activity occurs.
Trucking and Transport Activity	• Note location of hazardous materials (HazMat) cargo loading and unloading facilities. • Note vulnerable areas such as weigh stations and rest areas of cargo transit.
Waterways	• Map pipelines and process or treatment facilities (in addition to dams). • Note berths and ports for cruise ships, roll-on/roll-off cargo vessels, and container ships. • Note any international flagged vessels that conduct business in the area. ***NOTE:*** *The Harbor and Port Authorities, normally involved in emergency planning, should be able to facilitate obtaining information on the type of vessels and the containers they carry.*
Airports	• Note information on carriers, flight paths, airport layout, and types of aircraft that use this facility. • Annotate location of air-traffic control towers, runways, passenger terminals, and parking areas.
Trains/Subways	• Note location of rails and lines, interchanges, terminals, tunnels, and cargo/passenger terminals. • Note any HazMat material that may be transported via rail. • Note locations of subway stations and ventilation control systems.
Government Facilities	• Note location of federal, state, and local government offices. • Include locations of post office, law enforcement stations, fire/rescue, town/city hall, and local mayor/governor's residences. • Note judicial offices and courts. • Note locations of monuments, memorial structures, and prominent governmental symbols.
Recreation Facilities	• Map sports arenas, theatres, malls, special interest group facilities, and locations of special events.
Symbolic Buildings and Locations	• Note national monuments, internationally well-known facilities, and locations. • Note potential areas of congestion connected with such buildings and locations.
Other Facilities	• Map locations of financial institutions and the business district. • Make any notes in schedule that the business district may follow. • Determine whether shopping centers or heavily populated downtown areas are congested at certain periods. • Note location of special-events facilities that may have national importance. • Note location of prominent high-rise buildings.
Military Installations	• Note locations and type of military installations.
HazMat Facilities, Utilities, and Nuclear Facilities	• Map locations of facilities, such as electricity generating stations, oil refineries, spent nuclear fuel storage facilities, etc.
Water Supply Facilities	• Note the locations of water supply intakes from lakes or rivers. • Note the locations of water supply pipelines and holding areas, such as reservoirs and tanks. • Note the locations of water supply treatment plants.
Food and Agriculture	• Note the locations of key agricultural facilities, such as large grain elevators and livestock concentrations. • Note the locations of food processing and packaging facilities.
Computer Systems	• Identify governmental and business-related computer systems located within the jurisdiction and ascertain their level of protection against cyberterrorism.
NOTE: *Security and emergency personnel representing all of the above facilities should work closely with local and state personnel for planning and response.*	

RELEASE AREA

Standard models are available for estimating the effects of a nuclear, chemical, or biological release, including the area affected and consequences to population, resources, and infrastructure. Some of these models include databases on infrastructure that may prove useful when developing a terrorism annex.

Analogous to the area affected by a nuclear, biological, or chemical release is the area impacted by an explosive device. Models are also available for estimating the blast effects at various distances for various quantities of explosive materials. These models can be useful in preparing a terrorism annex, especially in regard to determining minimum setback distances from a potential vehicle bomb to a vulnerable facility or structure. If a specific minimum distance cannot be maintained, the planning effort may need to consider the cost and effectiveness of facility hardening to mitigate the effects of an assumed blast impact.

INVESTIGATION AND CONTAINMENT OF HAZARDS

Local first responders provide initial assessment or scene surveillance of a hazard caused by an act of WMD terrorism. The proper local, state, and federal authorities capable of dealing with and containing the hazard should be alerted to a suspected WMD attack as soon as first responders recognize symptoms that are unusual or of an unknown cause. Consequently, state and local emergency responders must be able to assess the situation and request assistance as quickly as possible.

ASSUMPTIONS ABOUT THE RESPONSE

Although situations vary, assumptions remain the same. These include the following:

- The first responder, health, or medical personnel will, in most cases, initially detect and evaluate the potential or actual incident, assess casualties, and determine whether assistance is required. If so, state support will be requested and provided. This assessment is based on warning or notification of a WMD incident that may be received from law enforcement, emergency response agencies, or the public.
- The incident may require federal support. To ensure that there is one overall lead federal agency (LFA), the Federal Emergency Management Agency (FEMA) is authorized to support the Department of Justice (DOJ). In addition, FEMA is designated as the lead agency for consequence management within the United States and its territories. FEMA retains authority and responsibility to act as the lead agency for consequence management throughout the federal response. In this capacity, FEMA will coordinate federal assistance requested through state authorities using standard procedures.
- Federal response will include experts in the identification, containment, and recovery of weapons of mass destruction (WMDs) that are chemical, biological, nuclear/radiological, or explosive in nature.
- Jurisdictional areas of responsibility and working perimeters defined by local, state, and federal departments and agencies may overlap. Perimeters may be used to control access to the affected area, target public information messages, assign operational sectors among responding organizations, and assess potential effects on the population and the environment. Control of these perimeters may be enforced by different authorities, which will impede the overall response if adequate coordination is not established.

- Response activities may continue for an extended period of days or weeks. Early emergency responders may be pushed to the limit of their capabilities, and regional and federal resources may be needed. The incident will be covered extensively by the media. There may be many volunteer responders and donations of food and supplies that will require management.

CONCEPT OF OPERATIONS

The terrorism annex should include a *concept of operations* section to explain the jurisdiction's overall concept for responding to a WMD incident. Topics should include division of local, state, federal, and any intermediate inter-jurisdictional responsibilities; activation of the EOP; and other elements set forth in a state's State and Local Guide (SLG) for responding to emergencies.

DIRECTION AND CONTROL

Local government response organizations will respond to critical incident scenes and make appropriate and rapid notifications to local and state authorities. Control of the incident scene is likely to be established by local first responders from either fire or police.

To ensure continuity of operations, an incident command post should be established at a safe location and at a distance appropriate for response to a suspected or known terrorist incident. In addition, in severe terrorist attacks, response operations may last for long periods, and there may be more leadership casualties because of secondary or tertiary attacks or events. Therefore, planning should provide for in-depth staffing of key leadership positions.

The Incident Command System (ICS) should be used by all responding local fire, police, and emergency management organizations, and all relevant responders should be trained in ICS use to prevent security and coordination problems in a multiorganizational response.

The ICS that was initially established likely will transition to a Unified Command System (UCS) as mutual-aid partners and state and federal responders arrive to augment the local responders. It is recommended that local, state, and federal regional law enforcement officials develop consensus on "rules of engagement" early in the planning process to smooth the transition from incident command to unified command system. The UCS structure facilitates crisis management and consequence management activities.

The site of a terrorist incident is a crime scene as well as a disaster scene, even though the lives, health, and safety remain top priority. Because of these considerations, as well as logistical control concerns, it is extremely important that a critical incident site and its perimeter be tightly controlled as soon as possible.

COMMUNICATIONS

In the event of a WMD incident, rapid and secure communications is important to ensure a prompt and coordinated response. Strengthening communications among first responders, clinicians, emergency rooms, hospitals, mass care providers, and emergency management personnel must be given top priority when planning. Planning should include adding 911 resources when an event requires an extraordinary response.

Terrorist attacks have been shown to overload nondedicated telephone lines and cellular telephones. In these instances, the Internet has proven more reliable for making necessary communications connections, although it should be recognized that computers may be vulnerable to cyber attacks in the form of viruses. It is recommended that response organizations establish relevant Internet connections with all coordinating emergency response organizations and have the use of these connections formalized in plans and practiced during training, drills, and exercises.

Responders with different functions within a jurisdiction, or from different jurisdictions, may use different radio frequencies, hindering communications. Use of 800 MHz radios alleviates this problem. Therefore, a backbone communications system to connect local, state, and federal responders is recommended. Also needed is the establishment of mutually agreed upon communications protocols during the planning and exercise stages so that all responding organizations understand each other's codes and terminology during response to real events.

Planning should also consider the need for an integrated communications system for all key state agencies and local emergency response organizations. The interoperability of such a system facilitates the integrated response to a terrorist incident. Planning also needs to consider the importance of reliable backup communications systems for emergency responders. Terrorist incidents may include the loss of radio transmission capabilities, and landlines and cellular phone connections will be overwhelmed. Satellite telephones, which can operate when cellular and nondedicated landlines are over loaded, are another option for backup telephone systems.

State and local planners should consider the distribution of priority emergency access telephone cards to their emergency workers. Planners should determine the existing status of their emergency communications systems and identify the funding needed to upgrade them.

WARNING

Every incident is different. There may or may not be warning of a potential WMD incident. Factors involved range from intelligence gathered from various law enforcement or intelligence agency sources to an actual notification from the terrorist organization or individual. The EOP should have hazardous materials (HazMat) facilities and transportation routes already mapped, along with emergency procedures necessary to respond. Typically, the FBI informs state and local law enforcement officials regarding potential threats, but the warning or notification of a potential WMD terrorist incident could come from many sources. Therefore, open but secure communication among local, state, and federal law enforcement agencies and emergency response officials is essential.

PRE-EVENT READINESS

The FBI operates with a four-tier threat level system that can be used as a basis for initiating precautionary actions when a WMD terrorist event is anticipated.

- Level 4 (Minimal Threat): Received threats do not warrant actions beyond normal liaison notifications or placing assets or resources on a heightened alert.
- Level 3 (Potential Threat): Intelligence or an articulated threat indicates the potential for a terrorist incident; however, this threat has not yet been assessed as credible.
- Level 2 (Credible Threat): A threat assessment indicates that a potential threat is credible and confirms the involvement of a WMD in a developing terrorist incident. The threat increases in significance when the presence of an explosive device or WMD capable of causing a significant destructive event.
- Level 1 (WMD Incident): A WMD terrorism incident resulting in mass casualties has occurred that requires immediate federal planning and preparation to provide support to state and local authorities. The federal response is primarily directed toward the safety and welfare of the public and the preservations of human life.

EMERGENCY PUBLIC INFORMATION

Terrorism is designed to be catastrophic. The intent of a terrorist attack is to cause maximum destruction of lives and property; create chaos, confusion, and public panic; and stress local, state, and federal response resources. Accurate and timely information, disseminated to the public and media immediately and often over the course of the response, is vital to minimize the accomplishment of these terrorist objectives.

Crisis research and case studies show that accurate, consistent, and expedited information calms anxieties and reduces problematic public responses, such as panic and spontaneous evacuations that terrorist hope will hamper response efforts.

The news media are the public's primary source of information, from both official sources and other non-official sources, during the course of the incident. Ensuring that the media receives accurate, consistent, and expedited official information from the outset and over what may be a rapidly changing and lengthy response, requires careful and considerable advanced planning and preparations. It is important to build and maintain a strong working relationship with the media. This relationship should include a clear commitment that government representatives will be immediately available to provide information over the course of the emergency.

Local plans should reflect responsibility for emergency information operations during the crucial initial response until state and federal personnel and resources can arrive to provide support. Planning should also reflect:

- a mechanism for sharing and coordinating information among all responders' agencies and organizations
- development and production of information materials
- dissemination of information through various methods
- monitoring and analysis of news media coverage with rapid response capabilities to address identified problems

A strong and ongoing public education program for terrorism response, built on outreach and awareness programs for other types of emergencies, can enhance the response organization's credibility and benefit the efforts of both members of the public and the first responder in the event of a terrorist attack.

PROTECTIVE ACTIONS

Evacuation may be required from inside the perimeter of the scene to guard against further casualties from contamination by release of a WMD agent, the possible release of additional WMDs, secondary devices, or additional attacks targeting emergency responders.

Temporary in-place shelters may be appropriate if there is a short-duration release of hazardous materials, or if it is determined to be safer for individuals to remain in place. Protection from biological threats may involve coercive or noncoercive protective actions, including isolation of individuals who pose an infection hazard, quarantine of affected locations, vaccinations, use of masks by the public, closing of public transportation, limiting public gatherings, and limiting inter-city travel.

As with any emergency, state and local officials are primarily responsible for making protective action decisions affecting the public. Protocols should be established to ensure that important decisions are made by people with the proper decision-making authority. A terrorism annex should include provisions for coordinating protective actions with other affected jurisdictions. Planning should also address ways of countering irrational public behavior that can hinder protective actions.

However, planning for evacuation should be flexible to account for difficult situations. After the September 11 attack on the Pentagon, federal buildings in Washington, DC were evacuated as a precautionary measure. Evacuation was hampered because roads in the federal area were under the jurisdiction of federal agencies, while roads in the rest of the District were under the authority of the District government. In addition, the Pentagon is located in Arlington County, Virginia, near a route people from Washington, DC were using to evacuate. This created a situation where some evacuation was moving toward the incident site rather than away from it. Because of heavy traffic in the vicinity of the Pentagon, fire vehicles had difficulty reaching the incident site.

MASS CARE

The location of mass care facilities is based partly on the hazardous agent. Decontamination, if necessary, may need to precede sheltering and other needs of victims to prevent further damage from a hazardous agent to either the victims themselves or to the care providers.

The American Red Cross (the primary agency for mass care), the Department of Health and Human Services (HHS), and the Department of Veteran Affairs (VA) should be actively involved in the planning process of a critical event to determine both in-place sheltering and mobile mass care systems for a terrorism annex.

A midpoint or intermediary station may be needed to move victims out of the way of immediate harm. This action would allow responders to provide critical attention (decontamination and medical services) and general lifesaving support, then evacuate victims to a mass location for further attention. Some general mass care issues to consider for inclusion in a terrorism annex include:

- location, setup, and equipment for decontamination stations, if any
- mobile triage support and qualified personnel
- supplies and personnel to support in-place sheltering
- evacuation to intermediary location to provide decontamination and medical attention
- determination of safety perimeters (based on agent)
- patient tracking/record keeping for augmentation of epidemiological services and support

HEALTH AND MEDICAL

A basic EOP should already contain a health and medical annex. Issues that may be unique during a terrorist incident that should be addressed in the terrorism annex include:

- decontamination
- safety of victims and responders
- in-place sheltering and quarantine versus evacuation
- multihazard and multiagent triage

Planning should anticipate the need to handle large numbers of people who may or may not be contaminated but who are fearful about their medical well-being. In addition, the terrorism annex should identify the locations and capacities of medical care facilities within the jurisdiction and in surrounding jurisdictions. The terrorism annex should also include a description of the capabilities of these medical care facilities, especially with regard to trauma care. Depending on the nature and extent of the terrorist attack, the most appropriate medical care facility may not necessarily be the closest facility.

In addition, first responders may encounter an environment rife with biological or chemical agents, radioactive materials, or hazardous pollutants from collapsed buildings, for example. In planning, address the need for first responders to perform a risk assessment and to modify standard protocols (establish plans for inoculating first responders) if necessary. The planning should also address how such assessments are made and what resources they may indicate are needed. The assessment may also indicate monitoring and sampling resources before federal resources can arrive. Responders may need appropriate PPE, including respirators. The planning process needs to address the availability of regional monitoring and sampling capability and PPE.

A bioterrorism incident may generate an influx of patients requiring specialized care. If an infectious agent is involved, it may be necessary to isolate the patients and use special precautions to avoid transmission of the disease to staff and other patients.

State planning should consider the need to obtain and integrate supplementary medical professionals and technicians who may be needed to respond to a terrorist incident. In addition to physiological health consideration, planners should take into account the need for mental health care in the consequence management planning. Support must be provided not only to those individuals directly affected by a terrorist attack, but also to those surviving family members experiencing emotional stress. Additional planning related to health and medical issues to consider includes:

- immunization and prophylaxis for biological agents
- notification to and receipt of information from doctors and clinics
- augmentation of medical facilities and personnel
- management of medical supplies and equipment
- patient tracking and record keeping for augmentation of epidemiological services and support
- analytical laboratory support, including memorandums of understanding (MOUs), specifying special considerations, as appropriate
- mental health support services, including clinical psychologists, psychiatrists, and social workers

RESOURCES MANAGEMENT

The following considerations are highly relevant to WMD incidents and should be addressed in a terrorsim annex:

- nuclear, biological, and chemical response resources that are available through inter-jurisdictional agreements
- unique resources that are available through state authorities (e.g., National Guard Units)
- unique resources that are available to state and local jurisdictions through federal authorities (e.g., the National Pharmaceutical Stockpile, a national asset providing delivery of antibiotics, antidotes, and medical supplies to the scene of a WMD incident)
- unique expertise that may be available through academic, research, or private organizations
- trained and untrained volunteer resources and unsolicited donated goods that arrive at the incident site

RECOVERY

The basic EOP should already contain a recovery annex. However, specific issues may arise during a WMD incident that should be addressed in the terrorism annex. Because a terrorist attack is a criminal act, its victims and their families may be eligible for assistance under a state's crime victim's assistance law, if one exists. Therefore, state crime victim's assistance staff should be included in the planning process, when appropriate. In addition, injured victims of a terrorist attack, those put at risk for injury, and their families may have suffered psychological trauma as a result of the attack and may be in need of crisis counseling.

In the event of an incident involving chemical or biological agents or radioactive materials, large areas or multiple locations may become contaminated. Decontamination may be required before buildings can be safely reoccupied. While decontamination is taking place, or until damaged areas are repaired or replaced, persons must be relocated from residences and offices, and office equipment must be relocated from office buildings. Recovery and relocation after a terrorist incident may be of longer duration and entail greater costs than recovery following a natural disaster.

URBAN SEARCH AND RESCUE

Urban search and rescue (US&R) involves a rapid deployment of task forces to provide specialized lifesaving support to state and local authorities, including locating and extricating, and providing on-site medical treatment to those trapped in collapsed structures.

There are several US&R task forces throughout the United States. They have the ability to deploy within six hours and to sustain themselves for 36 hours. Currently, deployment plans rely of commercial air transport. Consequently, in incidents where air traffic is curtailed, arrival of remote US&R task forces may be delayed.

Increasingly, the capabilities of these US&R task forces are being enhanced to operate in a collapsed building environment that is contaminated with biological or chemical agents or radioactive materials. These enhanced task forces have additional HazMat specialists, medical personnel, and monitoring and detection equipment.

ORGANIZATION AND ASSIGNMENT OF RESPONSIBILITIES

As with a hazard-specific emergency, the organization for management of local response will need to be tailored to address the special issues involved in managing the consequences of a terrorist incident. This organization should be defined in the terrorism annex or in the EOP.

The consequences of a terrorist attack can overwhelm local resources and require assistance from state or federal governments. The response by state and local governments, as well as the types of support from the federal government may be different from the response and support received for a natural disaster. Because of this, not only must the plans be upgraded to include response to a terrorist incident, but training and exercising must be expanded to ensure that the unique aspects of response to a terrorist incident can be carried out in a coordinated, effective manner.

Training should be planned for local, state, and federal staff involved in such a response. The plan should identify training needs, establish budgets for the training, and determine what funding is required to implement the training. Periodic integrated exercises must be conducted to ensure that the emergency response at all levels can adequately be coordinated. The following response roles and responsibilities should be addressed in the terrorism annex.

- Local emergency responders: Local fire departments, law enforcement personnel, HazMat teams, and EMS will be among the first to respond to terrorist incidents, especially those involving WMD. In incidents associated with public transportation, workers and officials from transportation organizations may be among the first responders. As response efforts escalate, the local emergency management agency and health department will help coordinate needed services.
- Inter-jurisdictional responsibilities: The formal arrangements and agreements for emergency response to a terrorist incident among neighboring jurisdictions, state, local, and neighboring states (and those jurisdiction physically located in those states) should be made prior to an incident. When coordinating and planning the risk assessment and risk area sections of the terrorism annex, inter-jurisdictional responsibilities should be readily identifiable.
- State emergency responders: If requested by the local officials, the state emergency management agency has the capabilities to support local emergency management authorities and the incident commander.
- State and local public health authorities: State laws grant the state and local public health authorities emergency powers to combat communicable disease. The powers available, diseases that trigger them, and procedures for enforcement vary from state to state. Typical powers include the power to isolate or quarantine persons and places and the power to compel vaccination and other preventive measure, such as wearing masks.
- Medical service providers: Hospitals generally perform emergency planning, both to protect their own facilities and patients and to respond to disasters in the community. State licensing and accreditation standards require hospitals to meet certain criteria for emergency preparedness, which often include participation in local or regional medical planning for disasters. Hospitals accredited by the Joint Commission on Accreditation of Healthcare Organizations (JCAHO) must be prepared for a variety of disaster scenarios, including facilities for biological, radioactive, or chemical isolation and decontamination, where appropriate.
- Local emergency planning committees and state emergency response commissions: These entities are established under the Superfund Amendment and Reauthorization Act of 1986 (SARA) Title III, and the implementing regulation of the Environmental Protection Agency (EPA). Local Emergency Planning Committees (LEPCs) develop and maintain local hazardous materials emergency plans and receive notifications of releases of hazardous materials. State Emergency Response Commissions (SERCs) supervise the operation of the LEPCs and administer the community right-to-know provisions of SARA Title III, including collection and distribution of information about facility inventories of hazardous materials, chemicals, and toxins. LEPCs will have detailed information about industrial chemicals within the community. It may be advisable for LEPCs and SERCs to establish memorandums of understanding (MOUs) with agencies and organization to provide specialized resources and capabilities for response to WMD incidents.
- Federal emergency responders: Upon determination of a credible terrorist threat, or if such an incident actually occurs, the federal government may respond through the appropriate departments and agencies. These departments and agencies may include FEMA, the DOJ and the FBI, the Department of Defense (DOD), the Department of Energy (DOE), the HHS, the EPA, the Department of Agriculture (USDA), the Nuclear Regulatory Commission (NRC), and possibly the American Red Cross and Department of Veteran Affairs (VA).

ADMINISTRATION AND LOGISTICS

There are many factors that make consequence management response to a terrorist incident unique. Unlike some natural disasters, the administration and logistics for response to a terrorist incident require special considerations. For example, there may be little or no warning and because the release of a WMD may not be immediately apparent, caregivers, emergency response personnel, and first responder are in imminent danger of becoming casualties before the actual identification of the crime can be made. Incidents can quickly escalate from one scene to multiple locations and jurisdictions.

Planning should address administrative protocols to ensure that proper purchasing procedures are followed and that duplicate purchases are avoided. On September 11, and during the days that immediately followed, commercial airlines were not allowed to operate. The shutdown delayed the arrival of supplies and federal responders from distant locations. To avoid the inefficiencies of ad hoc purchasing of supplies and of delays in the arrival supplies related to air traffic curtailments, consider regional warehousing of supplies and equipment for emergency responders, including equipment for use of US&R task forces.

One of the key logistical problems in the initial stages of emergency response to a terrorist incident is the establishment of an incident command post from which direction of response activities can be made. In a routine emergency, such as a fire, small hazardous materials releases, or police actions, the incident command post is established at a point that is close enough to observe the incident, but far enough away to maintain an overview perspective and a safe distance from the immediate hazards. Because of the unique nature of terrorist activity and the inherent unpredictability of such an incident, planners and emergency responders may need to rethink the protocol for locating the incident command post.

One of the key administrative and logistical challenges in managing the emergency response to a terrorist incident is the successful integration of the federal response into the initial response by local and state emergency response organizations.

Be aware that non-agricultural terrorist incidents are more likely to occur in urban areas than in rural settings. Urban centers have significantly higher numbers of emergency personnel and material resources, and they routinely deal with emergency response. Local emergency response organizations will likely want to maintain the direction and control of the emergency response to a terrorist incident.

As a planner, be aware of the potential logistical problems that may be caused by the unsolicited influx of volunteers and donated goods, as was experienced post–September 11. Site and perimeter control is important to avoid responder casualties and to prevent emergency operation from being disrupted by uncontrolled movement of such volunteers.

SUMMARY

When developing a critical incident response plan it is important to incorporate a test, training, and exercise program to evaluate every aspect of a critical incident such as a terrorist attack. Begin with orientations that describe the requirements for a response. Then, proceed to tabletop exercises, functional exercise, and full-scale exercises. Evaluate tests and exercises honestly, carefully, and thoroughly. Lessons learned from testing and exercises may point to EOP revisions or additional training needs. Stay focused on the jurisdiction's goals to:

- compare plans for various functions within an agency and between supportive agencies to improve the emergency reponse and coordination of efforts at all levels
- consider a broad range of incidents, including assaults on infrastructure and information that may affect life, health, and safety of the population

Preparing your jurisdiction for a terrorism response may seem like a daunting task, but by building on the systems and procedures that are already in place and testing, training, and exercising the plan, an effective response is attainable.

23

Workplace Violence Prevention

Nicholas Dillon

OBJECTIVES

The objectives of this chapter are as follows:

- define workplace violence and describe the different types of workplace violence
- identify the warning signs of workplace violence
- explain the steps to develop a violence prevention plan
- apply safety and security practices to prevent workplace violence
- assess threats in the workplace and implement effective disciplinary action when threats occur
- explore a zero-tolerance approach in the workplace as a positive tool for violence prevention

INTRODUCTION

The terrorist attack of September 11, 2001 was a tragic reminder to the United States of the threat posed by international terrorism. Many people and thousands of workplaces were affected by this incident, which ultimately changed the face of New York City and the lives of families around the world.

Among other things, the attacks of that day heightened awareness of workplace violence. Prior to 9/11, violence in the workplace was seen as extreme, isolated incidents perpetrated by angry or disgruntled employees or customers. After that day, businesses felt the need to have a plan not only for internal incidents of violence but also for external attacks—including the extremes of terrorism.

The information discussed in this chapter is designed to provide safety and security professionals with an understanding of workplace violence as a reality in our society and to provide information on taking a proactive approach to education and prevention.

WHAT IS WORKPLACE VIOLENCE?

Workplace violence is an act of aggression, physical assault, or threatening behavior that occurs in a work setting and causes physical or emotional harm to customers, co-workers, or managers. Broad definitions of workplace violence often include acts of sabotage on worksite property, including acts of terrorism.

Several of the most serious incidents of workplace violence occurred before the 1990s. On August 20, 1986, Patrick Henry Sherrill entered the United States Postal Service office in Edmond, Oklahoma. Having received a negative work report from his supervisor the previous day, Sherrill went to a National Guard armory, stole some weapons, and then went to the facility where he had worked for 18 months. He opened fire, and by the time he ended his killing rampage, 14 of his co-workers were dead and 6 were wounded.

On December 9, 1987, a recently fired employee of Pacific Southwest Airlines bypassed metal detectors and boarded a flight headed for San Francisco. Once the plane was airborne, he took out a pistol and killed the flight crew. The plane crashed near San Luis Obispo, killing all 43 people on board.

Although these incidents were among the most deadly ever reported in the media, they did very little to affect how companies and organizations in the United States conducted their business. It was not until the 1990s that things began to change. In the early 1990s several well-known companies and organizations experienced incidents of workplace violence involving multiple homicides. Among these were the Ford Motor Company, Luby's Cafeteria in Texas, and the United States Postal Service. These incidents made national headlines and were covered extensively by the news media.

Suddenly, workplace violence was in the consciousness of the American public and was the subject of considerable discussion and activity in America's boardrooms. The term "going postal" became popular, and it seemed that nearly every day there was a report of a new incident of workplace violence. As a result, some businesses and organizations began developing workplace violence prevention programs, usually with the help of a consultant with expertise in violence prevention, threat assessment, and incident management.

A decade into the 21st century, efforts to prevent violence in the workplace and in schools have changed since the 1990s. Some changes have been for the better, some have not.

Workplace violence is now recognized as a specific category of violent crime that calls for distinct responses from employers, law enforcement, and the community. This recognition is relatively recent. Prior to the Edmond, Oklahoma shootings, the few research and preventive efforts that existed were focused on particular issues—patient assaults on health care workers and the high robbery and murder risk facing taxi drivers, late-night retail workers, and convenience store clerks.

However, contrary to popular opinion, sensational multiple homicides represent a very small number of the workplace violence incidents. The majority of incidents that employees and managers deal with on a daily basis are lesser cases of assault, domestic violence, stalking, workplace bullying, harassment (including sexual harassment), and physical and/or emotional abuse that usually does not make media headlines. Many of the incidents are not reported to company officials or even to the police. Data on the exact extent of workplace violence "are somewhat spread out and a bit spotty," specialists acknowledged in a February 2001 report issued by the University of Iowa's Injury Prevention Research Center. Utilizing the responses to the National Crime Victimization Survey, a Justice Department report estimated that an average of 1.7 million "violent victimizations," 95% of them simple or aggravated assaults, occurred in the workplace each year from 1993 to 1999 (University of Iowa Injury Prevention Research Center 2001). Estimates of the cost from lost work time and wages, reduced productivity, medical costs, workers' compensation payments, and legal and security expenses are even less exact but clearly run into many billions of dollars, not to mention the indirect costs involved. See Table 23–A.

Table 23–A. Average Number, Rate, and Percent of Workplace Victimization by Crime Type, 1993-99

Crime Category	*Average Annual Workplace Victimization*	*Rate per 1,000 Persons in the Workforce*	*Percent of Workplace Victimization (%)*
All Violent Crime	1,744,300	12.5	100
Homicide	900	0.01	0.1
Rape/Sexual Assault	36,500	0.3	2.1
Robbery	70,100	0.5	4.0
Aggravated Assault	325,000	2.3	18.6
Simple Assault	1,311,700	9.4	75.2

Sources: Homicide data are obtained from the Bureau of Labor Statistics Census of Fatal Occupational Injuries. Rape and sexual assault, robbery, aggravated assault, and simple assault data are from the National Crime Victimization Survey.

According to the Bureau of Labor Statistics half of employers with 1,000 or more employees in the United States had an incident of workplace violence within the 12 months prior to completing a 2005 survey on workplace violence prevention. The Bureau of Labor Statistics conducted the survey for the National Institute for Occupational Safety and Health (NIOSH). The survey looks at the prevalence of security features, the risks facing employees, employer policies and training, and related topics associated with maintaining a safe work environment.

The survey asked whether an incident of workplace violence had occurred during the past year and, if so, how the incident had affected the staff and whether the employer had implemented changes to reduce the risk of further incidents.

While 5% of all establishments, including state and local governments, had a violent incident, half of the largest establishments (those employing 1,000 or more workers) reported an incident.

More than 34% of employers with 1,000 or more workers said they had a co-worker incident within the 12 months before completing the survey. More than 28% of large employers said they had a customer or client incident. More than 24% said they had a domestic violence incident. And more than 17% said they had a criminal incident.

More than 28% of respondents with 250 to 999 employees said they had an incident of workplace violence in the previous year.

Of all establishments reporting an incident of workplace violence in the previous 12 months, 21% reported that the incident affected the fear level of their employees and 21% indicated that the incident affected their employees' morale.

More than 70% of U.S. workplaces have no formal program or policy that addresses workplace violence. In private industry enterprises that reported having a workplace violence program or policy, 82% reported addressing co-worker violence. Customer or client violence was the next most frequent subject of private industry policies or programs (reported by 71%), followed by criminal violence (53%) and domestic violence (44%).

Twenty percent of establishments in private industry provided training on preventing workplace violence, while 32% of local government workplaces and 58% of state government workplaces provided this training. Only 4% of all establishments offered training on domestic violence and its impact on the workplace.

Forty-three percent of private industry establishments report tracking the cost of workplace injuries and illnesses, while fewer than half that amount (20%) report tracking costs related to incidents of workplace violence (Bureau of Labor Statistics 2006).

TYPES OF WORKPLACE VIOLENCE

As attention to the issue of workplace violence has grown, occupational safety and security professionals have broadly agreed that responding to workplace violence requires focusing on more than just actual physical attacks. Homicide and other physical assaults are on the continuum that also includes domestic violence, stalking, workplace bullying, threats, harassment, emotional abuse, intimidation, and other forms of workplace conduct that creates anxiety, fear, and a climate of distrust in the workplace. All of these are part of the workplace violence problem. Prevention programs that do not consider threats and harassment in all forms are unlikely to be effective. While agreeing on the broader definition of the problem, specialists have also come to a consensus that workplace violence falls into four categories, as shown in Table 23–B.

Table 23–B. Categories of Workplace Violence

Type and Description	*High-Risk Factors*
Type I—The Criminal The perpetrator has no legitimate relationship with the employer or the workplace. Crimes such as robbery and burglary occur along with violence against the employees. This is the main type of workplace violence and accounts for the most murders.	• Money handling • Night work • High-crime location • Working alone • Secluded workplaces
Type II—The Customer/Client The victimizer is receiving some product or service from the employer and is not an employee. This individual becomes violent because of some dissatisfaction with the product or service provided. This type can be clients, customers, medical patients, or students.	• Easy workplace access • Minimal exits • Customers with history of violence • Presence of few co-workers • Working alone
Type III—Employee-on-Employee This person has a relationship with the business or an employee. These perpetrators of violence or threats of violence are current or former employees or independent contractors. Fewer than 10% of workplace murders are committed by this type; however, they perpetrate most incidents of workplace violence.	• Employer allows or condones discriminatory treatment • Inadequate employee background check • High workplace stress and low morale • Incompetent, poorly trained managers and supervisors • Authoritarian type of management style • Atmosphere where work performance concerns go unnoticed
Type IV—Personal or Domestic This person has a relationship with an employee but not the workplace. This individual wreaks havoc on a job site to physically harm or threaten to harm an abused domestic partner, spouse, or other employees. This type is the fastest growing type of workplace violence.	• Slow electronic communication • Ineffective violence prevention program • Limited exits • Working alone • Perpetrator with history of violence • Easy access to workplace

Source: Frances E. Zollers and Elletta Sangrey Callahan, "Workplace Violence and Security: Are There Lessons for Peacemaking?" *Vanderbilt Journal of Transnational Law* 36 (2003).

Type I perpetrators—the criminal type—are otherwise unconnected to the workplace. They account for the vast majority—nearly 80%—of workplace homicides. In these incidents, the motive is usually theft, and in a great many cases, the criminal is carrying a gun or other weapon, increasing the likelihood that the victim will be killed or seriously wounded. This type of violence falls heavily on particular occupational groups whose jobs make them vulnerable: taxi drivers (the job that carries by far the highest risk of being murdered), late-night retail establishment or gas station clerks, and others who are on duty at night, who work in isolated locations or dangerous neighborhoods, and who carry or have access to cash.

Preventive strategies against Type I perpetrators include an emphasis on physical security measures, special employee policies, and employee training. In fact, it has been suggested that one of the reasons for the decline in workplace homicides since the early 1990s (see Figure 23–1) is the security measures put in place in businesses vulnerable to this type of activity. Surveillance cameras, night officers, and even security patrols have helped minimize these types of incidents.

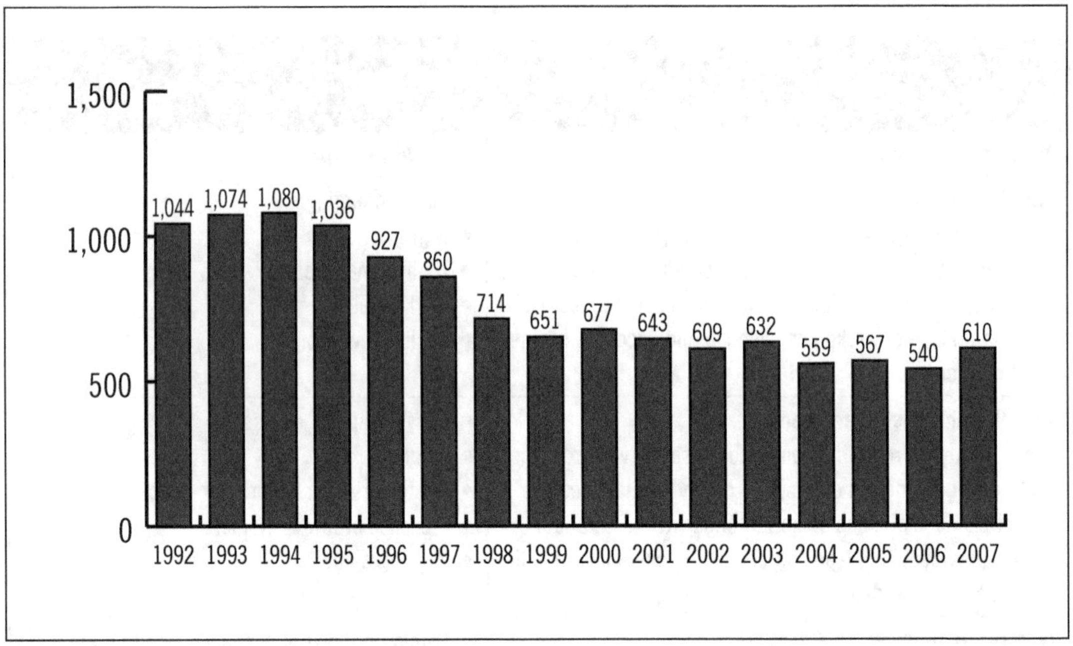

Figure 23–1. Workplace Homicides, 1992–2007. *Source: Bureau of Labor Statistics, U.S. Department of Labor.*

Because the outside criminal has no other contact with the workplace, the interpersonal aspects of violence prevention that apply to the other three categories are normally not relevant to Type I incidents. The response after a crime has occurred will involve conventional law enforcement procedures for investigating, finding, and arresting the suspect and collecting evidence for prosecution. For that reason, even though Type I events represent a large share of workplace violence (homicides in particular) and should in no way be minimized, the rest of this chapter will focus on the remaining types.

Type II cases typically involve assaults on an employee by a customer, patient, or someone else receiving service. In general, the violent acts occur as workers are performing their normal tasks. In some occupations, dealing with dangerous people is inherent to the job, as is the case of a police officer, correctional officer, security guard, or mental health worker. For other occupations, violent reactions by a customer or client are unpredictable, triggered by an argument, anger at the quality of service, denial of service, delays, or some other precipitating event.

Employees experiencing the largest number of Type II assaults are those in health care occupations—nurses in particular—as well as doctors and aides who deal with psychiatric patients; members of emergency medical response teams; and hospital employees working in admissions, emergency rooms, and crisis or acute care units.

Type III and Type IV violence—incidents involving violence by past or present employees and acts committed by domestic abusers or arising from other personal relationships that follow an employee into the workplace—will be the types most extensively discussed in this chapter. Violence in these categories is no less or more dangerous or damaging than any other violent act. But when the violence comes from an employee or someone close to an employee, there is a much greater chance that some warning sign will have reached the employer in the form of observable behavior. That knowledge, along with the appropriate prevention programs, can at the very least mitigate the potential from violence or, in the best-case scenario, prevent it altogether.

IDENTIFYING WARNING SIGNS OF WORKPLACE VIOLENCE

According to Northwestern National Life Insurance Company, 2,500 workers per 100,000 have been physically attacked on the job. Of these attacks:

- 44% were committed by customers or clients
- 24% were committed by strangers
- 20% were by co-workers
- 7% were by bosses
- 3% were by former employees (Northwestern National Life Insurance 1993)

Homicide is the most publicized form of workplace violence, but it is certainly not the most common type of behavior considered workplace violence. A broad range of conduct is considered workplace violence, including the following:

- sabotage of a co-worker or manager causing damage to her property
- violent confrontations in the workplace
- some forms of sexual harassment
- threats of violence by an employee
- assaults in the workplace
- armed robbery of employees
- suicide at the workplace
- hostage incidents in the workplace
- homicide

Research suggests there is almost always a warning sign before these types of violence occur. The first thing everyone in the workplace should be aware of is that it is not easy to know when someone is ready to commit a violent act. Some possible signs people will show before becoming violent in the workplace include:

- verbal abuse
- frequent arguments
- foul and insulting language
- threats of violent attacks

Workplace violence may begin with minor things and then explode into psychological and physical attacks. However, an employee may show one or two of these warning signs and never become violent. Since co-workers are sometimes completely surprised when an employee goes berserk, it is important for them and employers to look at a range of behaviors that could be the best warning signs of violence.

Just because someone is having a bad day or brings his personal problems to work with him does not mean he will resort to violence. But employees experiencing high levels of stress may be prime candidates for workplace violence. A violent incident is usually sparked by something that causes the employee in question to erupt. This is especially true if the individual is already feeling vulnerable or has low self-esteem.

Every employee who exhibits workplace violence warning signs will be different. Managers and co-workers should be sure to observe any negative changes in conduct. Because of the potential complexity of the issue, intervention and assistance for prevention may have to come from outside sources such as consultants or third-party specialists.

Many times the best warning sign of impending violence is what the person's body language says instead of what she actually says. Table 23–C lists examples of physical changes in conduct that could be warning signs of potential violent behavior.

Additional behavioral warning signs that could lead to potential workplace violence include the following:

- attendance and work performance problems
- requiring a significant amount of the supervisor's time
- inconsistent work patterns
- poor on-the-job relationships
- concentration problems
- unshakable depression
- poor health and hygiene
- safety issues
- unusual or changed behaviors
- fascination with guns or other weapons
- evidence of possible drug or alcohol abuse
- continual excuse making and blaming
- evidence of serious stress in the employee's personal life

As mentioned earlier, there is no one specific indicator of who will commit violence and when. The full range of workplace violence warning signs includes the manager's or employee's personality, motives, actions, and intentions. One must also consider the individual's state of mind in any given situation.

When employers and co-workers observe these warning signs prior to an actual violent act, they should (with appropriate training) be better equipped to keep the violence from occurring.

According to *A Comprehensive Guide for Employers and Employees,* published by Minnesota OSHA (with the Minnesota Department of Labor and Industry), there are five warning signs of escalating behaviors (MNOSHA 2009):

1. confusion: individual appears bewildered, distracted, or unsure what to do next
2. frustration: individual appears impatient or defeated
3. blame: individual holds others responsible for problems and is accusatory
4. anger: individual becomes visibly physical (pounding fists, pointing fingers, screaming, etc.)
5. hostility: individual becomes threatening and/or out of control

Table 23–C. Warning Signs for Potential Violent Behavior	
Clenched fists	Angry yelling
Getting too close	Sweating
Heavy or fast breathing	Foul language
Shaking or trembling	Frowning
Threatening movements	Finger pointing

What can we do when we see these escalating behaviors? Here are MNOSHA's suggested responses to these five warning signs:

1. confusion: Listen, ask questions, and give factual information.
2. frustration: Same as number one and relocate to private area and offer reassurance.
3. blame: Same as number one, then disengage and bring in a third party to the conversation. Restate the facts and ask probing questions.
4. anger: Use venting techniques; do not argue or offer solutions. Prepare to isolate or evacuate and contact supervisor or security.
5. hostility: Disengage and evacuate. Attempt to safely isolate individual and contact security immediately.

The key to identifying warning signs is that once they have been identified, they need to be brought to the attention of the employer and acted upon.

DEVELOPING A VIOLENCE PREVENTION PLAN

As with most other risks, prevention of workplace violence begins with planning. Also as with other risks, it is easier to persuade managers to focus on the problem after a violent act has taken place than it is to get them to act before something happens. If the decision to plan in advance is more difficult to make, however, it is also more logical. Any organization, large or small, will be far better able to spot potential dangers and diffuse them before violence develops, and better able to manage a crisis if one does occur, if its executives have considered the issue beforehand and have prepared policies, practices, and structures to deal with it.

When preparing a workplace violence plan, consider the following as part of your organizational assessment:

- The initiative and the support need to come from the top of the organizational structure.
- The plan should reflect the culture of your organization—it should incorporate needs and exposures that are particular to your place of business.
- The plan should be proactive and not reactive.
- The plan should follow a team approach in order to maximize the prevention of risk and increase its effectiveness.
- Management staff should take an active role in educating and training employees to identify the warning signs and provide instructions on what to do in case an incident occurs.
- Practice, monitor, and evaluate your plan on an ongoing basis. It is a good idea to survey the workplace every few years to identify any new risks or exposures that may have been missed initially. Revise your plan accordingly.

The following items should be included as part of your prevention plan:

- a policy statement that identifies the employer's zero-tolerance position on violence in the workplace
- a specific security assessment of the facilities or premises
- procedures for addressing possible threats and threatening behavior
- designation and training of an incident response team
- access to outside resources, such as threat assessment and debriefing/counseling professionals

- training at all levels of management that includes the different employee groups
- crisis response measures
- clear, consistent guidelines on how incidents will be handled, including effective disciplinary action and documentation

WRITTEN WORKPLACE VIOLENCE POLICY STATEMENT

An employer sets the standard for acceptable workplace behavior. A written policy statement should affirm the company's commitment to a safe workplace, employees' obligation to behave appropriately on the job, and the employer's commitment to take action on any employee's complaint regarding harassing, threatening, and violent behavior. The statement should be in writing and distributed to employees at all levels.

PREVENTIVE MEASURES

Preventive measures can include conducting preemployment screening, identifying problem situations and risk factors, and making security preparations. A thorough background check can be expensive and time-consuming. The depth of preemployment scrutiny will vary according to the level of sensitivity of the job being filled, the policies and resources of the prospective employer, and different states' legal requirements.

SECURITY SURVEY AND MEASURES

One important security tool is an employee questionnaire or survey to get ideas on the occurrence and potential for violent incidents and to identify or confirm the need for improved security measures. Surveys can be repeated at regular intervals, or when operations change or a violent incident has occurred, to help identify new or previously unnoticed risk factors. Responses can help identify jobs, locations, or work situations where the risk of violence appears highest.

Security personnel should be trained in techniques for handling threats or other confrontations without letting them escalate into violence. Security supervisors should have an up-to-date contact list for all employees, in case there is a need to advise workers of an emergency or distribute other information. They should also have a list of outside contacts, such as police, rescue, medical, social service, violence assessment, and employee assistance professionals. The security director should maintain a liaison with local law enforcement agencies, familiarizing them with the company's location and with evacuation and other emergency plans.

A plan for emergency evacuation should be designed and practiced. The evacuation plan should include not only procedures for getting workers out of the building, office, or facility but also some method for those evacuated to assemble or report in so that it can be accounted for. Evacuation plans should include provisions for workers with disabilities—for example, a way to make sure warnings are received by hearing-impaired employees and a system for safely evacuating anyone who uses a wheelchair.

ASSESSING AND INVESTIGATING WORKPLACE VIOLENCE INCIDENTS

Many times, a violent act is preceded by a threat. The threat may have been explicit or veiled, spoken or unspoken, specific or vague, but it occurred. In other instances, behavior that suggests the potential for some type of violent act may be observed by others. Yet in other cases, an offhanded remark or comments made to people close to the individual may suggest problematic behavior. Dealing with threats and threatening behavior—detecting them, evaluating them, and finding a way to address them—may be the single most important key to preventing violence in the workplace.

Any workplace violence strategy must include measures to detect, assess, and manage threats and behavior. In some ways, a physical assault or homicide is much easier to deal with than a threat. Normally there is no doubt that a homicide or assault has been committed because there is physical evidence. It is often harder to establish that a threat has been made or to assess that threat because often there is no physical evidence. Some threats are not criminal and, therefore, are not subject to law enforcement intervention or prosecution.

WHAT CONSTITUTES A THREAT?

Merriam-Webster's Collegiate Dictionary defines *threat* as "a statement or expression of intention to hurt, destroy, or punish, etc., as in retaliation or intimidation." This definition does not address the question of how legal authorities or employers frame and carry out a policy on threats. How do they determine when an intention to injure or harm someone has been expressed?

A purely subjective determination that whatever makes someone feel threatened is a threat is an uncertain guide for behavior, since different people can respond differently to the same words or acts. Employees who are required to observe "no threat" rules have a right to a reasonably clear statement on what is considered threatening behavior. That does not mean that subjective factors can or should be completely excluded from the definition. Employees can and should be held responsible for a reasonable regard for the feelings and concerns of co-workers and others in the workplace, and employers properly have an obligation to make sure employees do not feel frightened or intimidated.

A workplace violence prevention program addressing threats needs to include both a subjective and an objective component. It must set reasonably explicit standards of behavior so employees know how they are expected to act or not act; it must also make clear to employees that no one has the right to make anyone else feel intimidated. There should be a zero-tolerance approach to dealing with threats or threatening behavior in the workplace.

INVESTIGATING INCIDENTS OF WORKPLACE VIOLENCE

The best plans for threat assessment and response will be useless if employers or those assigned to respond to workplace violence do not know that a threat has been made. There must be procedures in place so that employees and the public know what to do and how to act. Detecting threats depends in large measure on the workplace culture. If employees are too afraid or too alienated from management to report violent or threatening behavior by co-workers, no violence prevention program will be effective. To encourage reporting, employers can create a climate in which safety is accepted as a common goal for workers and management, and all employees, including management, feel free to report disturbing incidents or possible danger signs.

Along with encouraging employees to report violence or threats, employers also have to inform them where to report and what to report. A designated office or person to whom complaints are directed, and perhaps a hotline number or suggestion box for employees who prefer to remain anonymous, can provide a concrete and clear venue for reporting.

To the extent that the employees feel comfortable reporting incidents to their immediate supervisors, the information may come through the normal management channels. However, having additional reporting channels can facilitate reporting where an employee finds it difficult to report through a supervisor. Whatever reporting system is adopted, publicizing it on bulletin boards, in employee newsletters and notes distributed with paychecks, or by other means will help ensure that all workers know how to report any behavior they consider troubling.

Just as important as knowing how to report incidents is knowing that reports will be heard and responded to. Feedback procedures, through which employees will be notified (subject to confidentiality rules) about how their reports were investigated and what actions were taken, will provide assurance and helpful closure to employees who make a report.

To further facilitate the identification of threats, employees, supervisors, and managers can receive training to help them detect out-of-bounds behavior or other warning signs. Training can also help educate workers and supervisors on how to respond to someone who seems troubled or potentially dangerous and how to report that behavior to managers. Training can also include a very clear statement to all employees on what to do if they see or become aware of a weapon. Any training program should be sensitive to cultural assumptions and stereotypes and emphasize focusing on an individual's manner, conduct, and behavior rather than ethnic or other group identity or a standardized "profile" of a dangerous person.

Disciplining an employee for abusive, threatening, or violent behavior serves two purposes. For the abusive or violent employee, the disciplinary action should serve as an appropriate penalty for past conduct and a deterrent against future offenses. For the rest of the work force, it should serve to reaffirm the employer's commitment to a workplace free from threats and violence and reinforce employees' confidence that their safety is protected by strong but fair measures.

To achieve those goals, penalties and the disciplinary process must be clearly proportionate, consistent, reasonable, and fair. Erratic or arbitrary discipline, favoritism, and a lack of respect for employees' dignity and rights are likely to undermine, not support, an employer's violence prevention efforts. Workers who perceive an employer's practices as unfair or unreasonable will nurse grievances and not report them with the expectation of a fair hearing and settlement. Grudges over unfair treatment will fester and may even erupt into further troublesome behavior.

Fairness in discipline begins with fairly and clearly spelling out what the rules are. Policies and workplace conduct should be written to clearly state the employer's standards and expectations. Penalties should be proportionate to the offense.

If there is a complaint or incident, the incident response team or designated security team should conduct or ensure a thorough investigation of the facts and, based on the results, determine appropriate disciplinary measures.

ZERO TOLERANCE AS A TOOL FOR PREVENTION

Historically, the phrase "zero tolerance" has referred to a standard, rather than a penalty. Zero tolerance on drugs meant that the standard of conduct would be no drug use. Zero tolerance on harmful substances in food or water supplies means that no amount of a particular toxic chemical or infectious agent would be considered safe.

In recent years, zero tolerance has taken on a new meaning: the application of an automatic penalty for a designated offense. In that sense, the policy has at times been criticized for overriding judgment and common sense, as when school administrators acting under a zero-tolerance drug or weapons policy expel a student for bringing a nail file to school or having a couple of aspirin tablets in a lunch box.

With regard to workplace violence, employers should make clear that zero tolerance in the original sense of the phrase applies; that is, no threatening or violent behavior is acceptable and no violent incidents will be ignored. Company violence prevention policies should require action on all reports of violence, without exception. That does not mean, however, that a rigid, one-size-fits-all policy of automatic penalties is appropriate, effective, or desirable. It may even be counterproductive, since employees may be more reluctant to report a fellow worker if she is subject to automatic termination regardless of the circumstances or seriousness of her offense.

Whether to use the phrase "zero tolerance" in its written workplace violence policy or find a different expression is a decision each employer will have to make. Whatever phrase is used, it should be made clear that the intent is to set a standard of conduct, not systematic penalties. Instead of warning of "automatic terminations," discipline policies should declare that violent workplace behavior will lead to penalties "up to and including termination." That leaves room for managers to consider circumstances and exercise penalties that are not imposed arbitrarily, but are consistent, proportionate, and fair.

SUMMARY

Although there is no one single way to prevent violence in the workplace, all organizations should recognize workplace violence as a specific category of violent crime that calls for a distinct response from employers, law enforcement, and the community. Violence in the workplace can take the form of an act of aggression, physical assault, or threatening behavior that occurs in a work setting and causes physical or emotional harm to customers, co-workers, or managers. It can also include acts of sabotage as well as terrorism. Preventive measures, such as pre-employment screening and security personnel staffing may help prevent violence in the workplace, but it is also important for an organization to develop a written policy on workplace violence and procedures for assessing and investigating incidents of violence in the workplace. When developing a plan for preventing workplace violence, an organization should consider the culture, history, industry, and size of its workforce to best accommodate the unique needs of the organization.

REFERENCES

Bureau of Labor Statistics 2006. *Survey of Workplace Violence Prevention, 2005.* U.S. Department of Labor, Bureau of Labor Statistics, Washington, DC. Accessed July 2009.

MNOSHA. 2009. *Workplace Violence Prevention: A Comprehensive Guide for Employers and Employees.* MNOSHA: http://www.dli.mn.gov/WSC/PDF/WorkplaceViolencePreventionGuide.pdf. Accessed July 2009.

Northwestern National Life Insurance. 1993. *Fear and Violence in the Workplace: A Survey Documenting the Experience of American Workers.* 1993.

University of Iowa Injury Prevention Research Center. 2001. *Workplace Violence: A Report to the Nation.* Violence in the Workplace, 1993-1999. Special Report, U.S. Department of Justice, Office of Justice Programs. Bureau of Justice Statistics, December 2001, NCJ 190076. Washington, DC.

24

Civil Unrest Preparedness

Ronald H. Beatty, CPP

CHAPTER 24

OBJECTIVE

The objective of this chapter is to consider ways to prepare for types of civil unrest that typically affect businesses and institutions, including demonstrations, activism, and civil disobedience.

INTRODUCTION

Civil unrest has been around as long as society and probably will be with us forever. It is an affliction endemic to the young and ideological, in particular, and to anyone else with grievances, real or imagined. It is found in those who advocate the causes of the underprivileged—either in genuine sympathy, out of expectation of power, or both. It is generally spawned by discontent with one's life or by a sense of injustice. It is a human condition driven by a desire to rectify an inequity, to obtain power, or to simply have what others have.

The framers of the U.S. Constitution also recognized this flaw in the human condition and in 1791 passed the First Amendment, which reads in part as follows:

> Congress shall pass no law respecting the establishment of religion or prohibiting the free exercise thereof; *or abridging the freedom of speech* or of the press, or the *right of the people peaceably to assemble, and partition the government for a redress of grievances.* (Emphasis added.)

In their wisdom, the founding fathers knew that if the oppressed have-nots could at least vent their frustration, they would be less likely to march up the hill with pitchforks—the principle being "Deny peaceful demonstration; incite revolution." Having just fought the Revolutionary War, they knew this principle well. This, then, became the means by which the young democracy would establish justice, ensure domestic tranquility, promote the general welfare, and secure the blessings of liberty for its new citizens and their posterity.

DEFINITIONS OF CIVIL UNREST

Forms of civil unrest include rebellion, revolution, uprisings, revolt, insurrection, mutiny, riots, and demonstrations. *Rebellion* implies an open, formidable resistance to civil authority that is often unsuccessful. *Revolution* implies a successful rebellion resulting in a major change in government. An *uprising* generally applies to a brief, limited, and often immediately ineffective rebellion. *Revolt* and *insurrection* imply an armed uprising that quickly fails or succeeds. And *mutiny* applies to group insubordination or insurrection generally against naval authority. The common elements of all these acts are that the target is generally a government body or civil authority and that the acts include the use of arms. There is no doubt that businesses and institutions could be affected by these conditions, but they are more likely to occur by accident than by design. In short, anything and everything in harm's way could be affected.

Merriam-Webster Dictionary defines a *riot* as "profligate behavior," which in turn is defined as "completely given up to dissipation and licentiousness." These definitions conjure up visions of the Roman Empire during its decline or a fraternity bash. They do not accurately depict mobs of citizens rampaging through the streets, throwing stones at cars, breaking store windows, looting, and burning everything in their path, but that is what some riots are.

The prevention and control of riots—like rebellion, revolution, uprisings, insurrection, revolt, and mutiny—rightfully belong to the civil authorities. However, in such situations, civil resources are often stretched thin, so most businesses are required to fend for themselves, at least at the incipient stages. There are things a company can and should do to prepare for major disasters. Many of the preparations made to survive an earthquake, a hurricane, or other natural catastrophes will serve a business well during a riot. The self-help principles are the same. However, unlike a natural disaster, a riot tends to focus on specific businesses in the riot location. Particularly vulnerable are pawn shops, liquor stores, and retail outlets. Factories and other income-producing businesses are less affected. This chapter focuses on the types of civil unrest that typically affect businesses and institutions in general (demonstrations, activism, and civil disobedience).

AD HOC DEMONSTRATIONS

An ad hoc demonstration is a public assembly of people exhibiting sympathy with or opposition to some specific social, political, or economic condition or movement. The aim of an ad hoc demonstration is to focus public attention on problems, people, and institutions against whom the action is directed. The intention is harassing or embarrassing the target and gaining publicity and support for the demonstrators' cause. The intended result is to apply sufficient social pressure on the target to force a confrontation over an issue and to obtain legitimacy for the action by having the target capitulate to the demands of the demonstrators. At the outset, demonstrators can be more interested in confrontation than in resolving problems. By the time they show up at the targeted institution, they may already have found it guilty of whatever had caused the offense.

THE ISSUES BEHIND THE ISSUES

As mentioned earlier, behind most movements are a small hard-core cadre of dedicated people who genuinely believe in their cause. Because the issues they champion are sometimes so diverse, there is some speculation that their primary motive is not the cause at hand but a much larger one—the acquisition of power. During the late 1970s and early 1980s a pacifist group, known as the Clergy and Laity Concerned, which had nationwide ties to several loosely connected groups, demonstrated against everything from nuclear power to infant formula. The avowed purpose of the organization was "to seek a new social order in the United States." Whether a new social order would benefit or harm the United States was irrelevant.

Much of the social and political change in our country in the past 100 years has been the result of activist groups. Women's suffrage at the turn of the century, the labor movement in the 1930s, and the Civil Rights movement in the 1960s, for example, all contributed to a new social order. During the cold war, many causes championed by activist groups seemed at variance with the established social order. However, the country became stronger, and some would argue that it was because public dissent was allowed.

DEMONSTRATORS

Most ad hoc demonstrators do not constitute a mob. Typically, demonstrations have identifiable leaders and are generally well-organized, orderly, and within the law. Individually, demonstrators can be persistent and view their causes with single mindedness. At times, the combined dedication and persistence of the demonstrators has turned a peaceful demonstration into a violent one. What begins as a peaceful protest can quickly be turned into civil disobedience.

Although the motivations of demonstrators can vary, in general, they seem to be idealistic, bright, articulate, and educated. However, in some instances, they may not be patient enough to use accepted channels to effect the changes they believe are necessary to achieve the results needed. Participating in the direct action of a demonstration can release a demonstrator's frustrations and offer a sense of identity.

In *Romeo and Juliet*, William Shakespeare wrote, "that which they call a rose by any other name would smell as sweet." This quote implies that the essence of a thing remains despite what it is called. That principle applies to the activist movement in the United States. It is interesting to note that in major national movements (for unilateral disarmament, against nuclear energy, antiwar, etc.) many of the groups are interactive. For example, on the East Coast, activists identified as the Atlantic Life Community, the Whales Tale, the Clamshell Alliance, and Jonah House are known in the Midwest as Clergy and Laity Concerned, the Draft Resisters League, and the St. Louis Economic Conversion Project. On the West Coast they have been known as the Concerned Citizens for National Security, the Alliance for Survival, the Community for Disarmament, and so forth. Many of the groups are interrelated, sharing the same address, the same office space, and the same resources. In the early 1980s the Trap Rock Peace Movement, of Massachusetts, set up an office in Missouri at an address shared by the St. Louis Economic Conversion Project, the National Alliance Against Racial and Political Oppression, the Woman's International League for Peace and Justice, and Clergy and Laity Concerned.

Organizations' names are chosen for a variety of reasons. At times, they try to identify with the mood of the noncadre workers (the Atlantic Life Community), or to reflect local issues (the Clayton Nine), or to emphasize a particular topical issue (First Strike Prevention Project), or simply to regroup in the event of stagnation or unfavorable publicity but always to solicit fresh donations and contributions.

ARE AD HOC ORGANIZATIONS LEGITIMATE?

The majority of "street" organizations exist in name only. They have mailing addresses but are not formally constituted or chartered. The cadre may belong to churches or social organizations, but when the group flies the banner of "Alliance for Survival" or some other organization, it is likely that the banner represents an ideal and not a registered organization. This approach obtains the most flexibility and the least liability. An organization that does not exist cannot be sued.

After the Seabrook New Hampshire nuclear power sit-in in 1977, the state of New Hampshire tried to recover damages and costs for the detention of hundreds arrested at the demonstration from the sponsoring organization, the Atlantic Life Community, only to find that the organization did not officially exist. Targeted industries and businesses should understand the ramifications of failure to register, not only from a recovery standpoint but also from the perspective of responding to the group's demands and giving legitimacy to a phantom organization. Activists are well aware of the difference between legitimacy and legality in popular movements. Their goals are to obtain legitimacy, to be recognized as the spokespeople for the issues and representatives of a political base, and to press their target to obtain that status.

HOW SPONTANEOUS ARE DEMONSTRATIONS?

The leaders of a demonstration would like the public to think they represent a groundswell of public opinion. Although a demonstration can appear to be spontaneous in nature, in reality there is generally a great deal of planning beforehand. During the initial or preparatory stage, core group members develop intelligence about the selected target based on what products it manufactures; what hazardous materials it produces or uses; and its record on industrial incidents, minority hiring, sexual discrimination, and other matters. Any of these issues could become the basis for a protest. Any issues that attract attention without much difficulty, have broad and accepted appeal, and give the aura of righteousness and respectability to the activists are fair game.

Often demonstrators will probe the target, handing out flyers, inviting employees of the targeted business or institution to attend a meeting at some public location. The response to the flyers may indicate whether there is community support for the organization or not. If the response is weak, the group may try to infiltrate the organization by getting on its payroll and developing support through one-on-one meetings at work. Having someone "inside" an organization helps ferret out employee attitudes and internal problems that can be useful in defining the issues.

Demonstrators may hold organized meetings at potluck dinners, where strategies are laid and consensus is built about planning the demonstration. Often, if civil disobedience is contemplated, potential demonstrators break up into smaller "affinity groups" to plan what to do and how to do it. These plans are kept secret from the larger organization to maintain the element of surprise and to limit the liability of the "parent" organization. The group may have legal support, advising members on what to expect, how to conduct themselves if arrested, how to post bail, and what to say at an arraignment, for example.

PUBLIC RELATIONS

Public relations is not left to chance. Activists are well aware of the power of the press and use it to their advantage by:

- drawing attention to their cause
- educating the public in general
- assisting indirectly in recruiting new members

Knowing this, a well-organized activist group uses the media in the same way a public relations organization would. It establishes a media contact list of reporters to cover the group's events for local newspapers, radio and television, and some periodicals. Members personally notify the press, send news releases, respond to letters to the editor, call press conferences, operate around deadlines, and establish clipping files. They also volunteer for talk shows, utilize public service announcements, and write articles.

ROLE OF THE SECURITY PROFESSIONAL

When a business is targeted for a demonstration, it is the responsibility of the safety or security manager to ascertain the reason for the demonstration. The manager should introduce himself to leaders of the protesting group and attempt to learn why the business has been targeted for demonstration. Because the group has a legal right to assemble and protest, the manager should inform them to exercise that right on the public sidewalk only and they may not impede traffic or enter company property.

By approaching the demonstrators, the security professional has accomplished a number of things. First, the target of attention is no longer a big, impersonal, unfeeling corporation. An employee, who is not afraid of them, has shown an interest in their protest. Second, notice has been served to the group that its members are welcome to exercise their constitutionally protected rights as long as those rights are not abused. This may prompt the group to seek a softer target. Some in the group, more interested in confrontation than problem solving, may resent the presence of the security officer.

A positive attitude is vital in situations involving activist groups. The best way to invite these groups to escalate a demonstration to civil disobedience or violence is to take no firm action from the beginning. A policy of sitting back and doing nothing at first and later taking action seems only to play into the hands of the militants who engage in lawlessness because it is easy and they think they can get away with it. It is more difficult to be militant if one thinks she will be arrested and prosecuted.

ESTABLISH A CLIMATE OF ACCEPTANCE

Equally important is establishing a climate of acceptance among the workforce, a climate that will discourage the development of antidemonstration sentiment. It is sometimes difficult for hardworking, law-abiding citizens who believe and work within a system to understand that others do not necessarily share their views. An activist's attack on the system is perceived as an attack on the ideals and values of law-abiding citizens. The fact that rule-following employees are working while malcontents are not does not sit well with some workers.

However, nothing seems to generate more publicity than a public clash between opposing sides, and publicity is what the activists seek. In addition to the possibility of injury to the participants, there is always the possibility of creating martyrs for the activists, and martyrs generally create more support for their cause. On some occasions the demonstrators may deliberately provoke strong reactions to set the stage for charges of brutality or to file civil rights lawsuits against the police or the company. Therefore, it is important that those responsible for safety and security work closely with communications, legal counsel, human resources, and the local police to educate employees and create a climate of acceptance.

JOIN THE GROUP

If it does not look as though the activists will go away, it may be necessary, to the extent possible, for the security manager to participate in the public planning and coordination of their meetings. Open participation, of course, is the Achilles' heel of many activist groups. Their first goals are always to increase public awareness and participation, which means they cannot selectively screen out passive observers, even those representing the business community.

In any case, contact with the organization's leaders should be maintained, and all flyers and publications should be carefully analyzed for clues concerning shifts in tactics, changes in leadership, and indications of what group or groups are providing funding and support to the group. Often the group's publications provide detailed demonstration plans, since this is one of the principal methods used by the organization to reach all members of the group.

DEVELOP INTELLIGENCE

Basic to any protective measures a business can take is learning all it can about the threat. The gathering of such information does not imply developing a sophisticated espionage net or using "undercover" operatives but suggests, rather, the gathering, collecting, and compiling for analysis information from overt sources (such as newspapers, magazines, and activist periodicals). Liaisons should be established with local, state, and federal law enforcement agencies. Networking with comparable professionals in other industries, trade associations, or professional societies will assist in keeping on top of the situation.

OBTAIN PHOTOGRAPHIC COVERAGE

In addition to normal television and news coverage provided by the media, the demonstrators often use their own video recorders and cameras to gather material for publicity after the demonstration is over. If scenes of violence or angry faces can be photographed, these can be used for future propaganda or in law suits. Consequently, it is essential for the company to obtain its own photographic record of the event to protect itself from adverse publicity and to use as evidence of any illegal actions on the part of the demonstrators. In addition, the knowledge that their actions are being photographed tends to discourage individuals from performing acts that could incite a crowd to violence. It diminishes the anonymity that is essential in encouraging massive, severe disturbances; and photographs are by far the best evidence in any subsequent court action. Ensure that the photographers are professional and well versed in the rules of identification and the preservation of evidence.

MEET WITH DEMONSTRATORS

One of the principal objectives of many demonstrations is to force a confrontation. The question then is: Should your company meet with the demonstrators? The answer often lies in the nature of the demands, which usually reflects the sincerity of the group. If the demands are outrageous or impractical, the leadership of the group is obviously more interested in confrontation than in problem solving. In this instance, a meeting would merely afford the group recognition and the opportunity to gain instant status or legitimacy.

If the demands are reasonable, modest, and within the company's ability to address, the demonstrators may be sincere and a private meeting may be desirable. Here, however, a judgment should be made on the basis of sound evidence that the group is not seeking entry just to enlarge and escalate demands or to conduct a sit-in. If an escalation occurs, the meeting will be aborted. As a general rule, a demonstration that has failed in its principal objective—confrontation—immediately loses its news value and financial support. Naturally, once the demonstrations have lost support, there should be no meeting except on the company's terms. In short, in most cases a business should resist meeting with demonstrators.

CIVIL DISOBEDIENCE

Civil disobedience has a long history in our country. Henry David Thoreau espoused it. It was a particularly effective tool during the Civil Rights movement of the 1950s and 1960s. Rosa Parks refused to move to the back of a bus, in violation of Southern segregation laws, and that simple defiant act sparked a movement that totally changed the character of racial relations in the United States. So successful were sit-ins, lie-ins, and chain-ins that they were quickly adopted by other groups with other causes and rationalized as a justifiable expression of free speech and therefore protected by the U.S. Constitution.

Civil disobedience is the deliberate violation of the law, with every expectation of arrest, as a protest to dramatize support of or opposition to an issue. It is the natural escalation of the simple demonstration. The purpose of a demonstration is to draw public attention to an issue, and this is generally done through media coverage. As more and more groups demonstrate to that end, the events lose their newsworthiness. Groups employ additional tactics to regain media attention and coverage, which may lead to more civil disobedience. Civil disobedience has the additional value of extending publicity from the incident into the courtroom and beyond.

VIOLATIONS OF LAW

Activists who participate in civil disobedience are often arrested for criminal trespassing or disorderly conduct. Criminal-trespass statutes vary from state to state, but most penalize people who unlawfully enter the property of another or who remain on the property after being asked to leave.

Criminal intent is an essential element of illegality; therefore, trespassers must have known that they were unlawfully on property or they must have known and understood the order to leave. In some states there is no criminal trespass if the trespasser thought that he had the right to be on the property (no criminal intent can be established) and used no force in the trespass.

Disorderly conduct is defined wholly by statute or municipal ordinance. Although some states specify certain conduct as constituting criminal disorderly conduct, most statutes define the crime generally as a disturbance of public peace and quiet, or conduct that could tend to provoke a breach of the peace or corrupt public morals. The presiding judge usually has the discretion of deciding whether the acts, in light of circumstances, constitute a crime. In most states, willful and unlawful intent is not essential to prove illegality. Where the statute does set out specific elements of disorderly conduct, however, the elements must all be met, or no crime has been committed in the eyes of the law.

KNOW THE LEGAL STEPS

It is essential to establish legal position against any possible illegal or extralegal actions by demonstrators. The company should research or have researched local ordinances on parade permits, loudspeaker systems, blocking of public thoroughfares, and other areas. Decisions should be made as to what extent the company will condone trespassing on its property. If arrests are to be made, know what sections of the penal code are applicable to possible offenses of demonstrators. Have all premises been properly posted with "No Trespassing" signs? Who will represent the company, and will the company sign complaints and prosecute vigorously? Once all of the issues regarding the legal position of the company have been established and responsibility assigned, it is essential to make the legal consequences of their activities known to the activists. Unless this is done, many of the demonstrators may be misled into believing the "moral right" of their cause protects them from normal civil and criminal prosecution. Most demonstrators do not consider themselves lawbreakers and often are shocked by the reality of criminal prosecution for their actions.

COUNTER LIE-INS

An implied threat in any demonstration is the use of direct-action tactics, such as lie-ins, sit-ins, and chain-ins, to dramatize the purpose of the protest. Such tactics are a surefire way to attract a larger turnout and increase media coverage. When these tactics are used, a special arrest procedure should be employed to remove those arrested from the scene. Any delay or vacillation by the company or law enforcement agents at that point could escalate the disturbance. Procedures should be worked out with law enforcement agencies well in advance of the demonstrations.

The security officer responsible for removing protesters from private property should approach protesters in a positive way. While on camera, he should introduce himself and state his position with the company. Inform trespassers that they are on private property, that they are not welcome, and that they should leave. If they make no effort to move, repeat the warning that they are on private property, are unwelcome, and that they should leave now. If, after an appropriate interval, they still do not move, repeat the warning again and ask whether they understand what was said. All of this should be recorded on camera. Do not use harsh language, threaten harm, or be rude; remain relaxed and positive but firm. If the demonstrators refuse to leave, turn the situation over to the local police and request the trespassers' arrest. Keep photographing any action taken by the police. Once the demonstrators have been removed, sign and file a police complaint. It is likely the trespassers will be booked and released on their own recognizance.

THE TRIAL

Typically there are four defenses that activists use, or try to use, in court: the First Amendment defense; defense of duress; international-law defense; and defense of necessity/choice of evils. *Defense of necessity/choice of evils* is usually reserved for cases in which the harm resulting from compliance with the law would exceed the harm resulting from breaking the law and where the harm was imminent, alternatives were not available, and the illegal action (trespass) was reasonable.

The *international-law defense* is a spin-off from the Nuremberg trials, at which the defendants were accused of committing crimes against humanity. Stopping the production of nuclear weapons might therefore be justified as preventing crimes against humanity.

The *defense of duress* excludes a person from criminal prosecution because she was forced to act under the threat of physical harm. However, this defense is rarely permitted. The *First Amendment defense*, however, has been used somewhat effectively, if not dramatically. This defense tries to prove that the facility trespassed was actually public or semipublic property and that what the trespassers did was only symbolic, designed to express their opinion to the public on public property, and therefore protected by the Constitution.

PRO SE

Many activists opt to represent themselves in court, or *pro se*, as it is termed in law. It is thought that direct personal appeal to the conscience of the jury members, the judge, the press, and all others present is the truest extent of their action and the means to preserve the integrity of their moral conviction. It is the activists' objective to indict the crime that caused their direct action rather than be diverted into the legal matters of the law as represented by the charges brought against them.

In their ignorance of proper and conventional courtroom processes, the *pro se* defendant enjoys a greater latitude in addressing what, for them are the real issues and do not wish to be bridled into compliance with the narrow concerns of the law. Most juries, like the defendant, are lost in the world of legal technicality. If the judge or prosecutor is harsh with the *pro se* defendant, sympathy may be evoked for the well-meaning defendant, who is as confused as they are. Consequently, the judge and prosecutor are likely to restrain energetic enforcement of their legal powers, giving the defendant more room to focus on the issues. The "ignorance" issue also provides the defendant with a great deal of latitude in cross-examination, which means being allowed to say things to a jury that ordinarily may not be said.

Sympathy is also aroused by the drama of a lone defendant in the face of overpowering intimidation by the state and court. He is often allowed to speak his case simply, without the fog of technical distractions. In that hush that comes when one speaks for oneself, the defendant hopes that the voice of conscience is heard. This drama also appeals to and intrigues the media, and the defendant's statements are often carried beyond the courtroom to the public forum. This, of course, is exactly what the defendant wants.

OTHER TACTICS USED BY ACTIVISTS

Activist groups can use other tactics against organizations. Activists legally can exercise their rights as shareholders in a corporation by submitting proposals and then arguing in favor of their cause at the shareowners meeting. Another route, which is illegal, is breaking and entering.

SHAREHOLDER PROPOSALS

In the late 1960s activists began using the shareowner proposal as a tool to gain access to and address their concerns to top management and board of directors of a company. According to the Securities and Exchange Commission's proxy rules, any owner of securities (beneficial or of record) in a public corporation may submit a proposal for consideration and vote at the company's shareholders' meeting. Activist groups purchased stocks in companies they targeted and began submitting proposals for inclusion in proxy statements sent to shareholders; then, at the meeting, representatives of the group would take the floor to argue their cause. Often the representatives would use this forum for demonstrations, and it was not uncommon for them to conduct sit-ins, spill animal blood on the floor, unfold banners, chant slogans, or otherwise disrupt and monopolize the meeting. Those responsible for security did everything possible to contain demonstrations and provide protection.

In the late 1980s two things happened: the Security and Exchange Commission tightened the rules on shareholder proposals, and the activists became less active. In fact, many groups admitted that they were doing quite well by their company investments. They continued submitting proposals, but the issues seemed to deal more with corporate governance than radical causes. For years, however, the proposal tactic became the uncomfortable bane of many chief executive officers.

BREAKING AND ENTERING

One of the more drastic tactics used by activists over the years was the penetration of defense facilities and military bases for the purpose of damaging military equipment. The activists' objective was to embarrass the government by underscoring the futility of spending billions of dollars on defense when the facilities are defenseless against unarmed pacifists. Dressed like commandos, they would enter the facilities under cover of darkness on holiday weekends, when staffing levels were low, strike quickly, and leave. These acts were clearly criminal, well beyond civil disobedience as typically practiced; and when those who committed the acts were captured, they received the full sanctions of law, including prison terms with restraining orders imposed after their release.

SUMMARY

Civil unrest is a fact of life. Our democracy has accommodated it by permitting freedom of speech and the right to assemble; these constitutional rights ensure equality under the law. Often many of citizens, either comfortable with the status quo or filled with patriotism, resent those who see things differently when they draw public attention to their causes. Similarly, those who draw public attention to their causes often become overzealous and abuse their rights of free speech and assembly. Good security and safety professionals must understand the feelings and emotions of both groups and establish an environment in which activists can exercise their rights peaceably but within the law and with a minimum amount of business disruption.

When activists begin pushing beyond justifiable limits, security and safety professionals must act positively to control them. As in labor disputes, the professionals should avoid becoming emotionally involved in the issues. They should be prepared for contingencies, monitor the activities, and act when action is required. Demonstrations and civil disturbances begin and end. The security professional must manage the situation with the least amount of interruption to the business.

BIBLIOGRAPHY

Anti-nuclear Legal Project. *The National No-Nukes Trial Handbook.* Haydenville, MA: Aldebaran.

Masotti, L. H. *Riots and Rebellion.* Thousand Oaks, GA: Sage Publications, 1968.

Momboisse, R.M. *Industrial Security for Strikes, Riots and Disaster.* Springfield, IL: Charles Thomas, 1968.

National Advisory Commission on Civil Disorders. Report of National Advisory Commission on Civil Disorders. Washington, DC: GPO, 1968.

Rosenthal, U. *Coping with Crisis.* Springfield, IL: Charles C. Thomas, 1989.

Schonborn, K.L. *Dealing with Violence.* Springfield, IL: Charles C. Thomas, 1975.

Security Management. American Society for Industrial Security. *Preparing for Unrest.* 3/94. *Domestic Extremists.* 12/94.

25

A Public–Private Partnership on Security

Robert Chapman

CHAPTER 25

Disclaimer: The opinions contained herein are those of the author and do not necessarily represent the official position of policies of the Office of Community Oriented Policing Services of the U.S. Department of Justice.

OBJECTIVE

The objectives of this chapter are as follows:

- to define community policing
- to understand the value of partnerships between private security and public law enforcement agencies
- to explore the goals of public–private partnerships and how they affect the community
- to examine the role of the federal government in promoting partnerships

INTRODUCTION

The policing profession and the private security industry each possess important histories, unique traditions, and talented and committed professionals. In addition, police practitioners and private security professionals hold significant expertise, often have access to impressive and state-of-the-art technology, and frequently have consistent and complimentary missions and goals. However, despite the many parallels that exist between these fields, the opportunities for leveraging their respective strengths through partnerships have not traditionally been fully maximized.

Fortunately, it appears that the lack of collaboration between these two vitally important professions is changing. As recently as 2000, a national study examined the extent to which partnerships existed between private security organizations and law enforcement agencies. The study identified only about 60 formal working partnerships (Connors, Cunningham, and Ohlhausen 2000). However, seven years later another federally funded examination exploring the number and nature of formal partnerships between private security and public law enforcement identified more than 450 partnerships (Law Enforcement-Private Security Consortium 2009). It is clear that as private security and police professionals become increasingly exposed to the value and benefits of collaborating with their colleagues to address issues of mutual concern, and see and hear examples of successful partnerships, they are much more likely to enter into formal arrangements in an attempt to resolve common issues.

The U.S. Department of Justice Office of Community Oriented Policing Services (COPS Office) has long promoted these partnerships by convening discussions among leaders from both fields to identify recommendations for policy and practice that would positively impact the depth and breadth of these partnerships. The COPS Office has also developed practical resources to support local collaborative efforts between private security and law enforcement, as well as publications and other materials on specific public safety and security topics that facilitate the work of these partnerships. Established as a component within the Department of Justice (DOJ) in 1994, the COPS Office advocates for partnerships because partnering aligns with the COPS Office's organizational mission to support policing and public safety fields through the advancement and implementation of community policing.

COMMUNITY POLICING

Community policing[1] comprises three primary elements, two of which are *problem-solving*:

1. to solve problems by reducing, eliminating, and preventing crime and disorder by addressing contributing underlying conditions
2. to bring about organizational changes so improvements in public safety practices are implemented, sustained, and ultimately institutionalized
3. to solve problems through partnerships by addressing priorities shared by private security and the police

Through partnerships, community policing encourages law enforcement agencies to engage and collaborate with a range of community segments, including other government agencies, community members and groups, nonprofits and service providers, the media, and private businesses. For the police, establishing partnerships serves two interrelated goals of *increasing trust* and *engaging in collaborative problem-solving efforts*.

INCREASING TRUST

In introducing the importance of developing trust through partnerships, Matthew Scheider from the COPS Office writes that "trust forms the basis for any effective relationship, whether it is between two friends, between customers and a company, or between a public service agency and citizens" (Scheider 2008). Executives from the public and private sector understand how critical trust is to the success of their respective ventures. In policing, citizens with little trust in their law enforcement agencies are less likely to report crime, as well as to participate in developing and implementing solutions to crime and disorder problems. In the private sector, trust can form a powerful foundation for the relationship between business enterprises and their customers.

The first goal of partnerships—increasing trust—is not always an exclusively desired outcome, although there are intrinsic benefits for public- and private-sector professionals (e.g., increasing citizen satisfaction with police services and elevating citizens' trust in the police, maintaining quality service standards in a corporate/client relationship, etc.), which can pay dividends over time. For example, some basic partnerships between private security and the police have as their goal: to foster relationships and share information that can be valuable in anticipation of the emergence of a problem. However, most effective and long-term partnerships have an identified purpose to solve an active, current problem that is shared by the partners, or one in which one or more of the partners have a stake in addressing the problem.

Successful partnerships are often those that are formed with a "compelling mission," an identified problem or condition that keeps the partners motivated to join and to stay actively involved as they drive towards developing and implementing a solution. During difficult economic times, police agencies must operate under reduced budgets with limited resources. Likewise, some corporate entities face the challenge of operating in an economic downturn. When this occurs, partnerships between the private and public sectors become crucial. Put simply, the purpose of the partnership must be meaningful and important to all partners to keep the partnership sustained.

[1]The COPS Office defines community policing as "a philosophy that promotes organizational strategies, which support the systematic use of partnerships and problem-solving techniques, to proactively address the immediate conditions that give rise to public safety issues such as crime, social disorder, and fear of crime." http://www.cops.usdoj.gov/Default.asp?Item=36

ENGAGING IN COLLABORATIVE PROBLEM SOLVING

Many police practitioners will freely admit that they are rarely able to solve crime and disorder problems alone, at least in the long-term. Comprehensive solutions that involve both traditional (i.e., reactive responses, such as making arrests and issuing citations) and nontraditional (i.e., proactive responses, such as altering the environment in which crime is taking place, "hardening" targets or victims, and deflecting offenders before they strike) responses must be developed in consultation with other stakeholders. Those stakeholders can bring unique tools, resources, and knowledge to bear and play a significant role in mitigating or solving the problem. Historically, the community policing philosophy has been operationalized and oriented from the police perspective, but these principles hold true and offer promise for problems that are identified by others, including those in the private sector. Consequently, the police may be engaged by private security professionals to establish a partnership.

Over the last several decades, community policing has become a predominant paradigm within the policing discipline. One would be hard-pressed to identify many law enforcement agencies that do not profess to engage in community policing in some form. Until recently, some have referred to private security as one of the last untapped partners touched by the spread of community policing practices. In response to this, and because community policing has proven to be an effective strategy for addressing a range of public safety problems, the COPS Office is committed to supporting the growth of these partnerships. And, as a result, it serves as the foundation for partnerships between the private sector and the police.

FEDERAL GOVERNMENT'S ROLE IN PROMOTING PARTNERSHIPS

Cooperation between law enforcement and private security dates back to the earliest periods of modern policing even though these arrangements were largely informal and limited in terms of the scope of cooperation. Consequently, major efforts to chronicle the private security field and its relationship to public law and order efforts, and to understand and improve the relationships between these two disparate but critical spheres with responsibility for safety and security, date back decades. The federal government has played an important role in developing a greater understanding about the strengths of private security and public law enforcement in preserving safety and security in the community, and examining ways to leverage efforts for a greater impact.

One of the first major studies on public–private partnerships was conducted in the mid-1970s by the Private Security Advisory Council to the Law Enforcement Assistance Administration (LEAA) within the Department of Justice (DOJ). LEAA, an early predecessor to DOJ components like the Office of Justice Programs and the COPS Office, was established to administer federal grant funding and promote criminal justice experimentation and innovation. The Private Security Advisory Council established a committee charged with improving the relationships between private security and law enforcement. In studying the state of collaborations, the committee identified several barriers to the creation of effective partnerships, including conflicts in roles, misunderstandings and negative perceptions about each other's capabilities and qualifications, and a lack of mutual respect.

Two related studies into the nature of the private security industry and activities related to public safety are commonly referred to as *The Hallcrest Reports*. Both studies were produced by the Department of Justice, National Institute of Justice. The first study, *Private Security and Police in America: The Hallcrest Report I,* was released in 1985 and included the seminal findings that private security was taking on increasing responsibilities in areas previously managed by public law enforcement, and employment and expenditures in private security were outstripping those in public policing as a result of:

- increasing workplace crime
- increasing fear of crime
- decreases in spending for public protection
- increasing awareness and use of private security as effective alternatives to protection

Five years later, a follow-up report titled *Private Security Trends 1997–2000, Hallcrest II* concluded that private security had definitely surpassed public policing as America's primary protective resource in terms of spending and employment, and affirmed the growing trends toward private security prominence in the area of public protection (Cunningham and Taylor 1990).

Another report that supported the creation of private security and law enforcement partnerships was *Operation Cooperation*, funded by the Department of Justice, Bureau of Justice Assistance. This 2000 report included information and guidance on how to establish partnerships, common types of partnerships, facilitators and inhibitors to success, and highlights and examples from prominent partnerships across the country (Connors et al. 2000).

In 2004, the COPS Office co-sponsored a National Policy Summit on Building Private Security and Police Partnerships during which 140 executives from the public and private sectors convened to identify gaps and challenges to collaboration, and policy recommendations that make up an action agenda.[2] These policy recommendations included:

1. Leaders of the major law enforcement and private security organizations should make a formal commitment to cooperation. In September 2009, ASIS International and the International Association of Chiefs of Police signed a formal memorandum of understanding (MOU) agreement to establish a formal partnership to develop educational and information-sharing programs and other activities with a goal of increasing collaboration resulting in improving security and public safety (Harwood 2010).
2. The Department of Homeland Security or the Department of Justice should fund research and training on relevant legislation, private security, and law enforcement–private security cooperation.
3. The Department of Homeland Security or Department of Justice should create an advisory council composed of nationally prominent law enforcement and private security professionals to oversee the day-to-day implementation issues of law enforcement–private security partnerships.
4. The Department of Homeland Security or Department of Justice, along with relevant membership organizations, should convene key practitioners to move this agenda forward in the future.
5. Local partnerships should set priorities and address key problems as identified by the summit (COPS Office 2004).

In response to the second recommendation, the COPS Office funded the Institute for Law and Justice (ILJ) to examine the state of public–private partnerships, to identify every formal partnership in existence across the country, and to learn about the circumstances of the partnerships.

[2]The International Association of Chiefs of Police (IACP), through an award from the COPS Office, organized this event. Along with IACP and COPS, the other co-sponsors were ASIS International, the International Security Management Association (ISMA), the National Association of Security Companies (NASCO), and the Security Industry Association (SIA).

Some circumstances of partnerships included how they are formed and organized and what has allowed them to be sustained over time. The resulting 2009 publication, *Operation Partnership: Trends and Practices in Law Enforcement and Private Security Collaborations*, identified the key forms that partnerships take and the trends and notable examples of effective partnerships.

Several benefits to partnering were identified by police and private security practitioners:

1. The *contacts* that are made through partnership activities have positive influences on personnel and pay other dividends over time.
2. Recognizing that private security firms have access to many more areas than their public-sector counterparts (by virtue of both their sheer numbers, as well as their unique responsibilities), they serve as a *force multiplier* for police.
3. Private security assets offer the potential to save costs for public safety services.
4. Law enforcement can benefit from the tremendous specialization and unique expertise and capacity within the private security industry, especially when addressing computer and high-tech crimes.
5. Law enforcement and private security can benefit from sharing information and resources in the investigation of complex financial crimes and intellectual property crimes.
6. Advanced technologies, like closed-circuit television (CCTV), electronic access control, alarm systems, gunshot recognition technology, and mapping and GPS, are being used by private security and law enforcement. Having access to these technologies, regardless of whether they are "owned" by the law enforcement or private security partner, can assist all partners.
7. Critical incident planning and response has proven critical to effectively responding to man-made and natural disasters. Knowing the answer to key questions, such as who are the "first responders," how will communication between responders take place, and how will order be restored, are all critical questions to be answered, planned, and practiced before an event or incident happens.
8. The increasing importance of sharing information and intelligence mandates the inclusion of not just law enforcement from various levels and jurisdictions, but also private security entities that also have access to critical information and intelligence.
9. Bringing private security into community policing efforts can only strengthen efforts to address and solve problems that are important to the partners and to the community.
10. There are many valuable training opportunities sponsored by both public and private sector partners.
11. Private security and law enforcement have many contacts in each other's field. Fostering these contacts through partnership activities offers tangible professional development opportunities and can lead to career opportunities and recognition for achievements and accomplishments (Law Enforcement-Private Security Consortium 2009).

For those who may be looking to establish a partnership, or improve and sustain one they are already involved in, the *Operation Partnership* report details key components of effective partnerships between the private security and the police. This work has informed the development of a forthcoming COPS Office sponsored web-based training resource that will help executives, practitioners, and others from both disciplines to understand the importance of these partnerships. It may also allow for development of a detailed, personalized action plan for partnership activities. This resource will be released in 2010, and can be accessed online at www.cops.usdoj.gov.

SUMMARY

A 2008 article in the COPS Office electronic newsletter, *The Community Policing Dispatch*, conveyed the rationale for law enforcement to think of its private security colleagues as important partners in identifying and addressing crime, disorder, and other problems:

> While numerous potential partners exist, one community in particular that holds much potential are the professionals included under the umbrella of private security. The *private security* profession is obviously extremely diverse, covering everything from armored guard services, to uniformed security, to corporate security departments, to facility security professionals. They often possess immense technology, skills, and extensive law enforcement and security experience. Additionally, by virtue of their unique responsibilities, they can serve as invaluable partners in the effort to secure communities and public and private assets (Chapman 2008).

The article also provided the largely police practitioner readership with a sense of the magnitude of the private security field, for those who were previously unaware of the tremendous scope of the profession.

> By some estimates, private security provides direct oversight of around 85 percent of the nation's critical infrastructure. Moreover, they outnumber their public sector counterparts by nearly 2 ½ to 1 (COPS Office 2004). Their access to both street-level knowledge and state-of-the-art security technology renders them an important—yet largely untapped—community partner (Chapman 2008).

Because the COPS Office understands the potential of these partnerships in addressing crimes and other problems, it continues to support and promote development of partnerships. There are vast and unique resources and expertise within the private security field. These valuable resources are ideal contributing partners for addressing safety or security issues of mutual concern or focus. The sheer numbers of private security professionals alone offers a substantial and tangible benefit to law enforcement. Similarly, so too does the private sector have much to gain by partnering with its public sector colleagues. The COPS Office remains committed to raising awareness about the potential these partnerships hold and providing resources to make them a reality.

REFERENCES

Chapman, Robert. 2008. *Community Partnerships: A Key Ingredient in an Effective Homeland Security Approach.* Community Policing Dispatch, Vol. 1, Issue 2. Washington, DC: U.S. Department of Justice, Office of Community Oriented Policing Services.

Connors, Edward, William Cunningham, and Peter Ohlhausen. 2000. *Operation Cooperation: Guidelines for Partnerships between Law Enforcement & Private Security Organizations*, Washington, DC: U.S. Department of Justice, Bureau of Justice Assistance.

COPS Office. 2004. *National Policy Summit: Building Private Security/Public Policing Partnerships to Prevent and Respond to Terrorism and Public Disorder.* Washington, DC: International Association of Chiefs of Police and U.S. Department of Justice, pp. 1–3.

Cunningham, William C. and Todd Taylor. 1990. *The Hallcrest Report I: Private Security and*

Police in America. Stoneham, MA: Butterworth-Heinemann.

Cunningham, William C., John J. Strauchs, and Clifford W. Van Meter. 1990. *The Hallcrest Report II: Private Security Trends 1970-2000*. Stoneham, MA: Butterworth-Heinemann.

Harwood, Matthew. 2010. *ASIS Signs Agreement with International Association of Chiefs of Police*. ASIS Dynamics, January/February.

Law Enforcement-Private Security Consortium. 2009. *Operation Partnership: Trends and Practices in Law Enforcement and Private Security Collaborations*, Washington, DC: U.S. Department of Justice, Office of Community Oriented Policing Services, pp. 9–17.

Scheider, Matthew. 2008. *The Purpose of Partnerships*. Community Policing Dispatch, Vol. 1, Issue 2. Washington, DC: U.S. Department of Justice, Office of Community Oriented Policing Services.

BIBLIOGRAPHY

NOTE: The COPS Office offers a number of resources for anyone who would like more information on how to develop productive and meaningful partnerships or address a crime or disorder issue. Some examples include those listed below. These publications and others are available at www.cops.usdoj.gov, or by calling the COPS Office Response Center at 1-800-421-6770.

Burglary of Retail Establishments. http://www.cops.usdoj.gov/RIC/ResourceDetail.aspx?RID=20

Community Policing Defined. http://www.cops.usdoj.gov/RIC/ResourceDetail.aspx?RID=513

Crimes Against Tourists. http://www.cops.usdoj.gov/RIC/ResourceDetail.aspx?RID=76

Law Enforcement Intelligence: A Guide for State, Local, and Tribal Law Enforcement Agencies, 2nd Ed. http://www.cops.usdoj.gov/RIC/ResourceDetail.aspx?RID=525

National Policy Summit: Building Private Security/Public Policing Partnerships to Prevent and Respond to Terrorism and Public Disorder. http://www.cops.usdoj.gov/RIC/ResourceDetail.aspx?RID=246

Operation Partnership: Trends and Practices in Law Enforcement and Private Security Collaboration. http://www.cops.usdoj.gov/RIC/ResourceDetail.aspx?RID=534

Partnering with Businesses to Addressing Public Safety Problems. http://www.cops.usdoj.gov/RIC/ResourceDetail.aspx?RID=147

Planning and Managing Security for Major Special Events: Guidelines for Law Enforcement. http://www.cops.usdoj.gov/RIC/ResourceDetail.aspx?RID=441

Policing Terrorism: An Executive's Guide. http://www.cops.usdoj.gov/RIC/ResourceDetail.aspx?RID=489

Shifting and Sharing Responsibility for Public Safety Problems. http://www.cops.usdoj.gov/RIC/ResourceDetail.aspx?RID=191

Understanding Risky Facilities. http://www.cops.usdoj.gov/RIC/ResourceDetail.aspx?RID=410

Using Crime Prevention through Environmental Design in Problem-Solving. http://www.cops.usdoj.gov/RIC/ResourceDetail.aspx?RID=440

Part 8

Special Topics

26

Terrorism Preparedness

Richard S. Post, PhD, CPP

CHAPTER 26

OBJECTIVE

The objectives of this chapter are as follows:
- explore the effects of terrorism on global business
- recognize what events or activities may trigger an act of terror against a company
- understand how a company can prepare for civil unrest when operating in a foreign country
- examine the importance of a company-specific political-risk report when doing business overseas

INTRODUCTION

Terrorism and what constitutes an "act of terrorism" is a concept that can mean different things to different people. Because one person's freedom fighter may be another person's terrorist, many governments have tried to establish common ground for defining what constitutes a terrorist, terrorist acts, and how these acts should be managed.

The U.S. Federal Criminal Code. 18 U.S.C. §2331 defines terrorism as:

> ...activities that involve violent... or life-threatening acts... that are a violation of the criminal laws of the United States or of any State and... appear to be intended (i) to intimidate or coerce a civilian population; (ii) to influence the policy of a government by intimidation or coercion; or (iii) to affect the conduct of a government by mass destruction, assassination, or kidnapping....

BACKGROUND

Most governments have standards they apply handle terrorist acts, including using diplomatic means and local law enforcement, engaging military forces, deploying intelligence agents to penetrate and destroy terrorist networks. Each of these approaches is based on the particular legislative or governmental authority in a country or groups of countries, such as the United Nations or European Union. When it comes to business operations, each company must find its own formula to protect its employees and its operations.

Dealing with terrorists and potential acts of terrorism should be addressed in corporate security planning and operations. The extent of preparations, programs, training, and risk analysis must be tailored to the specific risk profile of a company.

EFFECT OF TERRORISM ON GLOBAL BUSINESS

The global geopolitical scene is dominated by political, religious, criminal, and social activist groups using "terror" tactics to further their agendas or accomplish their goals. Every government, and some businesses, must deal with these groups, as well as their demands, actions and costs, on a regular basis.

The goal of some terror activities may be to further a religious cause or religious faction, such as the conflict between Sunni and Shiite Muslims. Other groups known to resort to terrorist acts include narco-terrorists operating in Colombia to control illegal drugs, Chechnyan

Muslims fighting Russians, Taliban fighters in Pakistan, Afghan warlords fighting for political and religious domination, or Al-Qaeda members against Western values and influences around the globe. The terrorist groups in the Middle East, Central Asia, and East Asia all have different historical roots but may cause the same degree of risk for businesses operating in these areas.

The reasons for the behaviors of some terrorist groups are lost in historical misunderstandings, misinterpretations of scripture or dogma, colonial atrocities committed by western powers, availability of social networks, political awakening without focus, or just garden variety criminals that use terrorist acts to obtain wealth. While each group has a specific set of guiding principles, methodologies, and aspirations, their weapons are similar: killing, maiming, extortion, armed fighting, kidnapping for ransom, and use of weapons of mass destruction.

There are numerous textbooks that deal with the causes of terrorism, how the guerrilla fighter is developed, how to fight terrorists, and how to develop both governmental and individual security programs to mitigate terrorist acts. David Kilcullen's book, *The Accidental Guerrilla*, and a report published by the National Academy of Sciences titled, *Making the Nation Safer,* are two useful resources for exploring the causes of terrorism.. The important thing to remember is most corporate involvement in terrorism occurs when:

- operations are located in a volatile or internally sensitive country
- the company is engaged in business or business activities that are targeted by a specific group
- the company may have employees engaged in personal activities that are targeted by a specific group

The safety of employees is paramount in security planning and risk analysis. If risks are known to a company, at-risk employees must be given information to assess specific risks and exposure while working on behalf of the company. In order to manage the safety expectations of employees a company has a duty to warn them of potential risks.

A domestic company, for example, that operates a public utility may have a nominal risk of a terrorist attack. Such a company should understand that damage to infrastructure is often a way to attack a society and cause wide scale disruptions and lack of confidence in a government to protect its citizens. The air transport system is aware of the risk from terrorists and has aggressively established safeguards. The establishment of the Transportation Security Administration (TSA) and recent changes in airline security is just one example.

In the United States, the Department of Homeland Security (DHS) has the lead role in counterterrorism by developing strategic initiatives to lower the risks to domestic activities. DHS operates much like corporate security in a company to pull together the best security practices, apply them to known risks, and develop appropriate countermeasures and operational programs to prevent incidents from occurring.

Sometimes identification or association of a company with a government can create problems. In the 1980s when terrorists planned attacks against U.S. interests, global companies like IBM, American Express, and Coca-Cola were targets. In the Middle East Exxon was viewed as an extension of U.S. interests. The employees and operations of these companies experienced higher risks globally. Facilities were attacked more often, and security efforts were enhanced and made more extensive.

Although there is no formula that will prevent attacks from occurring within a global company, being considered a good corporate citizen by the country in which a business operated can lower that company's profile and build goodwill in local populations. Cultural sensitivity by employees, marketing executives, and labor organizations also build rapport. Doing the morally right thing in local affairs and staying alert to changing political and social conditions can provide a local barrier to becoming a target of local terrorist groups.

Each type of business must conduct a risk review and develop appropriate plans to eliminate, deflect, redirect, or mitigate known or anticipated risks to operations and employees. A sample risk report at the end of this chapter looks at operations in a high-risk country.

Example 1. A U.S.-led multinational company had extensive agricultural operations in an Asian country. Twenty security guards were ambushed and killed by terrorists of a separatist guerilla group. The director of security was dispatched to negotiate with the group, which demanded a "revolutionary tax" be paid to stop further violence.

Offers to build hospitals and improve schools and housing for the group were rebuffed and U.S. dollars demanded instead. Company executives determined that payment of a tax would be a violation of U.S. and local laws, as well as a violation of the Foreign Corrupt Practices Act, and refused to pay. The terrorists threatened if payment was not made more killings would occur. Payment was refused and, as a result, two more guards were killed as several "company towns" were hit by mortar fire.

Using the local military forces of the country, 1,500 workers were evacuated to a secure location that was attacked a few days later.

Company officials determined that selling the business was the only recourse to stop the violence and protect its workers. A sale of the company was negotiated with a local business. The local company immediately paid the "tax" and stopped the violence. It was later learned that senior government officials made a deal with the terrorists to put pressure on the U.S. company so that they would sell the business to a local company. Was this a real terrorist threat, extortion, or both?

The Department of State, the DHS, and the CIA all have published lists of known terrorist groups. The following list includes groups currently designated as foreign terrorist organizations:

- Abu Nidal Organization (ANO) (International, Palestinian)
- Abu Sayyaf Group (ASG) (Philippines)
- Al-Aqsa Martyrs Brigade (Palestinian)
- Al-Shabaab (Somalia)
- Ansar al-Islam (Iraqi Kurdistan)
- Armed Islamic Group (GIA) (Algeria)
- Asbat an-Ansar (Lebanon)
- Aum Shinrikyo (Japan)
- Basque Fatherland and Liberty (ETA) (Spain, France)
- Communist Party of the Philippines/New People's Army (CPP/NPA) (Philippines)
- Continuity Irish Republican Army (CIRA) (Northern Ireland)
- Gama'a al-Islamiyya (Egypt)
- HAMAS (Islamic Resistance Movement) (Palestinian)
- Harakat ul-Jihad-i-Islami/Bangladesh (HUJI-B) (Bangladesh)
- Harakat ul-Mujahidin (HUM) (Pakistan)
- Hizballah (Party of God) (Lebanon)
- Hizbul Islam (Somalia)
- Islamic Jihad Group (Palestinian)

- Islamic Movement of Uzbekistan (IMU) (Uzbekistan)
- Jaish-e-Mohammed (Army of Mohammed) (JEM) (Pakistan)
- Jemaah Islamiya organization (JI) (South East Asia)
- al-Jihad (Egyptian Islamic Jihad) (Egyptian Islamic Jihad) (Egypt)
- Kahane Chai (Kach) (Israel)
- Kongra-Gel (formerly Kurdistan Workers' Party) (KGK, formerly PKK, KADEK, Kongra-Gel) (Turkey, Iraq, Iran, Syria)
- Lashkar-e Tayyiba (Army of the Righteous) (LT) (Muridke, Pakistan)
- Lashkar i Jhangvi (Pakistan)
- Liberation Tigers of Tamil Eelam (LTTE) (Sri Lanka)
- Libyan Islamic Fighting Group (LIFG) (Libya)
- Moroccan Islamic Combatant Group (GICM) (Morocco)
- Mujahedin-e Khalq Organization (MEK) (Iran)
- National Liberation Army (ELN) (Colombia)
- Palestine Liberation Front (PLF) (Palestinian)
- Palestinian Islamic Jihad (PIJ) (Palestinian)
- Popular Front for the Liberation of Palestine (PFLP) (Palestinian)
- PFLP-General Command (PFLP-GC) (Palestinian)
- Tanzim Qa'idat al-Jihad fi Bilad al-Rafidayn (QJBR) (al-Qaida in Iraq) (formerly Jama'at al-Tawhid wa'al-Jihad, JTJ, al-Zarqawi Network) (Iraq)
- al-Qa'ida (Global)
- al-Qa'ida in the Islamic Maghreb (formerly GSPC) (The Maghreb)
- Real IRA (Northern Ireland)
- Revolutionary Armed Forces of Colombia (FARC) (Colombia)
- Revolutionary Nuclei (formerly ELA) (Greece)
- Revolutionary Organization 17 November (Greece)
- Revolutionary People's Liberation Party/Front (DHKP/C) (Turkey)
- Shining Path (Sendero Luminoso, SL) (Peru)
- United Self-Defense Forces of Colombia (AUC) (Colombia)

In addition, there are countries determined by the Secretary of State to have repeatedly provided support for acts of international terrorism that are identified and sanctioned pursuant to three laws: section 6(j) of the Export Administration Act, section 40 of the Arms Export Control Act, and section 620A of the Foreign Assistance Act. Taken together, the four main categories of sanctions resulting from designation under these authorities include restrictions on U.S. foreign assistance, a ban on defense exports and sales, certain controls over exports of dual use items, and miscellaneous financial and other restrictions.

Designation by the U.S. Secretary of State under the above laws also implicates other sanctions that penalize people and countries engaging in certain trade with state sponsors from such countries. Currently there are four countries designated under these authorities (see Table 26–A).

The Overseas Security Advisory Council (OSAC) of the U.S. Department of State is the official government conduit to provide risk, threat, and protective advice to U.S. businesses operating abroad. Both the CIA and the FBI provides known-risk information to pertinent U.S. businesses if they are targets. However, a company operating overseas is primarily responsible for protecting itself. There are numerous sources and models that can be used to develop protec-

Table 26–A. Countries that Support International Acts of Terrorism as Designated by the U.S. Secretary of State

Country	*Designation Date*
Cuba	March 1, 1982
Iran	January 19, 1984
Sudan	August 12, 1993
Syria	December 29, 1979

tive measures within a company but they require the time, attention, and authority of company management to be meaningful.

Much like perpetrating a crime requires opportunity, desire, and ability for it to happen, terror attacks require a catalyst or target to trigger an attack. For example, being culturally insensitive in foreign operations, being viewed as taking advantage of local interests by the local population, or engaging in business activities targeted by a group are among the many triggers of a terrorist attack.

The type of business a company does can be a lightening rod for certain terror groups, but there are exceptions. A company may be a victim of terror because of what it does, or does not do in a foreign country. Groups like Al-Qaeda, Hammas, Fata, and FARC all have specific goals but anyone could be caught in their crossfire. A car bombing in Baghdad or Karachi could kill innocent visitors, damage hotels, and destroy vehicles without regard to those killed or injured. Such injuries may sometimes, but not always, be prevented when the act of terror involves suicide bombers and truck bombs.

In the past, the United States had little history of domestic terrorism. Two recent exceptions include the Oklahoma City federal building bombing and the Centennial Olympic Park bombing during the 1996 Olympic summer games in Atlanta, Georgia. In many countries domestic terrorism is a daily reality around the world, affecting both personal and business activities. The use of terror as a weapon has a long and varied history for political, religious, and economic interests.

Increasingly, terrorism directed against businesses has become a major security consideration in numerous Latin American, European, and Asian countries. Ensuring the protection of personnel and operations against various regional terrorist groups and religious or political movements continues to be a major activity for corporate security professionals with interests in those parts of the world. A great deal has been learned about how to identify vulnerabilities to terrorist groups, how to manage risks associated with a high-profile company, and how to manage the crisis that results from an attack against corporate interests. Corporate initiatives to reduce the risk of being victimized by terrorists, or lessen the impact of an attack, have been a part of multinational activities since the 1960s. Programs include kidnapping insurance, crisis management, political-risk reporting, physical security upgrades, travel security upgrades, intelligence collection, government liaison, and broader initiatives, such as the development of the U.S. Department of State's OSAC agency and travel advisories.

Organizations must take the initiative to protect their interests around the world, develop programs tailored to fit their specific requirements, and establish information resources to deal with a crisis situation wherever it might arise. A key element to protecting against terrorism is to have effective intelligence or political-risk information to assist in decision making. Target evaluation is a major factor in determining what must be done to protect company activities or interests.

THE POLITICAL RISK REPORT

Political-risk reporting seeks to provide an understanding of the threats, vulnerabilities, and risks to a company in a specific foreign country or region of the world. The collection of time-sensitive information related to economic, political, crime and terrorism issues, business polices, governmental stability, and corruption are elements that should be included in reports.

Reporting can be provided by:

- company employees
- travelers
- foreign operations
- contracted reporting services
- commissioned studies by consultants

These reports can be general in nature and prepared for a specific country. The report should cover all key economic indicators, political conditions, business and commercial factors, governmental stability, peace and order, and social factors. These reports require analysis for a specific company by internal staff to determine the implications of the information in the report. Consultant reports can also be tailored to a specific company and its operations in a country and the impact that current conditions in the country may have on the company. Sometimes a combination of company and purchased reports are used to crosscheck recommendations and conclusions.

A political-risk report often is based on the following types of specific criteria:

- crime and risk to personnel in the region
- corruption issues affecting business operations
- governmental stability and intervention in operations
- information protection and product/brand integrity
- operational integrity (protection of business operations)

PREPARING FOR CIVIL UNREST

Currently, the risk of civil unrest in the United States is relatively low compared to the 1960s and 1970s, when political activism centered on the Vietnam War, the civil-rights movement, and related issues resulted in open violence against the government. The existence and importance of civil unrest around the world, however, makes for a risky environment in which to live and work. Worldwide, risks may be country-specific. A risk in one country might not necessarily mean that same risk exists for the company in any other part of the world.

Preparations for dealing with civil unrest begin with understanding the political risks involved with being in the country in the first place. In many countries it is important to understand the attitudes toward foreigners and what might be expected if civil unrest occurs. Prior knowledge and preparation is important if risks are to be minimized. Before major unrest there may be warnings in the press about current events that should be closely watched. Typically, in countries with growing democratic movements, for example, as the democracy movement grows in strength, the likelihood for civil unrest will also grow. A political risk report for a country experiencing some civil unrest might include a statement similar to the following:

Because of the lengthy period that precedes the elections and the conflict that continues to be present, some precautionary measures are in order. Unless something unusual occurs to quell the current democracy movement without igniting full-scale antigovernment hostilities, the hit-and-run tactics by both sides in this contest are likely to continue and make the local living and work situations much more difficult.

Note that there may be an "anti-foreigner" sentiment in the minds of local citizens that may rise to the surface whenever tempers flare. It is recommended that whenever possible, expatriates avoid driving private vehicles and instead, use company or hired drivers. A vehicle accident or the death or injury of a local pedestrian could be the catalyst that sparks a riot or civil disobedience. When citizens cannot vent frustrations outsiders may become easy targets.

Business personnel should be advised to stock extra food and water supplies when operating in an environment of potential civil unrest. Emergency car pools should be set up in case evacuation to a clearing area is necessary. Primary and alternate assembly points should be designated, and staff should know where assembly points are before an incident occurs.

Terrorism and civil unrest often occur under similar circumstances. Reviewing local conditions for a company and developing adequate intelligence specific to a company's requirements is essential to prevent problems and to protect staff traveling into the area or on assignment in the area. The following example of a report on risk conditions in Pakistan, which, at the time, was one of the more difficult working environments for foreigners, describes the conditions that are problems and what should be done to lower risks.

Example 1. Risk for the operations of YOUR COMPANY in Pakistan in the mid-1990s. Each major city in the country has special risks associated with travel or business activities. These risks are reviewed for each city as well as for the key business issues.

Introduction

The entire country has almost endemic political violence, work stoppages, and regular strikes. Bombings are common as the local police have limited intelligence about the activities of local extremists. A strong undercurrent of civil unrest is in evidence across the country, with local "drug lords" and criminals in the local and national legislative bodies. Both Taliban and Al-Qaeda operatives operate throughout the country and are able to strike effectively almost at will.

There are several distinct regions in the country, each of which is associated with specific risks. Southern Pakistan, especially in the Karachi area, suffers political violence with its India neighbors. Northern Pakistan has violence related to Afghanistan issues. The Swat valley has an active Muslim separatist movement, embodied by the Taliban. The continuing tension between Pakistan and India, communal violence, Muslim extremists, a weak economy, an unemployment rate of more than 45%, a serious internal drug-abuse problem, and cumbersome bureaucracy makes doing business in the country difficult. Local businesses and political leaders resist foreign business because of a fear of losing power, money, and control of the country's economy. It is evident that making unofficial payments to governmental officials is the most effective way to operate a business and to successfully stay in business.

Lahore

Crime directed against foreigners is low in Lahore. However, the violent nature of attacks, such as the bombing on July 21, 1996, of the Lahore International Airport, which killed five and injured 62 people, makes travel to this area high-risk. The bomb was targeted to the immigration area, which strongly suggested its purpose was to kill foreigners. The inability of the local police to provide protection makes high-quality and well-supervised private security essential. Because airport security is lax, movement through the ticketing and departure areas should be handled as quickly as possible.

Travel into and out of the Lahore area should be considered on an essential-need basis only. Until the government can provide better intelligence and security against random violence, risk for travelers is higher than usual for this already high-risk environment.

In addition, because of an increase in kidnappings in this area, although not specifically directed against foreigners, high-profile visitors and visitors to locally prominent residents, should be discouraged.

YOUR COMPANY office facilities are protected at a level consistent with local risk. It is, however, essential that this protection is maintained with well-trained and supervised officers from a reputable commercial security company. Security plans for dealing with strikes, civil unrest, and violent employees should regularly be reviewed by local management.

Karachi

The overall crime rate of the metropolitan area is high, with more than 1 million of the 12 million residents addicted to heroin. Ongoing labor problems, including work strikes and labor-related violence also is high. Communal violence between various tribal groups living in close proximity is significant. Violence is prevalent and the risk of being caught in random violence is great even though violence is not generally directed against foreigners.

At the time of this writing, communal violence that required the use of army rangers, local police, mandatory curfews, and the arrest of thousands of people has ceased. Political pressure and police repression have managed to keep the most volatile situations under control. However, because the root causes of problems in this area (e.g., ethnic hostilities, unemployment, drug abuse, and poverty), have not changed, there likely will be outbreaks of violence in the future.

Local authorities attempted to lower the reported crime rate in Karachi by redrawing the reporting lines. They omitted the Karachi metropolitan area when reporting crime and accounted only for crime that occurred in the core central business district, which has a low population and virtually none of the ethnic disturbances taking place in other areas. Consequently, crime statistics in Karachi are deceptive and misleading.

There are no high-security residential areas in the metropolitan area. Currently, the three international class hotels afford the only level of consistent security for residents or visitors. Medical facilities, with the notable exception of the Aga Khan Hospital, are inadequate. The only other western standard hospital is in Islamabad.

Peshawar

The Northwest Khyber area of the country has specific risks for both travelers and expatriate residents. Peshawar is located adjacent to land designated as "tribal," which is administered differently than other parts of the country. Travel on three sides of the city is restricted to prevent encounters with local bandits, drug smugglers, or kidnappers. All have been historically prevalent in the area. There have been several cases reported in 1995 and 1996 involving western businessmen and travelers. Female travelers are particularly at risk in this area.

Tribal feuds, celebratory discharging of weapons, and the availability of weapons in this region make violence a common event. Violence is often common in job actions, discharging of employees, or in disciplinary situations in the workplace.

Historically, the tribal lands are a training area and "safe haven" for terrorists operating in Afghanistan. All factors contribute to the high-risk environment in the Peshawar area. This area is also used as a drug processing and transit area by both local drug lords and gangs from Afghanistan.

There is a large presence of "official and diplomatic" personnel in the area. The U.S. Consulate operates under a constant "yellow alert" because of local conditions. Local private security protection is essential for residents. As in most high-risk areas, life goes on with minor disruptions. One hotel has a conspicuously posted sign (as follows) to reduce risk for visitors and guests.

Hotel Policy
Arms cannot be brought inside the hotel premises.
Personal guards or gunmen are required to deposit their weapons with hotel security.
We seek your cooperation.

Islamabad and Rawalpindi

Unlike other parts of the country, in Islamabad crime rates are relatively low. It is possible to go anywhere within the city in relative safety. A high concentration of foreign diplomats, combined with the fact that it is a newly constructed city with none of communal violence that plagues other major cities in the country, makes for less violence in the capital city.

There is a large number of private security officers providing protection at all of the embassies and government facilities around the city. Although there has been occasional bombings, most notably an attack against the Egyptian Embassy and protests against the government, Islamabad is almost idyllic compared to its sister city of Rawalpindi. The local government strives to maintain the capital as a model of peace and order and has, in general, succeeded in this act of police efficiency. Major hotels are secure and risks for travelers are low compared to that of other major cities in the country.

Business Issues

Business issues specific to this region are detailed in the following sections.

Crime and Risk to Personnel in Pakistan

At the time of this writing the last U.S. State Department Travel Advisory of Jan. 10, 1996, is still in effect. Due to high levels of violence, travel in Baluchistan is restricted. The city of Quetta is also considered "off limits." Road travel anywhere in the country is considered risky, except in the company of armed local residents or security officers. Attacks against vehicles are common in rural areas.

Travelers and visitors are urged to be alert for possible surveillance and should not accept gifts or packages from unknown parties, as bombing attacks are common. Crime is relatively high throughout the country, but does not specifically target foreigners. High levels of alertness are called for by all travelers. Airport security and security service personnel is minimal.

SAMPLE POLITICAL-RISK REPORT

Using company-specific criteria when conducting a country analysis can produce a report that looks like the following example developed for the emerging market of Myanmar. You will note that there are both general and specific recommendations produced for the company that requested the report.

Key parts of an intelligence-gathering program are the mission statement (or strategic plan) and the task list. The mission statement outlines the information requirements, priority list, functional activities, and collection assignments.

The collection of whatever level and category of information is then determined by the sensitivity of the information required. Most of the information required by a company (90%

to 95%) is located in open sources that can be identified and collected using normal research methods and analytical techniques. The remaining 5% to 10% is sensitive information that can include intelligence provided from government sources and answers to specific questions related to sensitive corporate activities.

Collection of information is carried out through an internal organization and via consultants. The cost, time required, and methods of inquiry used in the investigation will be tailored to the degree of discretion required.

Example 2. Risk for the Operations of YOUR COMPANY in Myanmar in the mid-1990s.
The major cities in the country have no special risks associated with travel or business activities.

Introduction

A senior U.S. official described the business environment in Myanmar as "It's the Wild West." Doing business in this country has also been described as walking a high-wire without a safety net. The country has no effective legal structure, tort system, or civil code; insurance and consistent government policies do not exist. Corruption is endemic and unreliable, and what is promised is not always delivered after payments are made.

This business environment is, at the present time, for the high-risk investor or the short-term trader only. Most importantly, there is no protection available from the government unless the company is a major multinational company willing to pay the government for all of the services it receives.

The U.S. State Department Travel Advisory of May 23, 1996, indicated that nonessential travel to the country should be curtailed due to both political instability and the potential for violence. It is recommended that, if travel is essential, it should be to the major cities and not to the border areas with Thailand or China.

Travelers should keep a close watch on local situations because broadcasted news and information is unreliable. By law, foreign correspondents must be local citizens. Reporting is closely monitored by the State Law and Order Restoration Council. Networks like CNN and BBC are the only reliable foreign news services in the country. Travelers should stay only in a hotel that has access to these foreign news services. If there is indication of imminent trouble, travelers should try to leave the country as quickly as possible or remain in the hotel until the situation allows for an orderly and secure exit.

Crime and Risk to Personnel in Myanmar

Crime against foreigners is currently at low levels but increasing. There is no tradition of kidnapping in the country, except for in northern areas of the country where it is almost exclusively related to tribal or drug matters.

The single largest risk to travelers or resident expatriates is the potential for serious civil unrest due to the growth of the democracy movement in the country. The ruling military government controls the mostly young countryside population by force and coercion. If this control is loosened for any reason, serious civil unrest and reprisals could take place across the country. Because the military is poorly paid and poorly supervised, the potential for abuse is high.

Motor vehicle-accident fraud caused by locals stepping in front of a car owned or driven by foreigners is becoming commonplace. A few hundred dollars is often the price to settle these time-consuming crimes. There are no private security services in the country. All protection must be provided by the government.

Corruption Issues Affecting Business Operations

Corruption is widespread and blatant. Often generals who ask for money today will ask for more tomorrow without ever providing the service they agreed to make available. Loyalty to a company is determined by how much and how often payments are made. If a government official changes position, new arrangements must be made or governmental services will cease or contracts may be voided. Drug lords commonly channel drug-related funds into domestic development projects and commercial ventures. Some drug lords have reportedly built joint-venture hotels to launder money. Reportedly, Korean and Japanese companies are active in buying their way into the country. Major trading and infrastructure projects have allegedly been awarded to those firms that pay the most to State Law and Order Restoration Council officials.

Governmental Stability and Intervention in Operations

A military government, the State Law and Order Restoration Council, controls the country. The current democracy movement is gaining strength, and, if not properly handled, the State Law and Order Restoration Council could spark the democracy movement in an attempt to increase the pace of change in the country, setting off major civil unrest.

China is increasing pressure on the State Law and Order Restoration Council to hold on to power at any cost. It is clear that China has strategic designs on the country and has been active in funding its military, training its intelligence agencies and military officers, and providing foreign aid.

The local security services are inefficient in their surveillance of foreigners. They rely on following people rather than using electronics or listening devices. They will search rooms at will and are relatively well trained by the Chinese security service.

Although there have not been any reported attacks on computers, they are best left at home when traveling to Myanmar unless absolutely essential for the trip. If traveling with computers, take only the minimum files and no sensitive information.

Politically sensitive information should not be taken into the country to avoid potential embarrassment or expulsion. Politically sensitive discussions should also not be conducted because such discussions could prove damaging to your local contacts.

Protecting Information and Product-Brand Integrity

Myanmar is becoming a haven for counterfeiters of commercial and industrial products. With the necessary protection, it is possible for these operations to be carried out with little risk of criminal or civil sanctions.

All mail into and out of the country is opened and little is actually delivered. All international telephone calls are monitored. If encryption equipment is used, lines are cut off by the government.

YOUR COMPANY currently operates through distributors in the country. One of them has placed a YOUR COMPANY sign in a prominent location on the main route to the Yangon airport. The continued use of this sign should be reviewed to determine how much visibility is desirable for YOUR COMPANY's current activities in the country. Since many companies are coming under increasing public pressure in the United States for their business involvement in Myanmar, it is suggested that this sign be removed until the political situation stabilizes.

Protecting Business Operations

No mechanisms for handling commercial disputes exist. There are few private lawyers, and litigation is protracted and expensive. Even automobile accidents are handled by immediate cash payments to the victim.

Most companies with little business activity deal through agents or brokers. This will likely continue for the immediate future or until the political situation stabilizes.

The only reliable medical assistance in the country is the AEA Clinic in Yangon.

SUMMARY

Terrorism is a global problem and requires a broad-based global approach for each company. Protection of employees, operations, and products is the primary consideration for the security manager. Understanding the risk levels facing the company and its operations is the primary focus for all planning and programs.

Developing an understanding of global forces and the major terrorist groups and their ideologies and methods provides the background against which operational planning and actions will take place.

Developing and maintaining an active program of risk analysis and protective planning is a primary responsibility for the security manager. Making sure that the company through its global public profile maintains a low-risk profile is critical to risk reduction. Staying current with changing risk levels and adjusting security measures as needed is also essential.

Maintaining a carefully crafted employee protection program, physical security measures, and risk monitoring initiatives will provide the optimal level of protection against potential terrorist attacks. If, regardless of all efforts, attacks do occur, a disaster plan, a crisis management program, and law enforcement liaison must be in place.

BIBLIOGRAPHY

Ackerman, Gary and Jeremy Tamsett, eds. *Jihadists and Weapons of Mass Destruction.* Boca Raton, FL: CRC Press, an imprint of Taylor & Francis Group LLC. 2009.

Alexander, Dean C. and Yonah Alexander. *Terrorism and Business: The Impact of September 11.* Boston, MA: Hotei Publishing, an imprint of Brill. 2001.

Halibozek, Edward, Andy Jones, and Gerald L. Kovacich. *The Corporate Security Professional's Handbook on Terrorism.* Burlington, MA: Elsevier/Butterworth-Heinemann. 2008.

Harowitz, Sherry L. *Counterterrorism and Contingency Planning Guide*, 2nd ed. Alexandria, VA: ASIS International. 2003.

Jackson, Brian A., D. J. Peterson, James T. Bartis, Tom LaTourrette, Irene Brahmakulam, Ari Houser, and Jerry Sollinger. *Protecting Emergency Responders: Lessons Learned from Terrorist Attacks.* Santa Monica, CA: RAND Science and Technology Policy Institute. 2002.

Mannes, Aaron. *Profiles in Terror: A Guide to Middle East Terrorist Organizations.* Lanham, MD: Rowman & Littlefield Publishers, Inc. 2004.

Nance, Malcolm W. *Terrorist Recognition Handbook: A Practitioner's Manual for Predicting and Identifying Terrorist Activities.* 2nd ed. Boca Raton, FL: CRC Press, an imprint of Taylor & Francis Group, LLC. 2008.

Ness, Larry. *Securing Utility and Energy Infrastructures.* Hoboken, NJ: Wiley-Interscience, an imprint of John Wiley & Sons, Inc. 2006.

Purpura, Philip P. *Terrorism and Homeland Security: An Introduction with Applications.* Burlington, MA: Elsevier/Butterworth-Heinemann. 2007.

Ryan, Jeffrey R. and Jan F. Glarum. *Biosecurity and Bioterrorism: Containing and Preventing Biological Threats.* Burlington, MA: Elsevier/Butterworth-Heinemann. 2008.

27

Pandemic Preparedness

Michael J. Fagel, PhD, CEM
Douglas E. Himberger, PhD

CHAPTER 27

OBJECTIVES

The objectives of this chapter are as follows:

- define a pandemic and discuss its effects on a society
- discuss the preparedness requirements of a pandemic
- develop a pandemic preparedness plan
- explore recovery activities after a pandemic

INTRODUCTION

Pandemic preparedness is a topic of periodic and great concern, as are pandemics themselves. Often the public's awareness of these events is raised to a heightened level, although the same is not necessarily true for the public's overall preparedness. Enormous sums have been spent on such preparedness—billions of dollars in the United States alone between 2004 and 2009. However, many would say that preparedness for infectious disease events has changed little from a decade earlier. This is attributed to a combination of several effects: the nature of pandemics as compared to other crises; the unique preparedness requirements of pandemics; and the somewhat challenging or burdensome pandemic preparedness planning activities. This chapter discusses each of these factors, along with activities for during and after a pandemic, and the future of pandemic preparedness planning.

THE NATURE OF PANDEMICS

A *pandemic* is defined by Webster's Collegiate Dictionary as "...an outbreak of a disease...occurring over a wide geographic area and affecting an exceptionally high proportion of the population." (Webster's 1993). The *Collaborative International Dictionary of English* describes a pandemic as an "everywhere epidemic." Pandemics are often thought of in terms of notable historical events. (*Science Magazine* 2009). The Asiatic Cholera Pandemic of 1817; the Spanish Flu or the Great Pandemic of 1918; and the Severe Acute Respiratory Syndrome (SARS) Pandemic of 2003 (probably more appropriately deemed an "epidemic" because of its relatively limited geographic scope) are the most notable historical events. Table 27–A lists notable pandemics over the ages (this table excludes the 2005–2009 Avian Influenza type A/H5N1 variant and the 2009 Swine Influenza type A/H1N1 variant pandemics). Pandemic crises are recurring on average three times per century, and preparing for the next one should be continuous. Dr. David Nabarro, Senior United Nations System Coordinator for Avian and Human Influenza and a senior expert at the World Health Organization (WHO), said in 2005 that he believes a pandemic will erupt soon, and he predicted it might kill approximately 5 million to 150 million people (*CIDRAP News* 2005). Clearly, the threat is real, and preparedness is imperative. See Figure 27–1.

HEALTH CONCERNS

In each major pandemic, the most devastating impact was on health care. The scores of people sickened and killed by infectious disease affected families, communities, and entire countries or global regions. The diseases placed "sudden and intense demands on health systems," and overarching community structures also were often broken (World Health Organization 2009).

Illness and death caused by pandemics resulted in societal disruption. The psyche of populations was so severely degraded that the outlook of its citizens was not only depressed, but the pandemic also colored the future of their societies. As shown in Table 27–A, the toll can be devastating. Continuing cholera pandemics have killed staggering numbers of people worldwide. Although no reliable estimates are available, particularly for the early cholera pandemics, approximately 33,000 deaths occurred in a single day during the second global pandemic of 1829–1851 (Hoad 2009). The worst single pandemic, the Spanish Flu Pandemic[1] of 1918–1919, killed an estimated 30–50 million people globally (with some estimates as high as 50–100 million deaths), and 675,000 in the United States alone (U.S. Department of Health and Human Services (HHS) 2009). Numbers like these take a heavy toll on affected populations not only in terms of treating the sick and dying but also in dealing with mass burials in a culturally acceptable way and tending to the associated mental health impact.

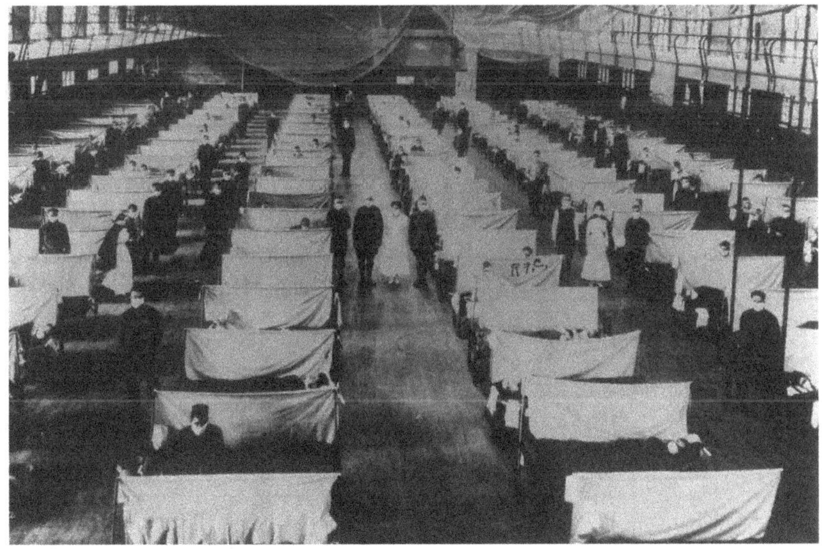

Figure 27–1. Dealing with pandemics has often been a seemingly insurmountable problem. Preparedness planning might change this outlook. *Source: Office of the Public Health Service Historian.*

In fact, in absolute terms of sickness and death (morbidity and mortality), the recent avian flu potential pandemic is relatively minor; only 413 cases resulted in a relatively high mortality of 256 deaths (WHO 2009). However, the effect that these cases have had on society worldwide is severe. The virus continued to tap health resources, and the fear generated by the disease spans many countries and continents, disproportionally affecting behavior.

The most troubling characteristic of these viruses may be their unpredictability. Public health academic Philip Alcabes, author of *Dread*, says that instead of looking to physicians to predict epidemics, "...we should leave the job of seeing the future to the mystics, prophets, and fortunetellers." This level of uncertainty breeds anxiety.

Nonetheless, there are certainties. Pandemics will recur, and they will have a great impact on society. In a prescient statement in 2005, a Congressional Budget Office (CBO) report stated, when referring to the threat of the H5N1 avian flu virus, "...[the H5N1 virus] could evolve in a way that rendered it harmless, and a pandemic could arise from an entirely different virus subtype" (CBO 2005).

COMMUNITY CONTINUITY CONCERNS

The impact of pandemics goes beyond health implications. The entire socioeconomic system is affected and felt well beyond the health community. "[Pandemics] expose existing weaknesses in these systems and, in addition to their morbidity and mortality, can disrupt economic activity and

[1] The pandemic was referred to as the "Spanish Flu" pandemic, which was also known as La Grippe Espagnole, or La Pesadilla (Science Daily website, accessed May 1, 2009), even though the disease is believed to have started at a military base in Kansas (note that other research indicates that it might have begun in the Far East or Austria) and been carried by WWI U.S. troops to other parts of the world. Because Spain was a neutral country during the conflict and had no wartime censorship in place, it was one of the few news outlets reporting on the pandemic (particularly when it moved from France to Spain)—hence, the term "Spanish Flu" might have been coined as a result. The variant of the influenza virus dominant in the Spanish Flu pandemic was H1N1.

Table 27–A. Pandemics Timeline	
Major Outbreaks of Disease Throughout History	
430 BC	First recorded instance of bubonic plague hits Athens, Greece.
542 AD	Mediterranean epidemic kills between 25 and 50% of the population of the Roman Empire. Emperor Justinian contracts the disease but survives.
1334–47	Black Death devastates Europe, spread partly by Crusaders from Constantinople. Three-quarters of the population die.
1563	Bubonic plague in London. One-third to one-half of the population dies.
1665	Great Plague of London. 20% of the population dies.
1729–30	Worldwide influenza pandemic. The first officially recorded "pandemic" (an epidemic of worldwide proportions) in history. Originating in Russia, this influenza virus spread to Europe and the United States.
1781	Worldwide influenza pandemic. Cited as among the worst in history, tens of millions of cases in Europe, North America, Spanish America, and the West Indies.
1817–23	First global cholera pandemic. This highly virulent outbreak spread from India across all of Asia.
1829–51	Second cholera pandemic. This cholera spread from Bengal, India, to Europe, Japan, and the United States. In Cairo and Alexander alone, 33,000 deaths were recorded in one 24-hour period.
1852–59	Third cholera pandemic. Multiple sporadic eruptions of the disease killed thousands world-wide.
1857–59	Worldwide influenza pandemic.
1863–79	Fourth cholera pandemic.
1881–96	Fifth cholera pandemic.
1889–90	"Russian flu" pandemic. An estimated 250,000 die in Europe, three times that worldwide.
1899–1923	Sixth cholera pandemic.
1918	"Spanish flu" pandemic. 20 to 40 million people die worldwide, more than were killed in the First World War. The worst pandemic in history.
1957–58	"Asian flu" epidemic. Affects 10 to 35% of the world population, though with proportionally low mortality rates (100,000 die).
1961–70	Seventh cholera pandemic. Some argue that this pandemic is ongoing. In 1991, nearly 4,000 deaths occurred in 16 countries.
1968–69	"Hong Kong" flu pandemic. Kills 700,000 worldwide.
1989–90	United Kingdom (UK) flu epidemic. Kills 29,000 in the United Kingdom.
1997	"Avian flu" outbreak in Hong Kong. Six people die.

Source: Phil Hoad. "Pandemics Timeline," accessed April 30, 2009. Guardian.co.uk

development" (WHO 2009). Nearly all facets of a community, or even a nation, can be affected gravely, from mundane daily tasks to broad strategic operations.

These operations are disrupted on several levels. The people required for any level of community activities—governance, education, health care, transportation, food distribution—are the parts of the infrastructure that are affected profoundly. Not only are there fewer people to conduct these vital activities, but also the people with appropriate skills are affected in unpredictable ways. Although most experts forecast high levels of absenteeism, the specific absentees themselves are not predictable; 30 to 40% of a population may be unable to function because of their own or a family member's sickness or death (HHS 2009). There is no way to predict exactly which individuals would be affected. Critical functions might be affected to a greater degree, and overall continuity of operations could be compromised severely.

Communities need a plan for pandemic preparedness, but how do they know when these plans must be put into practice? Several tools, or yardsticks, exist for measuring the pandemic threat. Two of the best known are WHO's pandemic alert level tool and the U.S. federal government's response stage tool from HHS. Figure 27–2 depicts these tools.

<table>
<tr><th colspan="2">WHO Phases</th><th colspan="2">Federal Government Response Stages</th></tr>
<tr><td colspan="4">INTER-PANDEMIC PERIOD</td></tr>
<tr><td>1</td><td>No new influenza virus subtypes have been detected in humans. An influenza virus subtype that has caused a human infection may be present in animals. If present in animals, the risk of human disease is considered to be low.</td><td rowspan="2">0</td><td rowspan="2">New domestic animal outbreak in at-risk country</td></tr>
<tr><td>2</td><td>No new influenza virus subtypes have been detected in humans. However, a circulating animal influenza subtype poses a substantial risk of human disease.</td></tr>
<tr><td colspan="4">PANDEMIC ALERT PERIOD</td></tr>
<tr><td rowspan="2">3</td><td rowspan="2">Human infection(s) with a new subtype, but no human-to-human spread, or at most rare instances of spread to a close contact.</td><td>0</td><td>New domestic animal outbreak in at-risk country</td></tr>
<tr><td>1</td><td>Suspected human outbreak overseas</td></tr>
<tr><td>4</td><td>Small cluster(s) with limited human-to-human transmission but spread is highly localized, suggesting that the virus is not well adapted to humans.</td><td rowspan="2">2</td><td rowspan="2">Confirmed human outbreak overseas</td></tr>
<tr><td>5</td><td>Larger cluster(s) but human-to-human spread still localized, suggesting that the virus is becoming increasingly better adapted to humans, but may not yet be fully transmissible (substantial pandemic risk).</td></tr>
<tr><td colspan="4">PANDEMIC PERIOD</td></tr>
<tr><td rowspan="4">6</td><td rowspan="4">Pandemic phase: increased and sustained transmission in general population.</td><td>3</td><td>Widespread human outbreaks in multiple locations overseas</td></tr>
<tr><td>4</td><td>First human case in North America</td></tr>
<tr><td>5</td><td>Spread throughout United States</td></tr>
<tr><td>6</td><td>Recovery and preparation for subsequent waves</td></tr>
</table>

Figure 27–2. World Health Organization and Federal Pandemic Response Levels. *Source: PandemicFlu.gov, accessed May 1, 2009.*

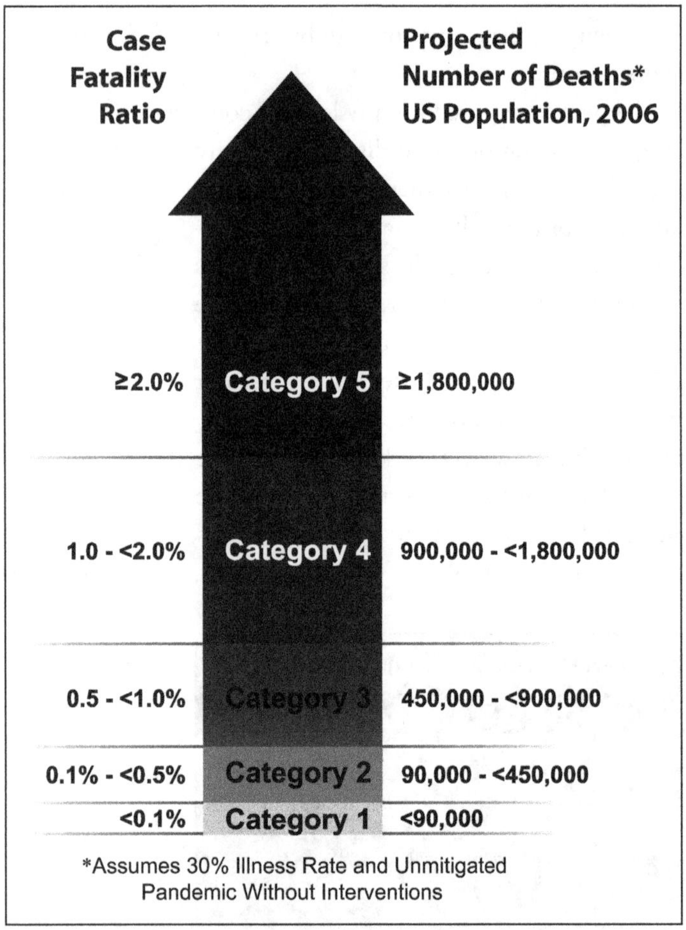

Figure 27–3. Pandemic Severity Index. *Source: U.S. Department of Health and Human Services, "Community Strategy for Pandemic Influenza Mitigation," February 2007.*

For each WHO phase or HHS stage, there are triggers that lead either organization to elevate the levels. These same triggers inform communities and populations about changes in the pandemic's nature. As Figure 27–2 illustrates, WHO and HHS monitor these triggers, but lower-level entities, including state and regional public health organizations, local community health authorities, and even businesses, also should monitor the triggers. The more aware these organizations are, the more prepared they will be to implement appropriate preparedness actions.

The WHO phases measure the *transmissibility* of a virus, not the *severity* of a resultant pandemic. This is a key difference because many pandemic plans are based on severity of the illness. Although WHO might measure a pandemic as being at a high level, it is indicating that triggers show the virus to be transmissible at the levels indicated in the chart (e.g., "widespread human infection" for levels 5 and 6), not that resulting illnesses are necessarily severe (although that might also be the case). WHO recently made clarifications to their phase structure,[2] but the phases themselves still relate to the transmissibility of the virus.

With respect to severity, at this writing, WHO has not yet developed a system for measuring this characteristic of a pandemic. Severity refers not only to the degree of virulence of a virus (e.g., number of severe illnesses and deaths, contagiousness of the virus, age distribution of cases, prevalence of chronic health problems and malnutrition of the population, viral mutations, number of waves of illness, and quality of health services) but also to the overall socioeconomic impact of the outbreak (*CIDRAP News* 2009).

Although WHO may be developing such a severity index, HHS already has one in place. Based on hurricane classifications, this system is shown in Figure 27–3. "HHS announced its Pandemic Severity Index in February 2007 as part of its guidance on community interventions for combating a pandemic. The index is based on case-fatality rates (CFR), with a CFR of 2% or greater signaling the most severe pandemic: category 5. The pandemics of 1957 and 1968 qualify as category 2 events, with CFRs between 0.1% and 0.5%" (HHS 2007). Note that there is potential for public confusion when using this index in combination with the WHO pandemic phases, which may have led to WHO's delay in presenting a similar index. HHS has presented numerous cross-references between their index and WHO's and has included characteristics and interventions for each severity level.

[2] "In the 2009 revision of the phase descriptions, WHO has retained the use of a six-phased approach for easy incorporation of new recommendations and approaches into existing national preparedness and response plans. The grouping and description of pandemic phases have been revised to make them easier to understand, more precise, and based on observable phenomena. Phases 1–3 correlate with preparedness, including capacity development and response planning activities, while Phases 4–6 clearly signal the need for response and mitigation efforts. Furthermore, periods after the first pandemic wave are elaborated to facilitate post pandemic recovery activities" (World Health Organization (WHO) *http://www.who.int/csr/disease/avian_influenza/phase/en/index.html*, accessed May 10, 2009.

The WHO and HHS tools depend on faithful and timely reporting of the relevant triggers. During the 2009 Swine Flu pandemic, it was reassuring to see that no apparent cover-up existed on the part of any nation with respect to reporting the illness (as some believed might have been the case for reporting the 2002–2003 SARS event). However, there was concern not only that some time had passed—perhaps three weeks or more—between the first cases in Mexico and the first WHO warnings, but also that some time had passed before the Centers for Disease Control and Prevention (CDC) identified the novel variant of the virus (*New York Times* 2009). These delays must be eliminated so that pandemic preparedness plans can be implemented when they can have the greatest impact.

PSYCHOSOCIAL CONCERNS

As stated earlier, the effects of a pandemic go well beyond health concerns. The cyclical nature of pandemics, as well as the rapid transmission of the diseases, generates fear. This "worried well" further overwhelms an already besieged health care infrastructure. Individuals whose family members or friends have been sickened or have died suffer from mental health issues that could disable the strongest among us. In 1918, during the Spanish Flu Pandemic, Philadelphia suffered such losses of adult citizens that the Bureau of Child Hygiene determined it was unable to care for the large number of orphans (HHS 2009). Even worse, the city was unable to cope with the numbers of dead people. If a pandemic of similar virulence were to occur today, there could be "1.7 million deaths in the United States and 180 to 360 million deaths globally" (Osterholm 2005), clearly overloading the health care system, as well as the funereal communities—taking these essential services beyond the breaking point.

One does not need to look to horrific pandemic data to sense the fragility of a community's health care infrastructure. In 2006, a two-week heat wave in California caused as many as 141 deaths, leading "coroners...[to deal with] the large jump in the number of bodies that were stuffed, some piled on top of others, into the freezers at the Fresno County morgue" (Maitre 2006).

Planning for pandemics must also include consideration of the nature of human behavior. It is well known that such behavior is nonlinear, and therefore might seem unpredictable. However, if sufficient information is known about a community's culture, historical actions, and motivations, behavior can be predicted with some fidelity. Unfortunately, a deep understanding of all these factors is seldom possible, particularly after a pandemic has started. Therefore, the more a community knows about itself and its underlying structure, and the more that knowledge is factored into preparedness planning, the more effective that planning will be.

When developing a plan that fits a community, it is important to consider all aspects of that population. There are five major aspects to consider (Sulek et al. 2008):

1. culture of the group
2. policies, strategies, and plans in place
3. economics, management, and budgeting of the entities involved
4. governance and operations
5. technology implemented to deal with crises (e.g., information databases, communication networks)

Planners must thoroughly understand all five aspects. For instance, if the culture of a group is to be prepared, but the policies for preparedness are not in place, or the technologies to enable the group to be thoroughly prepared are not implemented, preparedness might not be possible.

Similarly, developing plans for such preparedness must take into account all these elements, or the plans will not work.

ECONOMIC IMPACTS

The effects of pandemics go well beyond physical and psychosocial health care concerns. Pandemics can take a tremendous and far-reaching toll on the economic health and well-being of a community.

Economic costs occur at both ends of a pandemic. Costs to prepare the population for pandemics can be enormous, as can costs to respond to and recover from the health crisis itself. For example, within days of the first outbreaks of the Swine Flu in early 2009, the World Bank responded to the pandemic with "fast disbursing funds" ($25 million for drugs and supplies, and $180 million for epidemiologic, regulatory, institutional, and operational activities) and many other agencies and institutions followed course (World Bank 2009). The total amount of funding applied, that is, where those funds will be found, is unknown at this writing, but costs will be high.

As another example, in November 2005, the Bush administration requested $7.1 billion in emergency funding for pandemic preparedness. (Although less funding was ultimately made available to planners and responders, the majority of this funding request was allocated.) The activities supported included plan development, antiviral stockpiles, production of vaccines, and similar key actions. Long-term infrastructure would ultimately be strengthened, including source control and surveillance, vaccine research and development, antiviral drug research and development, and health care system readiness (CBO 2006).

Long-term infrastructure is of particular importance. When it comes to pandemics, our health care infrastructure is fragile and insufficient. During a severe pandemic, there would be far more demand on the U.S. health care system than the system could accommodate—as many as 5 to 10 million sick individuals (Lister 2005). This figure exceeds the roughly 970,000 staffed hospital beds and 100,000 ventilators, with 75% of those in use at any given time under normal, nonpandemic situations (CBO 2006). Consequently, building a stronger health care infrastructure *before* a pandemic is key. Even a mild pandemic, such as the 2009 Swine Flu (at the time of this writing, considered mild in severity, although transmissible at a relatively high level), demonstrated to communities that the stress on the health infrastructure could be enormous. Although a sizable percentage (about 25%), the federal stockpile of antiviral drugs was distributed quickly to states and localities; however, it was clear that local heath departments would have difficulty distributing them rapidly to the population as a result of budget cuts and layoffs at local levels. Treating a sick population would be difficult, and supporting the recovery of those who had been ill would add another level of stress to the health care system.

The indirect costs (e.g., supply and demand, such as a shrinking workforce, and a dramatic decline in demand for goods and services as people avoid public places) during and following a pandemic can be even more staggering. The economic impact of recent pandemics, even relatively minor ones such as the SARS pandemic of 2003, has been enormous. By some estimates, the indirect costs attributed to disruption of activity in business, civil, and governmental domains related to the SARS event were as much as $30 to $100 billion (note that CBO and others continue to debate these figures). In 2005, the World Bank estimated that a 2% loss of global gross domestic product (GDP) would result from a pandemic of similar severity to 1918, which translates into $800 billion in losses for a year (World Bank 2009). Recent estimates raise this expected impact to nearly 5% of global GDP, or about $3 trillion (Reuters 2009). This impact is "greater than in recent recessions and roughly the same size as the average postwar recession" (CBO 2006).

UNIQUE PREPAREDNESS REQUIREMENTS OF PANDEMICS

It is clear that we must prepare for pandemics. But what is the goal of such preparedness? This issue continues to be debated, but the CBO (drawing on WHO and others) has identified the following goals for the federal government (CBO 2006):

- support for the efforts of governments of other countries and international organizations to contain [virus] strains and control their evolution to diseases that are transmitted easily from person to person
- building stockpiles of vaccines, improving antiviral drugs, and putting in place new technologies that allow effective vaccines to be produced more rapidly and in large quantities
- improving the capacity of the health system to care for many people in all parts of the country who are sick simultaneously

Plans at other levels (states, regions, localities, and individual organizations and entities) have parallel goals that are appropriate for their own needs, but the last objective typically is the driving goal for planners.

PANDEMICS AT HAND: AVIAN AND SWINE INFLUENZAS

A pandemic of great recent concern is *avian flu*. This flu infects birds, such as domestic poultry (e.g., chickens) and wild fowl (e.g., ducks).

> Avian flu...is caused by influenza viruses that occur naturally among wild birds. Low pathogenic [avian flu] is common in birds and causes few problems [to humans]. Highly pathogenic H5N1 is deadly to domestic fowl, can be transmitted from birds to humans, and is deadly to humans. There is virtually no human immunity, and human vaccine availability is very limited (HHS 2009).

In fact, "viruses of the H5 subtype are not known to have ever circulated among the human population, which means that there would be little immunity to it" (CBO 2006). Therefore, it is this highly pathogenic version of avian flu that is a risk to communities and has the potential for pandemic impact.

Similarly, *swine flu* is caused by influenza viruses that occur in swine. However, although the immunity and human vaccine availability is like that of avian flu, the 2009 swine flu appeared to be a variant that was a mutation (or "re-assortment") of avian flu, swine flu, and human flu.[3] The resultant virus appeared to be transmitted with some efficiency from human to human, leading WHO to raise the pandemic alert phase to 5 (at this writing), with the potential of additional elevation possible. This phase carries with it the fact that the virus is being seen in "larger cluster(s), but human-to-human spread [is] still localized, suggesting that the virus is becoming increasingly better adapted to humans, but may not yet be fully transmissible" (WHO 2009).

[3] "According to virologists at the U.S. Centers for Disease Control and Prevention (CDC), the influenza A subtype H1N1 isolated from...patients in April [2009] was a genetic reassortment of four different influenza virus strains, including human influenza gene segments, swine influenza from North America and Eurasia, and avian gene segments from North America, never before reported among swine or human isolates from anywhere in the world." (Food and Agricultural Organization, *www.fao.org/ag/againfo/programmes/en/empres/AH1N1/Background.html*, accessed May 2, 2009.

As of this writing, a small but growing number of cases had been reported worldwide, with few deaths (though these deaths are at a rate that some have judged as alarming).[4]

Note that every community already suffers significant impact from *seasonal flu*, a common respiratory illness transmitted from person to person. Though many people have some immunity to this illness, and vaccines are available annually, each year in the United States alone has an average of 36,000 deaths occur as a result of seasonal influenza. These deaths, along with more than 200,000 U.S. hospitalizations and more than $10 billion in U.S. economic cost (Homeland Security Council 2005), accompany more than one-quarter to one-half million deaths worldwide annually (WHO 2009). This toll should lead to a heightened awareness of hygiene principles—principles that would be effective for not only a pandemic but also seasonal flu. Unfortunately, some say our culture has adopted a *laissez-faire* approach to this disease and to measures effective in preventing its transmission.

There are encouraging developments, however. Recently, public service announcements (PSAs) and other media communications have been emphasizing the importance of hygiene for limiting transmission of infectious disease (e.g., suggesting to citizens that they wash their hands regularly, taking the time necessary to sing the tune "Happy Birthday" twice). Although it has been suggested that the public had been suffering from "information fatigue" as a result of extensive coverage of the avian flu potential pandemic, and that subsequent media reports had waned, recent coverage has again risen, particularly as a result of the swine flu pandemic. This coverage can be helpful only in preparing the public for pandemics.

Numerous interesting new tools and techniques exist that are related to pandemics being introduced—not only to the health care community and emergency preparedness professionals but also to the public. For example, "Google's philanthropic branch, Google.org, ...launched a new U.S. public health initiative, *Google Flu Trends. Flu Trends* operates on the premise that an influenza outbreak is more likely to occur in a state where Google flu-related searches are increasing."[5] Given the popularity of Google web tools, including Google Earth (now being used to track avian flu outbreaks and mutations worldwide), the public will have techniques never before available to aid them in their awareness and sensitivity to preparedness activities. Pandemic preparedness plans must include these tools, and planners must use the tools in exercises.

PERSISTENCE AND PERVASIVENESS OF PANDEMICS

Pandemics are different from other crises. Events such as floods, tornadoes, hurricanes, and wildfires are typically limited to relatively small areas. By definition, pandemics are crises that affect wide areas, often covering entire countries, continents, or even the globe.

This geographic coverage greatly complicates preparedness planning. For most emergency scenarios, planners can depend on other resources and aid from areas not affected by the crisis. Fire fighters often travel to serious fires, first responders are often brought in to disaster scenes such as floods or hurricanes, and food and other resources are typically flown to stricken areas.

---

[4] As of May 1, 2009, 331 cases were reported in 11 countries, with 10 resultant deaths By May 10, 2009, only 9 days later, the reported cases had risen dramatically, to 4,379, with 49 deaths (World Health Organization, *http://www.who.int/csr/don/2009_05_10/en/index.html*, accessed May 1, 2009 and May 10, 2009).

[5] "The Centers for Disease Control and Prevention (CDC) helped Google validate and refine its *Flu Trends* software, and the CDC is looking forward to testing it [the 2009] flu season. According to Google, during the 2007–08 flu season, the *Flu Trends* model was able to accurately predict current influenza levels 1 to 2 weeks earlier than the data published in the CDC surveillance reports in each of the nine U.S. surveillance regions. Although the extra warning time will not prevent epidemic outbreaks, it might allow for hospitals and public health professionals to gear up preparedness for a surge in patients in a particular geographic area. ... Google's *Flu Trends* results are available for viewing at *www.google.org/flutrends/*" (DisasterPrepped.com, accessed April 23, 2009).

Planners for pandemics must assume the potential for the pandemic to hit *all* other areas, and people in other areas will be unable to assist. Pandemic plans must also build on locally available staff, food, water, power, transportation, and other resources. The plans must call on the affected areas to weather the pandemic organically.

TEMPORAL REQUIREMENTS OF PREPAREDNESS

Pandemic planners must not only assume that little or no outside help will exist but must also accommodate another unique aspect of pandemics: typically, pandemics are not bounded by a short, easily defined time frame. Rather, they often occur in "waves" (each with a several-week duration) separated by some time period. These waves can be individually devastating; collectively, they can drain a population's energy and resources as response, recovery, and planning phases begin to overlap.

These waves can be separated by many years, as has been the case for the global cholera pandemics (e.g., seven of which have occurred from 1817 to the present). Pandemics separated by such long time periods are more manageable in some ways. Preparedness for a successive wave can build on lessons learned from a previous wave, with sufficient time elapsing for replenishment of vital medical and other resources. However, if pandemic waves are separated by shorter periods (measured in weeks or months), the collective impact can be demoralizing and overwhelming. For example, the Spanish Flu pandemic of 1918–1919 attacked in three waves: the first occurring in the spring/summer of 1918; the second, in the fall of 1918; and the third and final wave, in the spring of 1919. Future influenza pandemics are expected to follow a similar pattern, causing great concern to planners and responders.

PANDEMIC PREPAREDNESS PLANNING

The notion of planning for a pandemic is not only a notion of practical or intuitive need but also one of policy. The highest level federal planning guidance addressing all crisis preparedness was the Department of Homeland Security's (DHS) *National Response Plan (NRP)* (issued December 2004), which was superseded by the *National Response Framework (NRF)* (issued March 22, 2008). The NRP, and subsequently the NRF, is an "all-discipline, all-hazards plan intended to establish a single, comprehensive framework for managing domestic incidents" (Department of Homeland Security 2004). The NRF describes federal support to be implemented through activation of 15 emergency support functions (ESF), including several relevant to pandemic planning: *Mass Care, Emergency Assistance, Housing, and Human Services* (ESF #6); *Public Health and Medical Services (ESF #8); Public Safety and Security* (ESF #13); and *Long-Term Community Recovery* (ESF #14). Each of the 15 ESFs, including those relevant to pandemics, includes details about crisis planning, implementation, training, and exercising (Department of Homeland Security 2008).

Further, the federal government has developed specific detailed guidance with respect to pandemics—the *National Strategy for Pandemic Influenza*—motivated largely by the threat of the type A/H5N1 virus (avian flu). Although the *National Strategy* was developed in response to the avian flu, it addresses the preparation for and response to any pandemic, and it is fully consistent with higher level policy and planning documents such as the NRP and the subsequent NRF. The *National Strategy*...

> ...guides our preparedness and response to an influenza pandemic, with the intent of (1) stopping, slowing, or otherwise limiting the spread of a pandemic to the United States; (2) limiting the domestic spread of a pandemic, and mitigating disease suffering and death; and (3) sustaining infrastructure and mitigating impact to the economy and the functioning of society (Homeland Security Council 2005).

There are three pillars in the *National Strategy*:

1. preparedness and communication
2. surveillance and detection
3. response and containment

Each has considerable detailed guidance for communities and the general population in business, civil, and governmental situations.

Following the *National Strategy*, the *National Strategy for Pandemic Influenza Implementation Plan* was released. This plan presents specific actions for more than 300 federal departments and agencies (e.g., Department of Defense [DoD], HHS, Veteran's Administration [VA]), and lower level implementation plans by these agencies have followed (Homeland Security Council 2006).

The following section addresses the process for developing pandemic preparedness plans, essential elements of not only the plans themselves but also the implementation of those plans.

DEVELOPING A PREPAREDNESS PLAN

The *National Strategy* includes key overarching goals (described previously) (Homeland Security Council 2007). Any lower-level pandemic preparedness plan must be in consonance with these high-level goals. Such lower-level plans may be developed carefully, and certainly well-intentioned, but many are "not legally or logistically feasible." Further, it has been said that "lessons from simulations had not been drawn on to revise plans" (Reuters 2009). In short, the plans must be realistic, appropriate to the entity or entities being addressed, and exercised thoroughly. Exercises must be carried out repeatedly so that participants will become trained and plans can be updated, based on exercise results. Situations for an affected population may change, and the plan must be agile enough to accommodate changes, but these can be appreciated only in the context of realistic exercises. All too often, a preparedness plan (for pandemics or for other crises) becomes "shelf-ware"—developed with care, but then relegated to a shelf until a crisis hits. At that point, it is too late to update the plan.

Numerous tools are in the pandemic preparedness plan toolbox, including the following:

- vaccines
- antiviral medications
- infection control measures
- community mitigation measures

Each is complex and could potentially have a broad effect on a pandemic; however, none is likely to completely address the severe impact of a pandemic. Most likely, all these tools (and perhaps others) will be needed. Lisa Koonin, Senior Advisor for the CDC's Influenza Coordination Unit, likens this issue to "Swiss cheese." A single tool is like a single slice of cheese—full of holes. However, when we lay one slice over another, and then another, and then another, we find that the holes may be partially—or even completely—filled. Similarly, these preparedness toolbox techniques are additive; as many as possible must be used to achieve the maximum mitigating effect.

Vaccines are a unique and troublesome element of a pandemic preparedness plan. Typically, vaccines can be quite effective, providing a measure of immunity to an individual. They are in widespread use for other viruses, including seasonal influenza, where they have been effective. Typically, pandemic preparedness plans specifically address the availability, distribution, and use of vaccines.

However, a vaccine must be matched precisely to the virus in circulation. The laborious process for developing this "match" typically takes months.[6] This time lag is crucial; if a highly transmissible virus were to cause a pandemic, many thousands (even millions) of people could be infected, many of whom could die before a vaccine were to become available. Furthermore, the time lag is not the only issue concerning vaccines. It is likely that the vaccines, once developed, would not be available in sufficient quantities to be distributed to all populations affected. Therefore, prioritization of recipients will be required (for instance, health care workers and emergency management responders would be first on such a list, and so on). Practical and ethical issues are certain to arise. For example, should those in failing health (unrelated to the influenza) be administered the vaccine rather than those who are younger and more vital? Should "key" staff be administered a vaccine before those deemed less critical for operations? Government and health care officials would be expected to resolve these issues. Other issues also would need to be considered, including topics such as use and efficacy of pre-pandemic vaccines (ones that might be available before a pandemic, but not necessarily perfect matches to the circulating virus).

Antivirals do not function like vaccines. Antivirals do not provide immunity, but they may make the illness less severe or shorten the course of the illness. They are not specific to a virus in circulation (although considerable differences exist in effectiveness for each variation of antiviral), and large stockpiles (i.e., millions of regimens in the United States alone) already have been developed and stored. Antivirals have demonstrated some effectiveness in prophylactic use, but they are not intended to immunize a recipient to a virus.

Antivirals also have shortcomings. Because they can be used prophylactically, there is fear that they will be overused or cause the virus strain to develop resistance. Resistance, whether a result of overuse or not, can be significant. For the 2008–2009 seasonal flu, virtually all flu cases have shown resistance to oseltamivir, a popular antiviral.[7] Further, although large quantities of antivirals are stockpiled, there still might be a need to prioritize who gets them and when.

Another tool is infection control. Many control measures are commonplace and intuitive, but others are less so. Encouraging standard hygiene processes (hand washing and covering coughs and sneezes) has a strong impact on limiting the transmission of an influenza virus. Other personal protective equipment (PPE) such as surgical masks and other respiratory devices (e.g., N95 masks, respirators) has been shown to be effective, although it can be burdensome, expensive, possibly misused, and effective only in certain situations. The key issue for all these techniques is compliance; without constant and vigilant use, these tools are ineffective. As dis-

[6] There is hope for great improvement in this area. New vaccine development techniques promise dramatic decreases in time required to develop the vaccine. For instance, the National Institutes of Standards and the University of Queensland in Australia have "successfully demonstrated ... findings that could reduce the time it takes to produce a vaccine from months to weeks ..." (*NIST Tech Beat*, December 8, 2008).

[7] Although this news report states specifically that only the H1N1 virus is showing this resistance (thus avoiding such resistance from the H5N1 avian flu variant), this may cause further concern over the H1N1 Influenza A (swine flu) variant posing a global threat in 2009 (L.A. Times, "Tamiflu No Longer Works for Dominant Flu Strain," February 4, 2009, *http://articles.latimes.com/2009/feb/04/science/sci-flu4*, accessed May 5, 2009).

cussed earlier, training is a key element in developing compliance; the more communities trained, the more likely that they will comply with these infection control approaches.

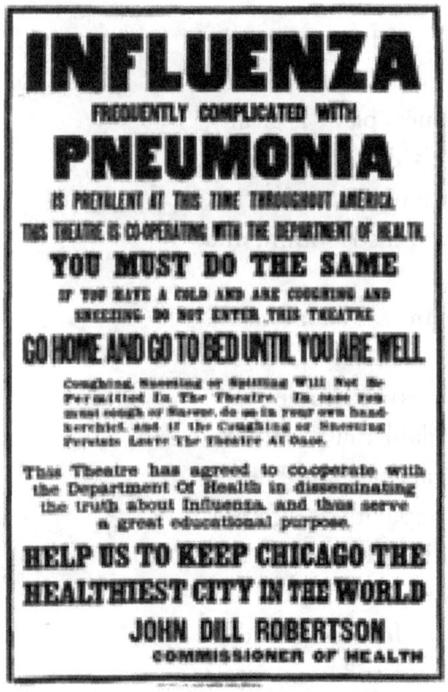

Figure 27–4. Infectious disease transmissibility mitigation techniques are long-standing and likely will be employed again as part of overall pandemic plans and their implementation. *Source: Office of the Public Health Service Historian.*

Finally, community mitigation techniques are an important part of pandemic preparedness planning. A key mitigation approach is social distancing, requiring that people maintain a minimum distance between one another to "reduce the duration and/or intimacy of social contacts and thereby limit the transmission of influenza" (Homeland Security Council 2008) either through shifts in operation times and/or locations, or individual behavior modification.[8] CDC and OPM published guidance about such behavior modification for the 2009 swine flu crisis: maintain 6 ft from each other unless PPE masks are employed. Other guidance for mitigation included using flexible work schedules to reduce face-to-face interactions (U.S. Office of Personnel Management (OPM) 2009). Other approaches could be used: minimizing mass gatherings (e.g., schools, churches, malls), practicing isolation (for people who are sick or tending to the sick), or even quarantining (a method not typically expected to be used extensively in a modern pandemic, but one that might be used in severe circumstances). These tools are available to planners and responders, and pandemic plans will almost certainly include several of these techniques. Figure 27–4 shows one technique used to mitigate the transmissibility of an infectious disease.

Many excellent examples of pandemic preparedness plans have been developed, most of which are available online at state emergency management sites, as well as others specific to businesses and civil entities. Each plan typically includes guidance regarding assumptions, scenario descriptions, roles and responsibilities, a concept of operations, incident management actions, and maintenance of the plan itself. An example of such a site is the Virginia Emergency Management website www.vaemergency.com.

TRAINING FOR AND EXERCISING PREPAREDNESS

Training is critical to the success of pandemic plans. It is often said that the military is successful largely because they "train like they fight, and fight like they train." Similarly, pandemic plans often may call for significant changes in behavior (e.g., social distancing). These changes will come about only with frequent training and exercising, particularly if behaviors are cultural in nature. For example, take the common practice of shaking hands when meeting someone. If not practiced often, *not* shaking hands might be perceived as rude rather than as a way to lessen the transmission of infectious diseases.

Training and exercising of preparedness plans is becoming more commonplace, even for pandemic plans. Setting the tone for such exercises is the series of federal exercises, Top Officials (TOPOFF). This series of exercises addressed emergencies such as the release of biological agents, radiological dispersal devices, and chemical agents. TOPOFF was intended to build an integrated federal, state, territorial, local, and private sector capability to prevent terrorist attacks, and

[8] These shifts can notably include "telework." OPM announced a new governmentwide telework policy in April 2009, which combined components of the Telework Improvements Act (H.R. 1722) and the Telework Enhancement Act (S. 707)—the U.S. Office of Personnel Management (OPM), *Memorandum for Heads of Executive Departments and Agencies—Advice to Federal Employees and Agencies on Preventing the Spread of the Current Flu and Maintaining Readiness to Use HR Flexibilities if Necessary,* April 26, 2009.

rapidly and effectively respond to, and recover from, any terrorist attack or major disaster that does occur. The full-scale exercise offers agencies and jurisdictions a way to not only test their plans and skills in a real-time, realistic environment but also gain in-depth knowledge that only experience can provide. Participants also exercise prevention and intelligence-gathering functions, which are critical to preventing terrorist attacks.

Lessons learned from TOPOFF will provide valuable insights to guide future planning for securing the nation against terrorist attacks, disasters, and other emergencies (Department of Homeland Security 2009). In particular, lessons learned from the exercises have helped pandemic planners design exercises at federal, regional, state, or local levels for pandemic scenarios: PANEX 07, a FEMA-hosted, joint federal–state exercise; and FBIIC/FSSCC Pandemic Flu Exercise of 2007, a Department of the Treasury-hosted exercise involving the banking and financial services sectors. The latter example has clear goals: "enhance the understanding of systemic risks...[on] sector[s]; provide an opportunity for firms to test their pandemic plans; and examine how the effect of a pandemic flu on other critical infrastructures will impact...sector[s]" (Department of Treasury 2009). These should be the goals for any exercise intended to validate a pandemic preparedness plan.

Other guidance for developing exercises intended to validate pandemic plans is available online; for example, WHO's "Exercise Development Guide for Validating Influenza Pandemic Preparedness Plans" can be found at the World Health Organization's website (WHO Interim Report 2006). The leadership and advice that these documents provide will increase the effectiveness of pandemic plans, but only if they are used to thoroughly and frequently exercise the plans themselves, and involve those who will be affected by a pandemic.

DYNAMIC REPLANNING FOR PREPAREDNESS

Any plan must be revisited regularly and adjusted for changes in the threat, environment, and participants. Therefore, plans must be reviewed and exercised not only periodically but also when circumstances necessitate review, such as a change in guiding policy and a change in the nature of a virus.

DURING AND AFTER A PANDEMIC

Even with diligent pre-pandemic planning, some issues need to be addressed when a population is in the midst of a pandemic. Pandemics may come in waves, and successive planning, taking into account lessons learned from a previous wave, can drastically reduce the socioeconomic impacts of subsequent waves (Figure 27–5). This section presents some issues that might be faced during and after a pandemic.

Figure 27–5. Pandemics have been part of our lives for centuries and will be again in the future. *Source: Office of the Public Health Service Historian.*

RESPONSE TO PANDEMIC INFECTION

One question that will be particularly relevant to any population is whether additional individuals are infected and who these people are. There is considerable risk in overreacting to this question (ranging from social shunning to illegal quarantining); therefore, screening, detection, and response will be necessary, and all should be considered and included in a holistic pandemic plan. Of interest is a set of new techniques for this screening and testing. For example, in early 2009, CDC developed a rapid diagnostic test kit to detect the type A/H1N1 virus and distributed these kits to all 50 states, the District of Columbia, Puerto Rico, and internationally. Although of lower sensitivity than viral culture or other similar traditional tests, these kits will increase rapid testing capacity and are likely to result in a more accurate, rapid picture of the impact of this disease.

Similarly, in April 2009, the U.S. Food and Drug Administration (FDA) cleared a new, more rapid test for the detection of type A/H5N1. "This test is an important tool for helping quickly identify emerging influenza A/H5N1 infections and reducing exposure to large populations," said Daniel G. Schultz, M.D., director of FDA's Center for Devices and Radiological Health. "The clearance of this test represents a major step toward protecting the public from the threat of pandemic flu."[9]

COMMUNICATING DURING A PANDEMIC

As with any emergency situation, effective, transparent, and complete communication is of utmost importance. Social networking tools are emerging as a powerful means of communication in a broad set of scenarios relevant for pandemic events.

Social networks and blogs are now the fourth most popular online activity, ahead of personal e-mail. Member communities are visited by 67% of the global online population, and time spent on social networking is growing at three times the overall Internet use rate, accounting for almost 10% of all internet time (Nielson Company 2009). Beyond the personal applications, business social networks also are growing. One of the fast-growing business networks, LinkedIn, is enjoying a surge of activity: "... the site's traffic is up in the recession. It hit 36 million members in April 2009, and is adding them at a rate of about one member per second" (Advertising Age 2009).

It is in this context of broad community participation that these networks are becoming effective tools for emergency communication. This is certainly true as it applies to pandemics. As examples, many such tools have set up specific sections for pandemic discussion and communication, including LinkedIn, MySpace, Facebook, and Twitter. These sites are in addition to the numerous, more traditional websites devoted to these topics; government sites (e.g., *PandemicFlu.gov; who.int/csr/en; cdc.gov/h1n1flu/guidance*); and many outstanding private or civic sites (e.g., *h1n1alliance.org; www.cidrap.umn.edu*). Add to these a wealth of other communication venues, such as PSAs, special public audio/video conferences and symposia,[10] and the paths to communicate the pandemic message (in the preparedness phase and in the response and recovery phases) are many and diverse.

One particular concern regarding pandemic communications is the notion of alarming the public unnecessarily. Certainly public awareness is important, but avoiding information fatigue (or even being perceived as "crying 'swine'" as one report quoted) is equally key (Associated Press 2009). Communication must be continually measured in terms of its intent, its process, and its effectiveness.

RECOVERY AFTER A PANDEMIC

Pandemics have a powerful effect on a population. Nearly every measure of a community or nation can be affected. For example, the average life span of U.S. citizens dropped nearly 12

[9] "The test, called AVantage A/H5N1 Flu Test, detects influenza A/H5N1 in throat or nose swabs collected from patients who have flu-like symptoms. The test identifies in less than 40 minutes a specific protein (NS1) that indicates the presence of the influenza A/H5N1 virus subtype. Previous tests that the FDA cleared to detect this influenza A virus subtype can take 3 to 4 hours to produce results." (FDA News release, April 7, 2009).

[10] A powerful example of this media communication is the Public Broadcast System (PBS) video, *Predicting Pandemics—How do we fight both the swine flu pandemic and our fear of it?* May 8, 2009.

years in the year following the Spanish Flu pandemic of 1918–1919. The way citizens behave, and the way they see themselves and each other, fundamentally changes when a pandemic devastates their way of life. Recovery, therefore, is neither an easy nor a quick process.

Specific guidance addresses recovery from pandemics. For example, DHS has prepared a document, *Pandemic Influenza Preparedness, Response, and Recovery Guide for Critical Infrastructure and Key Resources*. This document focuses on businesses (stating as its purpose to "stimulate the U.S. private sector to act now"), but it specifically includes consideration of recovery activities following a pandemic.

> Continuity of Operations–Essential (COP-E) is the central concept in the guidance, wherein COP-E is an extension and refinement of current business contingency and continuity of operations planning that fully exploits existing efforts and integrates...the suite of business disaster plans. The COP-E process assumes severe pandemic-specific impacts to enhance and complement existing business continuity plans (Department of Homeland Security 2009).

Relevant to recovery, COP-E incorporates an approach for "survival" under distinct COP-E scenarios, and it enhances business continuity planning to address other catastrophic disasters. This holistic approach to continuity of operations is at the heart of recovery for a nation, region, or community.

SUMMARY

Pandemic preparedness is imperative for all communities, as rampant infectious diseases can have broad and sometimes devastating consequences. Although the public's awareness of potential pandemic events may be raised through media coverage, the same may not be true for overall preparedness. Pandemics are unpredictable, but this much is known: they will happen again, and they will have great impact on society. The impact goes beyond the health implications; the entire socioeconomic system of a population will be affected.

Rapid transmission of disease generates enormous fear in a population. As mentioned, those who are ill stress the community's ability to respond; the "worried well," driven by that fear, overwhelms an already besieged health care infrastructure; and a large portion of the population can suffer from disabling mental health issues.

When developing a tailored pandemic preparedness plan for a community or large population group, five aspects must be considered: culture; policies, strategies, and plans; economics, management, and budgeting; governance and operations; and technology.

Planning for a pandemic must be consistent with existing culture and policy. The U.S. government has several goals for preparing for pandemics, and all lower-level plans should match these objectives: stopping, slowing, or otherwise limiting the spread of a pandemic; mitigating disease suffering and death; sustaining the infrastructure; and mitigating impact to the economy and functioning of society.

Training participants and exercising pandemic preparedness plans is critical to the ultimate success of implementing those plans. In the balance is the protection of citizens but ultimately the recovery of society.

Pandemic preparedness is not only a government construct but also a process that must involve nongovernmental entities (non-profits, faith-based organizations, and businesses) to reach a level of preparedness that enables resilience to be woven into the process of mitigating effects and accelerating recovery. Preparedness is neither an easy nor a quick process. Pandemics may pose one of the more difficult scenarios for planners, but also may be one of the most important—another pandemic is most certainly on its way.

REFERENCES

Advertising Age. March 2, 2009, http://adage.com/abstract.php?article_id=134962, available May 11, 2009.

Associated Press. "Swine Flu Warnings Totally Overblown, Some Say," May 7, 2009, MSNBC, *http://www.msnbc.msn.com/id/30627377/,* available May 11, 2009.

CIDRAP News, 29Sep05. http://www.cidrap.umn.edu/cidrap/content/influenza/swineflu/news/may1109severity.html available May 1, 2009.

CIDRAP News, *http://www.cidrap.umn.edu/cidrap/content/influenza/swineflu/news/may-1109severity.html* available May 11, 2009.

Congressional Budget Office. *A Potential Influenza Pandemic: Possible Macroeconomic Effects and Policy Issues,* December 8, 2005, revised July 27, 2006. CBO website *www.cbo.gov/ftpdocs/69xx/doc6946/12-08-BirdFlu.pdf*, available May 1, 2009. This report draws heavily on sources, including the World Health Organization (WHO), *Avian Influenza: Assessing the Pandemic Threat* (Geneva: WHO, January 2005); Howell, Pugh, *Pandemic, The Cost of Avian Influenza, Contingencies* (September/October 2005), pp. 22–27; and Garrett, Laurie, *The Next Pandemic, Foreign Affairs* (July/August 2005), pp. 3–23.

FBIIC FSSCC SIFMA and U.S. Department of the Treasury. *The FBIIC/FSSCC Pandemic Flu Exercise of 2007 After Action Report*, website *http://www.fspanfluexercise.com/Pandemic%20Flu%20AAR.pdf,* available May 13, 2009.

Hoad, Phil. *Pandemics Timeline.* www.guardian.co.uk website, available April 7, 2009.

Hoad, Phil. *Pandemics Timeline.* www.guardian.co.uk website, available April 30, 2009.

Homeland Security Council. *The National Strategy for Pandemic Influenza Implementation Plan*, online at *www.whitehouse.gov/homeland/nspi_implementation.pdf,* available June 4, 2008.

Homeland Security Council. *The National Strategy for Pandemic Influenza*, November 1, 2005.

Homeland Security Council. *The National Strategy for Pandemic Influenza Implementation Plan,* May 3, 2006.

Lister, Sarah. *Pandemic Influenza: Domestic Preparedness Efforts,* CRS Report for Congress RL33145 (Congressional Research Service, November 10, 2005), p. 10.

Maitre, Michelle. "Heat Linked to More Than 130 Deaths," *Oakland* (CA) *Tribune*, July 29, 2006.

Merriam-Webster's Collegiate Dictionary, Tenth Edition. 1993. Springfield, MA: Merriam-Webster, Inc.

Nielson Company. *"Global Faces and Networked Places—A Nielsen report on Social Networking's New Global Footprint,"* March 2009; Nielson website *http://server-uk.imrworldwide.com/pdcimages/Global_Faces_and_Networked_Places-A_Nielsen_Report_on_Social_Networkings_New_Global_Footprint.pdf,* available May 11, 2009.

New York Times. "A Spotty Response to the Flu Threat," *NYT* website, *www.nytimes.com/2009/05/02/opinion/02sat1.html?_r=1&emc=eta1*, available May 3, 2009.

Osterholm, Michael T. "Preparing for the Next Pandemic," *New England Journal of Medicine*. 352: 1839–42, May 5, 2005.

Reuters. Bird Flu Pushed Back, Pandemic Threat Remains—UN, *Reuter's* website *http://uk.reuters.com/article/healthNews/idUKTRE49K8BY20081021,* available May 8, 2009.

Science in Africa—Africa's First On-Line Science Magazine website, available April 7, 2009.

Sulek, D., R. Cowell, and M. Delurey. *Mission Integration—A Whole of Government Strategy for a New Century,* Booz Allen Hamilton white paper, December 2008.

U.S. Department of Health and Human Services. *"Community Strategy for Pandemic Influenza Mitigation,"* February 2007, website *http://www.pandemicflu.gov/plan/community/commitigation.html#I*, available May 13, 2009.

U.S. Department of Health and Human Services (HHS). *The Great Pandemic in the United States,* available April 15, 2009.

U.S. Department of Health and Human Services (HHS) website. *PandemicFlu.gov*; available May 1, 2009.

U.S. Department of Health and Human Services (HHS) website. *PandemicFlu.gov/plan/pandplan.html* (Pandemic Assumptions, Section 1.1), available May 11, 2009.

U.S. Department of Health and Human Services (HHS) website. *PandemicFlu.gov, "Pandemic Terms Defined,"* available April 15, 2009.

U.S. Department of Homeland Security website. *http://www.dhs.gov/xprepresp/training/gc_1179350946764.shtm,* available May 11, 2009.

U.S. Department of Homeland Security website. *http://www.pandemicflu.gov/plan/pdf/cikrpandemicinfluenzaguide.pdf,* available May 11, 2009.

U.S. Department of Homeland Security. *National Response Plan*, December 2004.

U.S. Department of Homeland Security. *National Response Framework*, March 22, 2008.

U.S. Office of Personnel Management (OPM). *Memorandum for Heads of Executive Departments and Agencies—Advice to Federal Employees and Agencies on Preventing the Spread of the Current Flu and Maintaining Readiness to Use HR Flexibilities if Necessary,* April 26, 2009.

World Bank.*"Influenza A (H1N1): Questions and Answers,"* available April 30, 2009.

World Bank. *http://web.worldbank.org/WBSITE/EXTERNAL/NEWS/ontentMDK:20715087~menuPK:3325365~pagePK:34370~piPK:42770~theSitePK:4607,00.html*, available May 5, 2009.

World Health Organization website. *http://www.who.int/mediacentre/factsheets/fs211/en/index.html*, available May 10, 2009.

World Health Organization Interim Report, February 2006; website. *http://www.wpro.who.int/NR/rdonlyres/DA340E3E-D27E-47A6-9833-452E7AAC9ED5/0/EDTedDRAFT1ExerciseDevelopmentGuide.pdf*, available May 11, 2009.

World Health Organization (WHO) website.*"Epidemic and Pandemic Alert & Response,"* available April 7, 2009.

World Health Organization (WHO) website. *"Cumulative Number of Confirmed Human Cases of Avian Influenza A/(H5N1),"* Reported to WHO, 30 March 2009, available April 7, 2009

28

Future of Security Management

Edward J. Krueger

CHAPTER 28

OBJECTIVES

The objective of this chapter is to reflect on the changes in security management as a result of the need for businesses and corporations to be secured from threats to personnel, property, and proprietary information. The ever-present dangers from global and domestic terrorism, workplace violence, corporate espionage, and cyber threats are new issues for many firms. Although it is not possible for today's security administrator to be an expert in all things, it is the intent of this chapter to emphasize the critical role security has come to play today's corporate environment.

INTRODUCTION

Often major events (positive or negative) require an immediate change in the way business is conducted; this is also the case with security. Almost all aspects of the security industry have changed since September 11, 2001. The terrorist event that occurred on that day was the impetus for immediate changes in the U.S. government and in non-governmental organizations. The impact on the U.S. government was so severe that the Department of Homeland Security (DHS) was created in 2002 and now has more than 230,000 employees. *Global terrorism* has become a term known and frequently used inside the board rooms of companies doing international business. Some security issues that followed the events of September 11, 2001 have included piracy, kidnappings, assassinations, human trafficking, biological threats, border gangs, drug wars, cyber crimes, and domestic terrorism—all of which have become headline news. For the U.S. government, and for most U.S.-based companies, these were incidents that, historically, occurred only in other countries. For corporate security, the importance of issues such as internal theft, workplace violence, legal issues, security audits, and internal investigations seemed to be in competition with these new security incidents.

CURRENT TRENDS

As noted in Chapter 6, Defining a Security Management Program, "the past 20 years have seen more changes in how security is viewed by society and corporate management than in any other time in history." The time has passed when providing security was viewed as a watchman or guard function. Although such fundamental services are still needed in today's environment, it is expected that those in this position be well trained, highly educated, and very knowledgeable about security functions. The words *watchman* and *security guard* have been replaced by the more descriptive title of *security officer*. The security officer has become the private security equivalent of the police officer. The word *officer* has an official and authoritative tone and infers that the person is a knowledgeable representative of his employer.

Because of the corporate need to address the challenges of security threats to a company, corporate appreciation for the security manager position has changed. The trend, which was slowly emerging prior to September 11, saw companies creating the position of security manager. For many years, whether propriety or contract, corporate security services were managed by the human resources director or by a facility manager. This was primarily for organizational convenience. The reporting responsibility was often not in the best interest of the organization because the potential for a security breach was not clearly understood or fully realized by non-security management. Today, however, because of the critical need for security to be an important part of the corporate structure, most companies have created an organizational position for the security manager.

LOOKING AHEAD

As previously discussed in this book, threats to corporations have drawn attention to the need for a sophisticated style of security management that parallels other aspects of corporate administrative management. Today, companies employ trained security professionals, such as former military personnel, former law enforcement officers (e.g., secret service agents, FBI agents, or police executives), and others with corporate security experience and training to manage the security functions of the organization. It is becoming typical to see organizational titles such as director of security, manager of security, loss prevention manager, or vice president of security. It is likely that these individuals not only oversee security functions but also assume management functions that include budgeting and staffing, as well as strategic planning to avoid, reduce, or mitigate corporate threats. In other words, being a visionary is an expectation of the successful security manager. As such, security should be considered an organizational business function. Given the state of affairs today, many firms are organized with security at the upper management level, reporting directly to the CEO. For many firms, security has become an integral part of corporate life. The growth and the importance of this position are recognized by the fact that the American Society for Industrial Security International (ASIS), the most prominent security organization in the world, has more than 37,000 members worldwide.

EDUCATION AND CAREERS

Many colleges and universities now offer an associate's, a bachelor's, a master's, or a PhD degree in security-related areas. Courses of study in many degree programs include topics such as computer security, terrorism, security management, risk management, as well as homeland security. An excellent resource for academic programs is the ASIS website: www.asisonline.org/education/academicresources.xml. As outlined within this book, the responsibilities of today's security manager are many and varied. Education, training, and certification programs offer individuals an opportunity to seek a degree in their field of interest, or take courses to upgrade their skill levels. Many courses are offered in a distance-learning format for practitioners already working in the field.

It is important to note that today's security professional is not expected to be a content expert in all facets of security. Instead, one must be knowledgeable in all facets of security. There are numerous experts offering services on specific security topics to support the security manager (see Chapter 29, Security Resources). For the career-minded individual, the previously mentioned ASIS website offers a wealth of information.

The nature of the security field today has security professionals agreeing that they should be part of the corporate business environment. Several chapters in this book help support this thesis, exploring such topics as professional liability, security legal issues, personnel security issues, global terrorism, computer security, facility design, workplace violence, civil unrest, critical incident management, and organizational management, as well as an understanding of security audits and the physical aspects of target hardening (lights, locks, alarms, etc.). Additionally, the security manager must have the capacity to manage personnel and to effectively communicate. The ability to communicate appropriate concerns to upper management is critical. This education of management contributes to the support of the overall security program initiatives.

Effective communication also requires an understanding of risk assessment and risk management and associating this understanding with the predictability of a possible threat. Essentially, the informed security manager must be a futurist in that he must be able to measure potential threats against the corporation and effectively communicate needs to get the appropriate resources to eliminate or mitigate such threats.

THE FUTURE SECURITY MANAGER

Far from his ancestors in security, the new security manager must be able to protect a company's resources (people, property, and proprietary information) by confronting threats. Companies today must consider incorporating several disciplines such as fire protection, environmental issues, risk management, and insurance into one department. This is somewhat similar to the organized effort of the Department of Homeland Security, which brings together 22 federal agencies under its umbrella (see Figure 28–1). This concept operates under the philosophy that it is easier to manage several departments under one manager. This fusion center of security is designed to capture crime data for analysis from several sources that is used as the foundation for security decision making.

SUMMARY

Security management, today and in the future, should be an integral part of an organization's management team. As important as finance, operations, and sales are to a company's growth, an effective security management department can contribute to the company's overall bottom line. Crime, or fear of crime, can negatively impact a firm's stability. Today, potential threats to a firm's operations are numerous, and increasingly, it is the security manager who is relied upon to give the company stability during times of security crises. Supporting this effort is the external relationship a company establishes with public agencies such as local law enforcement, as discussed in Chapter 25, A Public–Private Partnership on Security. The security manger in an organization can play a unique and vital role in the public–private partnership for security. It is the well-informed and well-prepared security manager who is in the best position to protect his company's assets—people, property, and proprietary information.

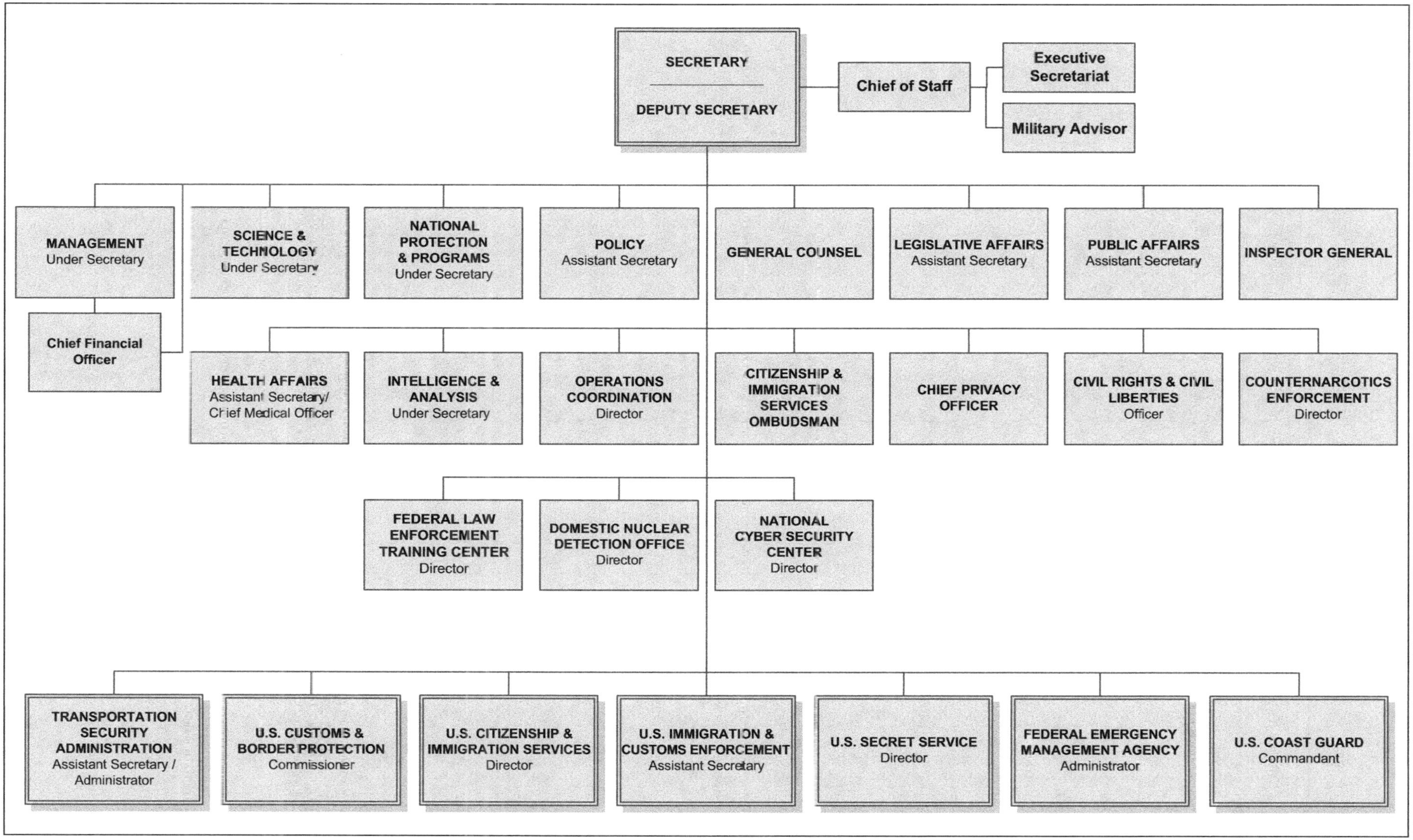

Figure 28–1. Organization Chart for the Department of Homeland Security

29

Security Resources

Mark C. Bach

CHAPTER 29

OBJECTIVE

With ever-changing technology, laws, standards, and practices, a security professional must maintain current ties to professional organizations and constantly review resources to stay abreast of the latest information.

This chapter provides a listing of some of the resources available to a security professional. Many can be utilized for free via the Internet, while some are offered with membership to a professional organization. Some publications have a yearly subscription cost or may be updated periodically for a fee. These resources provide a security manager with ways to stay current in the field and be able to research specific issues of concerns. The objective of this chapter is to provide a listing of some potential resources needed to manage a security mission.

SECURITY TRADE AND INTEREST ORGANIZATIONS

ACADEMY OF CRIMINAL JUSTICE SCIENCES (ACJS)
PO Box 960
Greenbelt, MD 20768-0960
(301) 446-6300 or (800) 757-2257;
Fax: (301) 446-2819
Officer: Mittie D. Sutherland, Executive Director
www.acjs.org

ACADEMY OF SECURITY EDUCATORS AND TRAINERS (ASET)
16 Penn Plaza, Suite 1570
New York, NY 10001
(212) 268-4555 or (800) 947-5827;
Fax: (212) 563-4783
Officer: Dr. Richard D. Kobetz, Director
www.personalprotection.com

AIR TRANSPORT ASSOCIATION OF AMERICA
1301 Pennsylvania Ave NW, Suite 1100
Washington, DC 20004
(202) 626-4000; Fax: (202) 626-4166
Officer: Mr. James C. May, President
www.airlines.org

AMERICAN ACADEMY OF FORENSIC SCIENCES
410 North 21st Street
Colorado Springs, CO 80904
(719) 636-1100; Fax: (719) 636-1993
Officer: Anne Warren, Executive Director
www.aafs.org

AMERICAN BANKERS ASSOCIATION (ABA)
1120 Connecticut Ave., NW
Washington, DC 20036
(202) 663-5000 or (800) 226-5377;
Fax: (202) 296-9258
Officer: Edward L. Yingling, President and CEO
www.aba.com

AMERICAN CORRECTIONAL ASSOCIATION (ACA)
206 N. Washington
Alexandria, VA 22314
(703) 224-0000 l; Fax: (703) 224-0179
Officer: James A. Gondles, Jr., Executive Director
www.aca.org

AMERICAN CRIMINAL JUSTICE ASSN (ACJA)
PO Box 601047
Sacramento, CA 95860
(916) 484-6553; Fax: (916) 488-2227
Officer: Karen K. Campbell, Executive Secretary
www.acjalae.org

AMERICAN FIRE SPRINKLER ASSOCIATION
12750 Merit Drive, Suite 350
Dallas, TX 75251
(214) 349-5965; Fax: (214) 343-8898
Officer: Steve Muncy, President
www.firesprinkler.org

AMERICAN HARDWARE MANUFACTURERS ASSOCIATION
801 N. Plaza Dr
Schaumburg, IL 60173-4977
(847) 605-1025; Fax: (847) 605-1030
Officer: Timothy S. Farrell,
President/Chief Executive Officer
www.ahma.org

AMERICAN INSTITUTE OF CERTIFIED PUBLIC ACCOUNTANTS (AICPA)
1211 Ave of the Americas
New York, NY 10036-8775
(212) 596-6200; Fax: (212) 596-6213
Officer: Barry C. Melancon, President
www.aicpa.org

AMERICAN LIGHTING ASSOCIATION
PO Box 420288
Dallas, TX 75342-0288
(214) 698-9898 or (800) 605-4448;
Fax: (214) 698-9899
Officer: Richard D. Upton
www.americanlightingassoc.com

AMERICAN MANAGEMENT ASSOCIATION (AMA)
135 West 50th Street
New York, NY 10020
(212) 586-8100; Fax: (212) 903-8168
Officer: Edward T. Reilly,
President/Chief Executive Officer
www.amanet.org

AMERICAN POLYGRAPH ASSOCIATION (APA)
PO Box 8037
Chattanooga, TN 37414-0037
(423) 892-3992 or (800) 272-8037;
Fax: (423) 894-5435
Officer: Robbie Bennett, Manager
www.polygraph.org

AMERICAN SECURITY EDUCATORS
PO Box 1337
Downey, CA 90240
(562) 928-1847; Fax: (562) 806-2461
Officer: James W. Gonos, President
www.americansecurityeducators.com

AMERICAN SOCIETY FOR AMUSEMENT PARK SECURITY & SAFETY (ASAPSS)
c/o Director of Security
Sea World Ontario
7007 Sea World Drive
Orlando, FL 32821
(407) 363-2161; Fax: (407) 354-6747
Officer: Paul Volkerson
www.asapss.com

AMERICAN SOCIETY FOR INDUSTRIAL SECURITY (ASIS)
1625 Prince Street
Arlington, VA 22314-2818
(703) 519-6200; Fax: (703) 519-6299
Officer: Michael J. Stack, President
www.asisonline.org

AMERICAN SOCIETY FOR TESTING AND MATERIALS INTERNATIONAL (ASTM)
100 Barr Harbor Dr.
West Conshohocken, PA 19428-2959
(610) 832-9500; Fax: (610) 832-9555
Officer: James A. Thomas, President
www.astm.org

AMERICAN SOCIETY FOR TRAINING AND DEVELOPMENT (ASTD)
1640 King St., Box 1443
Alexandria, VA 22313
(703) 683-8100; Fax: (703) 683-8103
Officer: Tony Bingham, President
www.astd.org

AMERICAN TRUCKING ASSOCIATION (ATA)
950 North Glebe Road, Suite 210
Arlington, VA 22203-4181
(703) 838-1700; Fax: (703) 684-5751
Officer: Bill Graves,
President/Chief Executive Officer
www.trucklinc.com

ARMED FORCES COMMUNICATIONS & ELECTRONICS ASSOCIATION (AFCEA)
4400 Fair Lakes Ct.
Fairfax, VA 22033-3899
(703) 631-6100 or (800) 336-4583;
Fax: (703) 631-4693
President: Kent Schneider
www.afcea.org

ASSOCIATED CREDIT BUREAUS
1090 Vermont Ave., NW, Suite 200
Washington, DC 20005-4905
(202) 371-0134; Fax: (202) 371-0134
Officer: D. Barry Connelly, Pres.
www.acb-credit.com

ASSOCIATED LOCKSMITHS OF AMERICA
3500 Easy St.
Dallas, TX 75247
(214) 819-9733 or (800) 532-2562;
Fax: (214) 827-1810
Officer: Hans Mejlshede, President
www.aloa.org

ASSOCIATION OF CERTIFIED FRAUD EXAMINERS
716 West Avenue
Austin, TX 78701-2727
(512) 478-9070 or (800) 245-3321;
Fax: (512) 478-9297
Officer: Joseph T. Wells, Chairman
www.acfe.org

ASSOCIATION OF FORMER AGENTS OF THE U.S. SECRET SERVICE
525 SW 5th St., Suite A
Des Moines, IA 50309-4501
(515) 282-8192; Fax (515) 282-9117
Officer: Kathy Rinkenberger, Executive Director
www.oldstar.org

ASSOCIATION OF FORMER INTELLIGENCE OFFICERS
6723 Whittier Ave., Suite 200
McLean, VA 22101-4533
(703) 790-0320; Fax (703) 991-1278
Officer: Elizabeth Bancroft, Executive Director
www.afio.org

AIT GLOBAL (AIT)
9 Byrd Court
Kings Park, NY 11754
(631) 921-8850; Fax: (516) 269-3881
Officer: Carol Lackey, Vice President
www.aitglobal.com

ASSOCIATION OF INFORMATION TECHNOLOGY PROFESSIONALS
401 N. Michigan Ave., Suite 2400
Chicago, IL 60611-4267
(312) 245-1070 or (800) 224-9371;
Fax: (312) 673-6659
Officer: Eric Hawkinson,
Association Executive Director
www.aitp.org

ASSOCIATION OF OLD CROWS
The AOC Bldg, 1000 N. Payne St., Suite 300
Alexandria, VA 22314-1696
(703) 549-1600; Fax: (703) 549-2589
Officer: Kermit Quick, President
www.crows.org

ASSOCIATION OF RECORDS MANAGERS & ADMINISTRATORS (ARMA)
13725 W. 109th Street, Suite 101
Lenexa, KS 66215
(913) 341-3808 or (800) 422-2762;
Fax: (913) 341-3742
Officer: Marilyn Bier, Executive Director
www.arma.org

ASSOCIATION OF STATE CORRECTIONAL ADMINISTRATORS
213 Court Street, Suite 606
Middletown, CT 06457
(860) 704-6410; FAX (860) 704-6420
Officer: George and Camille Camp,
Executive Directors
www.asca.net

AUTOMATIC IDENTIFICATION MANUFACTURERS NORTH AMERICA (AIM USA)
124 Warrendale-Bayne Road, Suite 100
Warrendale, PA 15086
(724) 934-5688; Fax: (724) 934-4495
Officer: Dan Mullen, President
www.aim-na.org

BANK ADMINISTRATION INSTITUTE (BAI)
115 S. LaSalle Street, Suite 3300
Chicago, IL 60603-3801
(312) 683-2464 or (888) 284-4078;
Fax: (312) 683-2373
Officer: Deborah Bianucci, President
www.bai.org

BUREAU OF JUSTICE STATISTICS
810 Seventh Street, NW
Washington, DC 20531
(202) 307-0765 or (800) 851-3420;
Fax: (301) 519-5212
www.ojp.usdoj.gov/bjs

BUSINESS ESPIONAGE CONTROLS AND COUNTERMEASURES ASSOCIATION (BECCA)
PO Box 55582
Shoreline, WA 98155-0582
(206) 364-4672; Fax: (206) 367-3316
Officer: William M. Johnson, Executive Director
www.becca-online.org

BUSINESS SOFTWARE ALLIANCE (BSA)
1150 18th Street NW, Suite 700
Washington, DC 20036
(202) 872-5500; Fax: (202) 872-5501
Officer: Robert W. Holleyman II, President
www.bsa.org

CENTRAL STATION ALARM ASSOCIATION (CSAA)
8150 Leesburg Pike, Suite 700
Vienna, VA 22182
(703) 242-4670; Fax: (703) 242-4675
Officer: Stephen Paul Doyle,
Executive Vice President
www.csaaul.org

COMMUNICATIONS FRAUD CONTROL ASSOCIATION (CFCA)
4 Becker Farm Road, 4th Floor
Roseland, NJ 07068
(973) 871-4032; Fax: (973) 871-4075
Officer: Roberta Aronoff, Executive Director
www.cfca.org

COMPUTER & COMMUNICATIONS INDUSTRY ASSOCIATION
900 17th Street, NW, Suite 1100
Washington, DC 20006
(202) 783-0070; Fax: (202) 783-0534
Officer: Edward Block, President
www.ccianet.org

COMPUTER SECURITY INSTITUTE
600 Harrison St.
San Francisco, CA 94107
(415) 947-6769 or (866) 535-9002;
Fax: (415) 947-6023
Officer: Robert Richardson, Director
www.gocsi.com

COUNCIL OF INTERNATIONAL INVESTIGATORS (CII)
2150 N. 107th St., Suite 205
Seattle, WA 98133-9009
(206) 361-8869 or (888) 759-8884;
Fax: (206) 367-8777
Officer: Nancy Barber, President
www.cii2.org

DOOR AND HARDWARE INSTITUTE
14150 Newbrook Dr.
Chantilly, VA 20151
(703) 222-2010; Fax: (703) 222-2410
Officer: Jerry Heppes, Chief Executive Officer
www.dhi.org

ELECTRONIC FUNDS TRANSFER ASSOCIATION (EFTA)
11350 Random Hills Road, Suite 800
Fairfax, VA 22030
(703) 934-6052; Fax: (703) 934-6058
Officer: Kurt Helwig, President
www.efta.org

ELECTRONIC INDUSTRIES ALLIANCE (EIA)
2500 Wilson Blvd.
Arlington, VA 22201
(703) 907-7500; Fax: (703) 907-7501
Officer: Matthew Flanigan, Interim President
www.eia.org

ETHICS RESOURCE CENTER, INC.
2345 Crystal Drive, Suite 201
Arlington, VA 22202
(703) 647-2185; Fax: (703) 647-2180
Officer: Patricia J. Harned, President
www.ethics.org

FEDERAL EMERGENCY MANAGEMENT AGENCY (FEMA)
500 C Street SW
Washington, DC 20472
(202) 646-2500 or (800) 621-3362;
Fax: (800) 827-8112
Officer: Craig Fugate, Administrator
www.fema.gov

FINANCIAL & SECURITY PRODUCTS ASSOCIATION
5300 Sequoia Road, NW, Suite 205
Albuquerque, NM 87120
(505) 839-7958 or (800) 843-6082;
Fax: (505) 839-0017
Officer: John and Liz Vrabec, Managers
www.fspa1.com

FIRE EQUIPMENT MANUFACTURERS ASSOCIATION
1300 Sumner Ave.
Cleveland, OH 44115-2851
(216) 241-7333; Fax: (216) 241-0105
Officer: Jim Widmer, President
www.femalifesafety.org

FIRE SUPPRESSION SYSTEMS ASSOCIATION
5024-R Campbell Blvd.
Baltimore, MD 21236-5974
(410) 931-8100; Fax: (410) 931-8111
Officer: Crista LeGrand, Executive Director
www.fssa.net

INDEPENDENT ARMORED CAR OPERATORS ASSOCIATION (IACOA)
8000 Research Forest Dr., Suite 115-155
The Woodlands, TX 77382
(281) 292-8208
Officer: John Margaritis, Administrator
www.iacoa.com

INFORMATION SYSTEMS AUDIT & CONTROL ASSOCIATION
3701 Algonquin Road, Suite 1010
Rolling Meadows, IL 60008
(847) 253-1545; Fax: (847) 253-1443
Officer: Lynn Lawton, President
www.isaca.org

INFORMATION TECHNOLOGY INDUSTRY COUNCIL (ITI)
1101 K Street, NW, Suite 610
Washington, DC 20005
(202) 737-8888; Fax: (202) 638-4922
Officer: Dean C. Garfield, President
www.itic.org

INSTITUTE OF ELECTRICAL & ELECTRONICS ENGINEERS (IEEE)
3 Park Avenue, 17th Floor
New York, NY 10016-5997
(212) 419-7900; Fax: (212) 752-4929
Officer: Dr. James Prendergast, Executive Director
www.ieee.org

INSTITUTE OF INTERNAL AUDITORS
247 Maitland Ave.
Altamonte Springs, FL 32701-4201
(407) 937-1100; Fax: (407) 937-1101
Officer: Richard F. Chambers, President and CEO
www.theiia.org

INSTITUTE OF NUCLEAR MATERIALS MANAGEMENT
111 Deer Lake Road, Suite 100
Deerfield, IL 60015
(847) 480-9573; Fax: (847) 480-9282
Officer: Leah McCrackin, Executive Director
www.inmm.org

INTERNATIONAL ASSOCIATION FOR COMPUTER SYSTEMS SECURITY (IACSS)
6 Swarthmore Lane
Dix Hills, NY 11746
(631) 499-1616; Fax: (631) 462-9178
Officer: Robert J. Wilk, President
www.iacss.com

INTERNATIONAL ASSOCIATION FOR HEALTHCARE SECURITY AND SAFETY (IAHSS)
PO Box 5038
Glendale Heights, IL 60139
(630) 529-3913 or (888) 353-0990;
Fax: (630) 529-4139
Officer: Evelyn F. Meserve, President
www.iahss.org

INTERNATIONAL ASSOCIATION OF CAMPUS LAW ENFORCEMENT ADMINISTRATORS (IACLEA)
342 N. Main Street
West Hartford, CT 06117-2507
(860) 586-7517; Fax: (860) 586-7550
Officer: Peter Berry, Chief Staff Officer
www.iaclea.org

INTERNATIONAL ASSOCIATION OF CHIEFS OF POLICE (IACP)
515 N Washington St.
Alexandria, VA 22314
(703) 836-6767 or (800) 843-4227;
Fax: (703) 836-4543
Officer: Daniel N. Rosenblatt, Executive Director
www.theiacp.org

INTERNATIONAL ASSOCIATION OF FINANCIAL CRIME INVESTIGATORS (IAFCI)
1020 Suncast Lane, Suite 102
El Dorado Hills, CA 95762
(916) 939-5000; Fax: (916) 939-0395
Officer: Janis Moffett, Executive Director
www.iafci.org

INTERNATIONAL ASSOCIATION OF EMERGENCY MANAGERS
201 Park Washington Court
Falls Church, VA 22046-4527
(703) 538-1795; Fax: (703) 241-5603
Officer: Elizabeth Armstrong, Chief Executive Officer
www.iaem.com

INTERNATIONAL ASSOCIATION OF LAW ENFORCMENT PLANNERS (IALEP)
Box 11437
Torrance, CA 90510-1437
(301) 225-5148
Officer: Denise Thurston, Manager
www.ialep.org

INTERNATIONAL ASSOCIATION OF PROFESSIONAL SECURITY CONSULTANTS (IAPSC)
525 SW 5th St., Suite A
Des Moines, IA 50309-4501
(515) 282-8192; Fax: (515) 282-9117
Officer: Kathy Rinkenberger, Executive Director
www.iapsc.org

THE INTERNATIONAL FOUNDATION FOR PROTECTION OFFICERS
1250 Tamiami Trail, Suite 305
Naples, FL 34102
(239) 430-0534; Fax: (239) 430-0533
Officer: Sandi Davies, Executive Director
www.ifpo.com

INTERNATIONAL GRAPHOANALYSIS SOCIETY
842 Fifth Avenue
New Kensington, PA 15068
(724) 472-9701; Fax: (509) 271-1149
Officer: Greg Greco, President
www.igas.com

INTERNATIONAL NARCOTIC ENFORCEMENT OFFICERS ASSOCIATION
112 State St., Suite 1200
Albany, NY 12207-2079
(518) 463-6232; Fax: (518) 432-3378
Officer: John J. Ryan, Executive Director
www.ineoa.org

INTERNATIONAL PERSONNEL MANAGEMENT ASSOCIATION FOR HUMAN RESOURCES
1617 Duke St.
Alexandria, VA 22314
(703) 549-7100; Fax: (703) 684-0948
Officer: Neil Reichenberg, Executive Director
www.ipma-hr.org

INTERNATIONAL SECURITY MANAGEMENT ASSOCIATION (ISMA)
PO Box 623
Buffalo, IA 52728
(563) 381-4008; Fax: (563) 381-4283
Officer: Susan W. Pohlmann,
Consulting Business Manager
www.isma.com

INTERNATIONAL SOCIETY OF CRIME PREVENTION PRACTICIONERS
Box 476
Simpsonville, SC 29681
(864) 884-8466
Officer: Richard Cannady, Executive Director
www.iscpp.org

INTERNATIONAL SOCIETY OF FIRE SERVICE INSTRUCTORS
2425 Highway 49 East
Pleasant View, TX 37146
(800) 435-0005; Fax: (615) 746-1170
Officer: Eddie Buchanan, President
www.isfsi.org

ISCA LABS
1000 Bent Creek Blvd, Suite 200
Mechanicsburg, PA 17050
(717) 790-8100; Fax: (717) 790-8170
Officer: George Japak, Managing Director
www.icsalabs.com

JEWELERS SECURITY ALLIANCE OF UNITED STATES
6 E 45th St.
New York, NY 10017
(212) 687-0328 or (800) 537-006;
Fax: (212) 808-9168
Officer: John J. Kennedy
www.jewlerssecurity.org

JUSTICE RESEARCH & STATISTICS ASSOCIATION
4777 N Capitol St., NE, Suite 801
Washington, DC 20002
(202) 842-9330; Fax: (202) 842-9329
Officer: Joan C. Weiss, Executive Director
www.jrsainfo.org

NATIONAL ALARM ASSOCIATION OF AMERICA
Box 3409
Dayton, OH 45401
(800) 283-6285; Fax (937) 866-5098
Officer: Gene Riddlebaugh, President
www.naaa.org

NATIONAL ARMORED CAR ASSOCIATION, INC (NACA)
9532 Stevebrook Road
Fairfax, VA 22032
(703) 426-1976; Fax: (703) 666-9085
Officer: Larry Sabbath, Executive Director
www.nationalarmoredcar.com

NATIONAL ASSOCIATION OF FLEET ADMINISTRATORS
125 Village Blvd., Suite 200
Princeton Forrestal Village
Princeton, NJ 08540
(609) 720-0882; Fax (609) 452-8004
www.nafa.org

NATIONAL ASSOCIATION OF LEGAL INVESTIGATORS (NALI)
235 N. Pine Street
Lansing, MI 48933
(517) 702-9835 or (866) 520-6254;
Fax (517) 372-1501
Officer: Julian Vail, Association Manager
www.nalionline.org

NATIONAL ASSOCIATION OF SECURITY COMPANIES (NASCO)
444 N. Capitol Street, NW, Suite 345
Washington, DC 20001
(202) 347-3257; Fax: (202) 393-7006
Officer: Jeff Flint, Executive Director
www.nasco.org

NATIONAL ASSOCIATION OF SHOPLIFTING PREVENTION (NASP)
380 N. Broadway, Suite 306
Jericho, NY 11753
(516) 932-0165 or (800) 848-9595;
Fax: (516) 932-9393
www.shopligtingprevention.org

NATIONAL ASSOCIATION OF SURETY BOND PRODUCERS
1828 L Street, NW, Suite 720
Washington, DC 20036-5104
(202) 686-3700; Fax: (202) 686-3656
Officer: Richard Foss, Executive Vice President
www.nasbp.org

NATIONAL BURGLAR AND FIRE ALARM ASSOCIATION (NBFAA)
2300 Valley View Lane, Suite 230
Irving, TX 75062
(214) 260-5970 or (888) 447-1689;
Fax: (214) 260 5979
Officer: Merlin Guilbeau, Executive Director
www.alarm.org

NATIONAL CABLE & TELECOMMUNICATIONS ASSOCIATION
25 Massachusetts Ave. NW, Suite 100
Washington, DC 20001
(202) 222-2300; Fax: (202) 222-2514
Officer: Kyle McSlarrow, President
www.ncta.com

NATIONAL CONSUMERS LEAGUE
1701 K St., Suite 1200
Washington, DC 20006
(202) 835-3323; Fax: (202) 835-0747
Officer: Sally Greenberg, Executive Director
www.nclnet.org

NATIONAL COUNCIL OF INVESTIGATION & SECURITY SERVICES (NCISS)
7501 Sparrows Point Blvd.
Baltimore, MD 21219-1927
(800) 445-8408; Fax: (410) 388-9746
Officer: Carolyn Ward, Executive Director
www.nciss.com

NATIONAL CRIME PREVENTION COUNCIL
2345 Crystal Drive, Suite 500
Arlington, VA 22202
(202) 466-6272; FAX: (202) 296-1356
Contact: Ann Harkins, President and CEO
www.ncpc.org

NATIONAL CRIME PREVENTION INSTITUTE (NCPI)
University of Louisville
Louisville, KY 40292
(502) 852-5577 or (800) 334-8635;
Fax: (502) 852-6990
Officer: Marianna A. Perry, Director
www.louisville.edu/ncpi

NATIONAL DEFENSE INDUSTRIAL ASSOCIATION
2111 Wilson Blvd, Suite 400
Arlington, VA 22201
(703) 522-1820; Fax: (703) 522-1885
President, Lawrence P. Farrell
www.ndia.org

NATIONAL FIRE PROTECTION ASSOCIATION (NFPA)
1 Batterymarch Park
Quincy, MA 02169-7471
(617) 770-3000; Fax: (617) 770-0700
Officer: James M. Shannon, President/CEO
www.nfpa.org

NATIONAL FIRE SPRINKLER ASSOCIATION
40 Jon Barrett Road
Patterson, NY 12563
(914) 878-4200; Fax: (914) 878-4215
Officer: John A. Viniello, President
www.nfsa.org

NATIONAL INSTITUTE OF STANDARDS & TECHNOLOGY
100 Bureau Drive, Stop 1070
Gaithersburg, MD 20899-1070
(301) 975-6478; Fax: (301) 948-0279
Officer: Patrick Gallagher, Deputy Director
www.nist.gov

NATIONAL FORENSIC CENTER
Box 270529
San Diego, CA 92198-2529
(800) 735-6660; Fax: (858) 487-7747
Officer: Mark Haslam, Director
www.national-experts.com

NATIONAL INSURANCE CRIME BUREAU
1111 E. Touhy Ave., Suite 400
Des Plaines, IL 60018
(847) 544-7000 or (800) 447-6282;
Fax: (708) 430-2446
Officer: Joseph H. Wehrle, Jr., President
www.nicb.org

NATIONAL RETAIL FEDERATION
325 7TH Street, NW, Suite 1100
Washington, DC 20004
(202) 783-7971 or (800) 673-4692;
Fax: (202) 737-2849
Officer: Tracy Mullin, President
www.nrf.com

NATIONAL SAFETY COUNCIL
1121 Spring Lake Dr.
Itasca, IL 60143-3201
(630) 285-1121; Fax: (630) 285-1315
Officer: Janet Froetscher, President
www.nsc.org

NATIONAL DEFENSE INDUSTRIAL ASSOCIATION
2111 Wilson Blvd., Suite 400
Arlington, VA 22201
(703) 522-1820; Fax: (703) 522-1885
Officer: Lawrence P. Farrell, President & CEO
www.ndia.org

RETAIL INDUSTRY LEADERS ASSOCIATION (RILA)
1700 N. Moore St., Suite 2250
Arlington, VA 22209
(703) 841-2300; Fax: (703) 841-1184
Officer: Sandy Kennedy, President
www.rila.org

RISK & INSURANCE MANAGEMENT SOCIETY (RIMS)
1065 Avenue of the Americas, 13th Floor
New York, NY 10018
(212) 286-9292; Fax: (212) 922-9716
Officer: Mary Roth, Executive Director
www.rims.org

SECURITY INDUSTRY ASSOCIATION (SIA)
635 Slaters Lane, Suite 110
Alexandria, VA 22314
(703) 683-2075 or (866) 817-8888;
Fax: (703) 683-2469
Officer: Richard Chase, CEO
www.siaonline.org

SOCIETY FOR HUMAN RESOURCE MANAGEMENT
1800 Duke Street
Alexandria, VA 22314
(703) 548-3440 or (800) 283-7476;
Fax: (703) 836-0367
Officer: Laurence O'Neil, President & CEO
www.shrm.org

SOCIETY OF COMPETITIVE INTELLIGENCE PROFESSIONALS (SCIP)
1700 Diagonal Rd., Suite 600
Alexandria, VA 22314
(703) 739-0696; Fax: (703) 739-2524
Officer: Ken Garrison, Chief Executive Officer
www.scip.org

SOCIETY OF FIRE PROTECTION ENGINEERS
7315 Wisconsin Ave., Suite 620 E
Bethesda, MD 20814
(301) 718-2910; Fax: (301) 718-2242
Officer: Allan Freedman, Executive Director
www.sfpe.org

TECHAMERICA
5201 Great America Pkwy, Suite 400
Santa Clara, CA 95054
(408) 987-4200 or (800) 284-4232;
Fax: (408) 987-4298
Officer: Christopher Hansen, President
www.aeanet.org

SECURITY PERIODICALS

ANNUALS/BIENNIALS

BEST'S LOSS CONTROL MANUAL
AM Best Company
Ambest Road
Oldwick, NJ 08858
(908) 439-2200; Fax: (908) 439-3296

INSURANCE FACT BOOK
Insurance Information Institute
110 William St., 24th Floor
New York, NY 10038
(212) 346-5500 or (800) 331-9146
www.iii.org

LEGAL NEWSLETTERS IN PRINT
Infosources Publishing
140 Norma Rd.
Teaneck, NJ 07666
(201) 836-7072; Fax (201) 836-9591
www.infosourcepub.com

NATIONAL DIRECTORY OF EXPERT WITNESSES
NATIONAL FORENSIC CENTER
Box 270529
San Diego, CA 92198-2529
(800) 735-6660; Fax: (858) 487-7747
Officer: Mark Haslam, Director
www.national-experts.com

NATIONAL DIRECTORY OF LAW ENFORCEMENT ADMINISTRATORS
National Public Safety Information Bureau
PO Box 365
Stevens Point, WI 54481
(800) 647-7579; Fax (715) 345-7288
www.safetysource.com

JOURNALS

Some of these publications are offered as part of an association membership or can be purchased as a subscription.

BUSINESS INSURANCE
Crain Communications
711 Third Ave.
New York, NY 10017-4036
(212) 210-0425 or (888) 446-1422
Editor: Regis Coccia
Weekly

CAMPUS LAW ENFORCEMENT JOURNAL
International Association of Campus Law Enforcement Administrators
342 N. Main Street
West Hartford, CT 06117-2507
(860) 586-7517; Fax: (860) 586-7550
Officer: Peter Berry, Chief Staff Officer
www.iaclca.org

CONSTRUCTION RISK MANAGEMENT
International Risk Management Institute, Inc
12222 Merit Dr., Suite 1450
Dallas, TX 75251-2276
(972) 960-7693; Fax: (972) 371-5120
www.irmi.com
Editors: Ann R. Hickman and Jack P. Gibson
Updated Quarterly

CRYPTOLOGIA
Taylor and Francis
325 Chestnut Street, Suite 800
Philadelphia, PA 19106
(800) 354-1420
www.informaworld.com
Editor: Craig Bauer
Subscription: Quarterly

DISASTER RECOVERY JOURNAL
1862 Old Lemay Ferry
Arnold, MO 63010
(636) 282-580; Fax (636) 282-5802
www.drj.com
Editor: Jon Seals
Subscription: Quarterly

FBI LAW ENFORCEMENT BULLETIN
Federal Bureau of Investigation
936 Pennsylvania Ave., NW
Washington, DC 20535-0001
(202) 512-1800; Fax: (202) 512-2250
Subscription: Monthly
Professional journal for civilian and law enforcement community

THE LEGAL INVESTIGATOR
NATIONAL ASSOCIATION OF LEGAL INVESTIGATORS (NALI)
235 N. Pine Street
Lansing, MI 48933
(517) 702-9835 or (866) 520-6254;
Fax (517) 372-1501
Officer: Julian Vail, Association Manager
www.nalionline.org
Subscription: Six times a year

LIBRARY & ARCHIVAL SECURITY
Taylor and Francis
325 Chestnut Street, Suite 800
Philadelphia, PA 19106
(800) 354-1420
www.informaworld.com
Editor: Christopher Brown-Syed
Subscription: Two issues per year

MILITARY POLICE
464 Manscen Loop, Suite 2662
Fort Leonard Wood, MO 65473-8926
(573) 563-4137
Biannual

SECURITY TECHNOLOGY EXECUTIVE
12735 Morris Road
Deerfield Point, Building 200, Suite 180
Alpharetta, GA 30004
(800) 547-7377
www.securityinfowatch.com
Editor: Steve Lasky
Subscription: Monthly

NEWSLETTERS/NEWS SERVICES

AMERICAN POLYGRAPH ASSOCIATION NEWSLETTER
POLYGRAPH JOURNAL QUARTERLY-Polygraph
AMERICAN POLYGRAPH ASSOCIATION (APA)
PO Box 8037
Chattanooga, TN 37414-0037
(423) 892-3992 or (800) 272-8037;
Fax: (423) 894-5435
Officer: Robbie Bennett, Manager
www.polygraph.org
Subscription: Quarterly

CARD TECHNOLOGY TODAY
Elsevier Ltd.
PO Box 150
Kidlington, 0X5 1AS UK
+44 (0) 1865 843000; Fax: +44 (0) 1865 843971
Subscription: Ten times a year
Editor: Mark Lockie
www.elsevierscitech.com

CKC REPORT
The Hotel Technology Newsletter
Chervenak, Keane & Co
307 E. 44th Street
New York, NY 10017
(212) 966-8230; Fax: (212) 983-5275
Publisher: Larry Chervenak
Subscription: 10 times a year

CODES AND STANDARDS
Kelly P. Reynolds & Assoc, Inc
337 Shore Dr.
Burr Ridge, IL 60527-5821
(800) 950-2633; Fax: (866) 814-2633
www.codexperts.net
Editor: Kelly Reynolds
Subscription: Monthly

COMPUTER SECURITY ALERT
COMPUTER SECURITY INSTITUTE
600 Harrison St.
San Francisco, CA 94107
(415) 947-6769 or (866) 535-9002;
Fax: (415) 947-6023
Officer: Robert Richardson, Director
www.gocsi.com
Subscription: Part of membership services

EDPACS
The EDP Audit Control and Security Newsletter
Taylor and Francis
325 Chestnut Street, Suite 800
Philadelphia, PA 19106
(800) 354-1420
www.informaworld.com
Subscription: Monthly

THE JOURNAL OF POLYGRAPH SCIENCE
National Training Center of Polygraph Science
200 W. 57th St., Suite 1400
New York, NY 10019
(212) 755-5241
Editor: Richard O. Arther
Subscription: Bimonthly

LIPMAN REPORT
Guardsmark, Inc
22 S. Second St
Memphis, TN 38103-2695
(901) 522-6000; Fax: (901) 522-6013
Editor: Ira A. Lipman
Subscription: Monthly

NATIONAL CRIMINAL JUSTICE REFERENCE SERVICE
Box 6000
Rockville, MD 28049-6000
(301) 519-5500 or (800) 851-3420;
Fax: (301) 519-5212
www.ncjrs.gov

PRATT'S BANK SECURITY REPORT
A. S. Pratt & Sons
807 Las Climas Pkwy., Suite 300
Austin, TX 78746-4223
(512) 472-2244; Fax: (512) 305-6575
www.aspratt.com
Editor: Jo Taylor
Subscription: Monthly

THE RISK REPORT
International Risk Management Institute, Inc
12222 Merit Dr., Suite 1450
Dallas, TX 75251-2276
(972) 960-7693; Fax: (972) 371-5120
www.irmi.com
Editor: Jack P. Gibson and Bonnie Rogers
Subscription: Monthly

SECURITY INDUSTRY ASSOCIATION (SIA) QUARTERLY RESEARCH UPDATE
635 Slaters Lane, Suite 110
Alexandria, VA 22314
(703) 683-2075 or (866) 817-8888;
Fax: (703) 683-2469
Officer: Richard Chase, CEO
www.siaonline.org
Monthly for the members of SIA

SECURITY LAW NEWSLETTER
Strafford Publications, Inc
590 Dutch Valley Rd.
Atlanta, GA 30324
(404) 881-1141 or (800) 926-7926;
Fax: (404) 881-0074
Editor: Jennifer Vaughan
Subscription: 10 issues

TRADE PUBLICATIONS

LAW AND ORDER
Hendon, Inc.
130 N. Waukegan Rd., #202
Deerfield, IL 60015-5652
(800) 843-9764; Fax: (847) 444-3333
Editor: Ed Sanow
Subscription: Monthly
www.hendonpub.com

LAW ENFORCEMENT PRODUCT NEWS
Cygnus International
1233 Janesville Avenue
Fort Atkinson, WI 53538
(800) 547-7377; Fax (920) 563-1702
Publisher: Michael L. George
www.law-enforcement.com
Subscription: Seven Times a year

LAW ENFORCEMENT TECHNOLOGY
Cygnus International
1233 Janesville Avenue
Fort Atkinson, WI 53538
(920) 563-6388 or (800) 547-7377;
Fax (920) 563-1702
Contributing Editor: Ronnie Garrett
www.letonline.com
Subscription : Monthly

LOCKSMITH LEDGER INTERNATIONAL
Cyngnus International
3030 Salt Creek Lane, Suite 200
Arlington Heights, IL 6005
(847) 454-2702; Fax: (847) 454-2759
Editor in Chief: Gale Johnson
Subscription: 15 issues

THE NATIONAL LOCKSMITH (monthly)
THE INSTITUTIONAL LOCKSMITH & SECURITY (quarterly)
National Publishing Company
1533 Burgundy Pkwy.
Streamwood, IL 60107
(630) 837-2044; Fax: (630) 837-1210
Publisher: Marc Goldberg
www.thenationallocksmith.com

POLICE/THE LAW OFFICER MAGAZINE
Elsevier Public Safety
525 B Street
San Diego, CA 92101
(800) 266-5367; Fax: (619) 699-6396
Editor in Chief: Dale Stockton
Subscription: Monthly
www.lawofficer.com

POLICE & SECURITY NEWS
Days Communications, Inc.
1208 Juniper Street
Quakertown, PA 18951-1208
(215) 538-1240; Fax (215) 538-1208
Publisher: David A. Yaw
Subscription: Bimonthly
www.policeandsecuritynew.com

SECURITY DEALER & INTEGRATOR
Cygnus Publishing
3 Huntington Quadrangle, Suite 301N
Melville, NY 11747
(516) 845-2700 or (800) 547-7377;
Fax: (516) 845-7109
Publisher: Carol Enman
www.securityinfowatch.com
Subscription: Monthly

SECURITY DISTRIBUTING & MARKETING (SDM)
BNP Media
1050 IL Route 83, Suite 200
Bensenville, IL 60106
(630) 694-4027; Fax (630) 227-0214
Editor: Laura Stepanek
ww.sdmmag.com
Subscription: Monthly

SECURITY MANAGEMENT
AMERICAN SOCIETY FOR INDUSTRIAL SECURITY (ASIS)
1625 Prince Street
Arlington, VA 22314-2818
(703) 519-6200 or (800) 368-5685;
Fax: (703) 518-1518
www.securitymanagement.com
Editor in Chief: Sherry Harowitz
Subscription: Monthly

SECURITY SYSTEM NEWS
United Publications
106 Lafayette Street
Yarmouth, ME 04096
(207) 846-0600; Fax (207) 846-0657
Editor: L. Samuel Pfeifle
www.securitysystemnews.com
Subscription: Monthly

SECURITY SALES & INTEGRATION
Bobit Publishing Co
3520 Challenger Street
Torrance, CA 90503
(310) 533-2400; Fax: (310) 533-2502
Editor in Chief: Scott Goldfine
www.securitysales.com
Subscription: Monthly

SECURITY TECHNOLOGY EXECUTIVE
Cygnus Publishing
12735 Morris Road
Deerfield Point 200, Suite180
Alpharetta, GA 30004
(800) 547-7377; Fax (404) 601-5120
www.securityinfowatch.com
Publisher/Editor: Steven Lasky
Subscription: Monthly

SOUND & VIDEO CONTRACTOR
Penton Media
249 W. 17th Street
New York, NY 10011
(818) 236-3667; Fax: (913) 967-1905
Editorial Director: Cynthia Wisehart
www.svconline.com
Subscription: Monthly

STORES
National Retail Federation
325 7th Street, NW, Suite 1100
Washington, DC 20004
(2020) 626-8101; Fax (202) 661-3042
Editor: Susan Reda
www.stores.com
Subscription: Monthly

SECURITY LITERATURE/BOOKS

HIGHWAY LOSS REDUCTION STATUS REPORT
Insurance Institute for Highway Safety
1005 N Glebe Rd.
Arlington, VA 22201
(703) 247-1500; Fax (703) 247-1595
www.iihs.org
Subscription: Periodically

Charles C. Thomas Publishers
2600 South First Street
Springfield, IL 62704
(217) 789-8980 or (800) 258-8980;
Fax: (217) 789-9130
Editor in Chief: Michael Payne Thomas

TRAINING PROGRAMS AND RESOURCES

AMERICAN MANAGEMENT ASSOCIATION (AMA)
135 West 50th Street
New York, NY 10020
(212) 586-8100; Fax: (212) 903-8168
Officer: Edward T. Reilly,
President/Chief Executive Officer
www.amanet.org

AMERICAN SECURITY EDUCATORS
PO Box 1337
Downey, CA 90240
(562) 928-1847; Fax: (562) 869-6906
Contact: James W. Gonos, President
www.americansecurityeducators.com

AMERICAN SOCIETY FOR INDUSTRIAL SECURITY (ASIS)
1625 Prince Street
Arlington, VA 22314-2818
(703) 519-6200; Fax: (703) 519-6299
Officer: Michael J. Stack, President
www.asisonline.org

ANSUL FIRE SCHOOL
One Stanton St.
Marinette, WI 54143
(715) 735-7411; Fax: (715) 732-7973
Contact: Eric Calderaro
www.ansul.com

BANK ADMINISTRATION INSTITUTE (BAI)
115 S. LaSalle Street, Suite 3300
Chicago, IL 60603-3801
(312) 683-2464 or (888) 284-4078;
Fax: (312) 683-2373
Officer: Deborah Bianucci, President
www.bai.org

COMPUTER SECURITY INSTITUTE
600 Harrison St.
San Francisco, CA 94107
(415) 947-6769 or (866) 535-9002;
Fax: (415) 947-6023
Officer: Robert Richardson, Director
www.gocsi.com

CRIMINAL JUSTICE CENTER
John Jay College of Criminal Justice
555 West 57th Street, Room 603BMW
New York, NY 10019-1029
(212) 237-8638; Fax: (212) 237-8637
www.jjay.cuny.edu
See also: Security Management Institute
Officer: Herbert A. Johnson, Director

CRISIS PREVENTION INSTITUTE
3315-K N 124th St.
Brookfield, WI 53005
(262) 783-5787 or (800) 558-8976;
Fax: (262) 783-5906
Contact: Robert Rettmann
www.crisisprevention.com

DIEBOLD, INC.
5995 Mayfair Road
North Canton, OH 44720
(330) 490-4000 or (800) 999-3600;
Fax: (330) 490-3794
Contact: Thomas Swidarski, President and CEO
www.diebold.com

EXECUTIVE PROTECTION INSTITUTE
16 Penn Plaza, Suite 1570
New York, NY 10001
(212) 268-4555 or (800) 947-5827;
Fax: (212) 563-4783
Contact: Richard W. Kobetz, Executive Director

FEDERAL EMERGENCY MANAGEMENT AGENCY (FEMA)
500 C St., SW
Washington, DC 20472
(202) 646-2500 or (800) 621-3362;
Fax: (800) 827-8112

INSTITUTE OF PUBLIC SAFETY
BROWARD COLLEGE
3501 Davie Rd
Building #21 and #22
Ft Lauderdale, FL 33314
(954) 201-6801; Fax: (954) 201-6305
Contact: Ken Shives

INTERNATIONAL ASSOCIATION FOR HEALTHCARE SECURITY & SAFETY (IAHSS)
PO Box 5038
Glendale Heights, IL 60139
(630) 529-3913 or (888) 353-0990;
Fax: (630) 529-4139
Contact: Evelyn F. Meserve, Executive Director
www.iahss.org

INTERNATIONAL ASSOCIATION OF CHIEFS OF POLICE (IACP)
515 N. Washington St.
Alexandria, VA 22314
(703) 836-6767 or (800) 843-4227;
Fax: (703) 836-4543
Officer: Daniel N. Rosenblatt, Executive Director
www.theiacp.org

INVESTIGATION TRAINING INSTITUTE
PO Box 770579
Orlando, FL 32877-0579
(407) 816-7273; Fax: (407) 816-7232
Contact: Edward R. Burke
www.investigationtraining.com

JARVIS INTELLIGENCE SOLUTIONS
11720 E. 21st St.
Tulsa, OK 74129-1824
(918) 437-1100; Fax: (918) 437-1191
Contact: Ray Jarvis, President
www.jarvisinternational.com

JUSTICE RESEARCH & STATISTICS ASSOCIATION
4777 N. Capitol St. NE, Suite 801
Washington, DC 20002
(202) 842-9330; Fax: (202) 842-9329
Officer: Joan C. Weiss, Executive Director
www.jrsainfo.org

MIS TRAINING INSTITUTE, INC.
498 Concord St.
Framingham, MA 01701-2357
(508) 879-7999; Fax: (508) 872-1153
Contact: Linda Burton, President and CEO
www.misti.com

NATIONAL BURGLAR AND FIRE ALARM ASSOCIATION (NBFAA)
2300 Valley View Lane, Suite 230
Irving, TX 75062
(214) 260-5970 or (888) 447-1689;
Fax: (214) 260-5979
Officer: Merlin Guilbeau, Executive Director
www.alarm.org

NATIONAL CRIME PREVENTION COUNCIL
2345 Crystal Drive, Suite 500
Arlington, VA 22202
(202) 466-6272; Fax: (202) 296-1356
Contact: Ann Harkins, President and CEO
www.ncpc.org

NATIONAL CRIME PREVENTION INSTITUTE (NCPI)
University of Louisville
Louisville, KY 40292
(502) 852-5577 or (800) 334-8635;
Fax: (502) 852-6990
Officer: Marianna A. Perry, Director
www.louisville.edu/ncpi

NATIONAL FIRE PROTECTION ASSOCIATION (NFPA)
1 Batterymarch Park
Quincy, MA 02169-7471
(617) 770 3000; Fax: (617) 770 0700
Officer: James M. Shannon, President/CEO
www.nfpa.org

NATIONAL RIFLE ASSOCIATION OF AMERICA (NRA)
11250 Waples Mill Rd.
Fairfax, VA 22030
(703) 267-1640 or (800) 672-3888;
Fax: (703) 267-3957
Contact: Wayne LaPierre, Executive Vice President

NATIONAL SAFETY COUNCIL
1121 Spring Lake Dr.
Itasca, IL 60143-3201
(630) 285-1121; Fax: (630) 285-1315
Officer: Janet Froetscher, President
www.nsc.org

NOR-CAL TRAINING ACADEMY
2016 Oakdale Ave.
San Francisco, CA 94124-2098
(415) 550-8282; Fax: (415) 647-0300
Contact: Moura Borisova, Owner
www.nor-caltrainingacademy.com

NATIONAL SEMINARS TRAINING
PO Box 419107
Kansas City, MO 64141-6107
(913) 432-7755 or (800) 258-7246;
Fax (913) 432-0824
www.nationalseminarstraining.com

PINKERTON CONSULTING & INVESTIGATIONS
120 Howard Street
San Francisco, CA 94105
(415) 808-1730; Fax: (415) 520-0109
Contact: Michael Delamere, Managing Director
www.securitas.com

PROFESSIONAL SECURITY TELEVISION NETWORK (PSTN)
4101 International Parkway
Carrollton, TX 75007
(800) 624-2272; Fax: (972) 309-5352
Contact: Norm Wilcox, CEO
www.pstn.com

PROPERTY CASUALTY INSURERS ASSOCIATION OF AMERICA
2600 S. River Road
Des Plaines, IL 60018-3286
(847) 297-7800; Fax: (847) 297-5064
Contact: David A. Sampson, President and CEO
www.pciaa.net

SECURITY MANAGEMENT INSTITUTE
John Jay College of Criminal Justice
555 West 57th Street, Room 603BMW, Rm. 636
New York, NY 10019-1029
(212) 237-8638; Fax: (212) 237-8637
Officer: Herbert A. Johnson, Director
www.jjay.cuny.edu

SMITH & WESSON ACADEMY
299 Page Blvd.
Springfield, MA 01104
(413) 846-6461; Fax: (413) 736-0776
Contact: Barbara Grissom
www.smith-wesson.com

WACKENHUT EDUCATION SERVICES
4200 Wackenhut Dr.
Palm Beach Gardens, FL 33410-4243
(561) 622-5656 or (800) 922-6488;
Fax: (561) 691-6700
Contact: Grahame Gibson, COO
www.g4s.com

AUDIOVISUAL DISTRIBUTORS

ADVANCED DRIVING AND SECURITY
210 Airport, Suite 10
North Kingstown, RI 02852
(401) 294-1600 or (888) 394-2374;
Fax: (401) 294-7600
www.1adsi.com

ALTERNATIVE COMPUTER TECHNOLOGY
7908 Cin-Day Rd., Suite W
West Chester, OH 45069
(513) 755-1957; Fax: (513) 755-1958
www.altcomp.com

AMERICAN HOSPITAL ASSOCIATION
1 N. Franklin St.
Chicago, IL 60606-3421
(312) 422-3000; Fax: (312) 422-4796
www.aha.org

AMERICAN MEDIA, INC
4900 University Ave.
West Des Moines, IA 50266
(515) 224-0919 or (800) 262-2557;
Fax: (515) 224-0256
www.ammedia.com

AMERICAN POLYGRAPH ASSOCIATION (APA)
PO Box 8037
Chattanooga, TN 37414-0037
(423) 892-3992 or (800) 272-8037;
Fax: (423) 894-5435
Officer: Robbie Bennett, Manager
www.polygraph.org

AMERICAN SOCIETY FOR INDUSTRIAL SECURITY (ASIS)
1625 Prince Street
Arlington, VA 22314-2818
(703) 519-6200; Fax: (703) 519-6299
Officer: Michael J. Stack, President
www.asisonline.org

AMERICAN TRUCKING ASSOCIATION (ATA)
950 North Glebe Road, Suite 210
Arlington, VA 22203-4181
(703) 838-1700; Fax: (703) 684-5751
Officer: Bill Graves, President/Chief Executive Officer
www.truckline.com

ARMA INTERNATIONAL
13725 W. 109th Street, Suite 101
Lenaxa, KS 66215
(913) 341-3808 or (888) 301-3324;
Fax: (913) 341-3742
www.arma.org

ASSOCIATED GENERAL CONTRACTORS OF AMERICA
2300 Wilson Blvd., Suite 400
Arlington, VA 22201
(703) 548-3118 or (800) 242-1767;
Fax: (703) 548-3119
www.agc.org

ASSOCIATION OF CERTIFIED FRAUD EXAMINERS
716 West Avenue
Austin, TX 78701-2727
(512) 478-9070 or (800) 245-332;
Fax: (512) 478-9297
Officer: Joseph T. Wells, Chairman
www.acfe.org

BANK ADMINISTRATION INSTITUTE (BAI)
115 S. LaSalle Street, Suite 3300
Chicago, IL 60603-3801
(312) 683-2464 or (888) 284-4078;
Fax: (312) 683-2373
Officer: Deborah Bianucci, President
www.bai.org

BNA COMMUNICATIONS, INC
1801 S. Bell Street
Arlington, VA 22202
(703) 341-3500 or (800) 372-1033;
Fax: (800) 253-0332
www.bna.com

BUSINESS & LEGAL REPORTS
141 Mill Rock Road E
Old Saybrook, CT 06475
(860) 510-0100 or (800) 727-5257;
Fax: (860) 510-7225
www.blr.com

COMMONWEALTH FILMS, INC
223 Commonwealth Ave.
Boston, MA 02116
(617) 262-5634; Fax: (617) 262-6948
www.commonwealthfilms.com

corVISION MEDIA
872 S. Milwaukee Ave. #295
Libertyville, IL 60048
(847) 549-7860 or (877) 364-7485;
Fax: (866) 440-2614
www.corvision.com

CRISIS PREVENTION INSTITUTE
3315-K N. 124th St.
Brookfield, WI 53005
(262) 783-5787 or (800) 558-8976;
Fax: (262) 783-5906
Contact: Robert Rettmann
www.crisisprevention.com

CRM LEARNING
2215 Faraday Ave., Suite 110
Carlsbad, CA 92008
(760) 431-9800 or (800) 421-0833;
Fax: (760) 931-5792
www.crmlearning.com

DIEBOLD, INC.
5995 Mayfair Road
North Canton, OH 44720
(330) 490-4000 or (800) 999-3600;
Fax: (330) 490-3794
Contact: Thomas Swidarski, President and CEO
www.diebold.com

EDUCATIONAL INSTITUTE OF THE AMERICAN HOTEL & MOTEL ASSOCIATION
2113 N. High Street
East Lansing, MI 48906
(517) 372-8800 or (800) 752-4567;
Fax: (517) 372-5141
www.ei-ahla.org

EDUCATIONAL RESOURCES, INC
7500 W. 160th Street
Stilwell, KS 66085
(913) 239-2658 or (800) 667-7531;
Fax: (913) 685-2381
www.eriworld.com

FOOD MARKETING INSTITUTE
2345 Crystal Drive, Suite 800
Arlington, VA 22202
(202) 452-8444; Fax: (202) 429-4519
www.fmi.org

GOAL PRODUCTIONS, INC
2623 E Foothill Blvd., #101
Pasadena, CA 91107-3466
(626) 584-9515; Fax: (626) 792-2709
www.goalproductions.com

GRIGGS PRODUCTIONS, INC
220 30th Ave.
San Francisco, CA 94121
(415) 750-5100 or (800) 210-4200;
Fax: (415) 750-5145
www.griggs.com

HOME CENTER INSTITUTE
5822 W. 74th St.
Indianapolis, IN 46278
(317) 299-0339; Fax: (317) 328-4354

JOHN E. REID & ASSOCIATES
209 W. Jackson Blvd., Suite 400
Chicago, IL 60606
(312) 583-0700 or (800) 255-5747;
Fax: (312) 583-0701
www.reid.com

LIBRARY VIDEO NETWORK
320 York Rd.
Towson, MD 21204
(410) 887-2090 or (800)441-8273;
Fax: (410) 887-2091
Officer: Jim Fish
www.lvn.org

MIS TRAINING INSTITUTE, INC
498 Concord St.
Framingham, MA 01701-2357
(508) 879-7999; Fax: (508) 872-1153
Contact: Linda Burton, President and CEO
www.misti.com

NATIONAL ASSOCIATION OF CONVENIENCE STORES
1600 Duke Street
Alexandria, VA 22314
(703) 684-3600 or (800) 966-6227;
Fax: (703) 836-4564
www.nacsonline.com

NATIONAL BURGLAR AND FIRE ALARM ASSOCIATION (NBFAA)
2300 Valley View Lane, Suite 230
Irving, TX 75062
(214) 260-5970 or (888) 447-1689;
Fax: (214) 260-5979
Officer: Merlin Guilbeau, Executive Director
www.alarm.org

NATIONAL COMPUTER SECURITY ASSOCIATION
1200 Walnut Bottom Rd.
Carlisle, PA 17013
(717) 258-1816 and (888) 627-2281;
Fax: (717) 243-8642
www.icsa.net

NATIONAL CRIME PREVENTION COUNCIL
2345 Crystal Drive, Suite 500
Arlington, VA 22202
(202) 466-6272; Fax: (202) 296-1356
Contact: Ann Harkins, President and CEO
www.ncpc.org

NATIONAL CRIMINAL JUSTICE REFERENCE SERVICE
Box 6000
Rockville, MD 28049-6000
(301) 519-5500 or (800) 851-3420;
Fax: (301) 519-5212
www.ncjrs.gov

NATIONAL FIRE PROTECTION ASSOCIATION (NFPA)
1 Batterymarch Park
Quincy, MA 02169-7471
(617) 770-3000; Fax: (617) 770-0700
Officer: James M. Shannon, President/CEO
www.nfpa.org

NATIONAL RETAIL FEDERATION
325 7TH Street, NW, Suite 1100
Washington, DC 20004
(202) 783-7971 or (800) 673-4692;
Fax: (202) 737-2849
Officer: Tracy Mullin, President
www.nrf.com

NATIONAL SAFETY COUNCIL
1121 Spring Lake Dr.
Itasca, IL 60143-3201
(630) 285-1121; Fax: (630) 285-1315
Officer: Janet Froetscher, President
www.nsc.org

NATIONAL TECHNICAL INFORMATION SERVICES
5301 Shawnee Road
Alexandria, VA 22312
(703) 605-6050 or (888) 584-8332;
Fax: (703) 605-6700
www.ntis.gov

PHOENIX LEARNING GROUP
2349 Chafee Dr.
St. Louis, MO 63146-3306
(314) 569-0211 or (800) 221-1274;
Fax: (314) 569-2834
www.phoenixlearninggroup.com

PROFESSIONAL SECURITY TELEVISION NETWORK (PSTN)
4101 International Parkway
Carrollton, TX 75007
(800) 624-2272; Fax: (972) 309-5352
Contact: Norm Wilcox, CEO
www.pstn.com

PSYCHEMEDICS CORPORATION
125 Nagog Park
Acton, MA 01720
(978) 206-8220 and (800) 628-8073;
Fax: (978) 264-9236
www.psychemedics.com

PUBLIC RISK MANAGEMENT ASSOCIATION
500 Montgomery Street, Suite 750
Arlington, VA 22314
(703) 528-7701; Fax: (703) 739-0200
www.primacentral.org

RESOURCE NETWORK
Box 247
Harrisonburg, VA 22801
(540) 433-5750 or (800) 296-5750;
Fax: (540) 433-5640
www.resourcenetwork.com

SECURITY TRAINING CORP
PO Box 12244
Longview, TX 75607-2244
(903) 759-6619 or (888) 284-7928;
Fax: (903) 759-9208
www.jackhenry.net

VIDEO TRAINING, INC
4242 22nd Ave. W
Seattle, WA 98199
(800) 600-1555; Fax: (206) 682-1865
www.videotraining.com

COMPUTER RESOURCES ON THE INTERNET

Internet resources change on nearly a daily basis. Firms merge, publications cease to exist, and associations wither away. Many of the publications and membership associations listed in the sections above offer digital versions of publications (with some limitations), often at no cost to visitors and members. Likewise, some firms offer electronic newsletters or updates periodically that offer timely news articles. Search and develop your own set of favorites or bookmarks that you can check frequently, and subscribe to some blogs so you can stay current with ever-changing technology.

Index